Microbiology and Chemistry for Environmental Scientists and Engineers

Microbiology and Chemistry for Environmental Scientists and Engineers

Second edition

J. N. Lester and J. W. Birkett

London and New York

First published 1988 by E & FN Spon,
an imprint of Chapman & Hall

Second edition published 1999 by E & FN Spon
11 New Fetter Lane, London EC4P 4EE

Simultaneously published in the USA and Canada
by Routledge
29 West 35th Street, New York, NY 10001

E & FN Spon is an imprint of the Taylor & Francis Group

© 1999 J. N. Lester and J. W. Birkett

Typeset in Times by Mathematical Composition Setters Ltd, Salisbury, Wiltshire
Printed and bound in Great Britain by TJ International Ltd, Padstow, Cornwall

British Library Cataloguing in Publication Data
A catalogue record for this book is available from the British Library

Library of Congress Cataloging in Publication Data

Lester, J. N. (John Norman). 1949–
   Microbiology and chemistry for environmental scientists and
engineers / J.N. Lester and J.W. Birkett.
      p.   cm.
   Includes bibliographical references and index.
   1. Sanitary microbiology. 2. Drinking water–Microbiology.
3. Microbiology. 4. Environmental chemistry. 5. Water chemistry.
6. Chemistry. I. Birkett, J. W. (Jason W.), 1970– . II. Title.
QR48.L47 1999
579–dc21   98-49280
                                                                 CIP

ISBN 0-419-22680-X

Contents

Preface to the second edition

Environmental science and engineering students come from a wide variety of disciplines. As such, they may not be well versed in the fundamentals of chemistry and microbiology that are essential to these types of courses. This second edition has been produced to give environmental scientists and engineers a sound understanding of the chemical and biological principles involved in water and wastewater treatment processes.

The first edition found widespread use on both undergraduate and postgraduate courses in presenting the fundamentals of microbiology in these processes. Owing to the need for a chemistry element, the second edition now incorporates chemistry sections that cover the fundamental principles of inorganic, organic and environmental chemistry relating to water quality. A new chapter on water pollution discusses the types of chemicals encountered in aquatic systems with particular emphasis on the pollutants listed in the EC Dangerous Substances Directive. Previous sections on biological processes have also been revised and contain up to date references from the literature.

The second edition is a unique text that encompasses all aspects of microbiology, chemistry and water and wastewater treatment processes, that are relevant to environmental scientists and engineers. Owing to its novel approach, this book is also suitable for non-biological and non-chemical professionals who require a background knowledge of the principles of water and wastewater treatment.

Microbiology, chemistry and the treatment of water have been topics of teaching and research at Imperial College for over a century. In 1885 Percy Frankland, a lecturer in chemistry at the Royal School of Mines (now a part of Imperial College) introduced Koch's gelatine process of water examination to the UK. This was applied to examining the effect of a slow sand filter on viable bacteria concentrations in treated and untreated water, and was utilized on London's water supplies. In 1869, Percy's father, Sir Edward Frankland, a professor of chemistry at the Royal School of Mines, made advances in biological wastewater treatment processes by developing the intermittent downward filtration technique, the precursor to the percolating filter.

Professor Lester has over 25 years experience dealing with environmental microbiology and biological wastewater treatment. In particular the biological degradation and transformation of toxic organic micropollutants and heavy metals has been a major area of research as well as industrial wastewater treatment. These activities have extended to sewage sludge disposal, contamination of lowland rivers, water re-use and contamination of estuarine and coastal environments. Dr Birkett has several years experience of the behaviour of metals and binding to humic substances, and effects on bioavailability and metal speciation. He has recently been appointed as lecturer in environmental chemistry and pollution at Imperial College.

John N. Lester
Jason W. Birkett
London

Acknowledgements

The authors would like to thank the co-author of the first edition, Dr Robert Sterritt, for his invaluable contributions to this edition. They also wish to acknowledge the contributions made by Dr T. Rudd to Chapters 2 and 19, and Dr G. Shaw to Chapter 6. They would particularly like to thank Richard Whitby and the staff at E & FN Spon Ltd for their help and guidance given during the production of this book.

Chapter 1

Introduction to microbiology

1.1 Origins of microbiology

The existence of the microbial world was first recognized by Antonie van Leeuwenhoek (1632–1723), a Dutch merchant working in the town of Delft. An essential prerequisite for the discovery of micro-organisms was the microscope. The first microscopes were developed at the beginning of the seventeenth century. Two types of instrument were available. These were the simple microscope with a single lens of short focal length utilizing the same optical principle as the magnifying glass and the compound microscope with a double lens system from which contemporary instruments have been developed. It was Leeuwenhoek's excellence in the manufacture of the former type of instrument, shown diagrammatically in Fig. 1.1, which allowed him to make his extraordinarily accurate observations which marked the origin of microbiology. Although simple in concept and construction and almost entirely dependent for their effectiveness upon the quality of the nearly spherical lens, Leeuwenhoek's microscopes could achieve magnifications of 50 to 300 diameters. Leeuwenhoek constructed many such instruments since no variation in the magnification of each instrument was possible, this being predetermined by the lens selected. A range of magnification was dependent upon the availability of a selection of microscopes. These microscopes were used in the horizontal plane rather than the vertical plane in which contemporary instruments function. The lens was mounted in a plate and holding the plate close to the eye the observer squinted through the lens. To focus the instrument the material under examination was mounted on a movable pin which permitted its position to be changed with respect to the lens. With the aid of the simple microscope Leeuwenhoek was able to identify the principal groups of unicellular micro-organisms, protozoa, algae, yeasts and bacteria.

Following Leeuwenhoek's observations there occurred an interregnum, during which the existence of "microbes" was generally accepted. Then the simple microscope was surpassed by the compound microscope, which became prominent between 1820 and 1850, and the birth of experimental

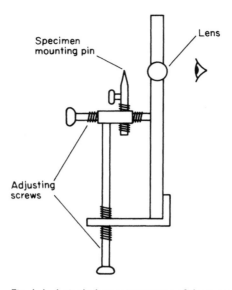

Fig. 1.1 A single-lens microscope of the type used by Leeuwenhoek.

bacteriology occurred, conceived in the experiments of Lazzaro Spallanzani and brought to fruition by Louis Pasteur (1822–95). Spallanzani demonstrated that prolonged heating prevented the putrefaction of various infusions and that subsequent microbial growth in these sterilized infusions was dependent upon airborne contamination since when hermetically sealed no putrefaction was observed.

The presence of micro-organisms in air was first demonstrated by Pasteur who filtered a large volume of air through a guncotton plug. When the guncotton was dissolved in alcohol and ether the residue contained small bodies identical to micro-organisms. Subsequently Pasteur demonstrated that guncotton containing large numbers of micro-organisms as a consequence of filtration was able to induce growth in previously sterile infusions. As a result of these experiments Pasteur undertook further studies using swan-necked flasks (see Fig. 1.2) for which he remains famous. Infusions sterilized in these flasks remain uncontaminated despite contact with air through the neck. Although the micro-organisms could travel along the descending limb of the neck they were unable to pass through the ascending part and were thus prevented from contaminating the infusion. If, however, the neck of the flask was broken or the flask tilted so that the infusion was allowed to enter the neck and then returned, contamination occurred. Although not the first to propose that fermentation was a microbial process Pasteur finally persuaded the scientific community of this fact with the aid of experimental evidence. He also identified microbial growth in the absence of oxygen

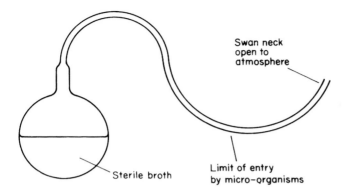

Swan neck
open to
atmosphere

Sterile broth

Limit of entry
by micro-organisms

Fig. 1.2 Pasteur's swan-necked flasks.

which he termed anaerobic to distinguish it from growth in the presence of oxygen which he called aerobic.

In subsequent experiments he demonstrated the existence of facultative anaerobes which have two energy yielding mechanisms available to them. These organisms in the presence of oxygen obtain energy for maintenance and growth by aerobic respiration. In the absence of oxygen, however, they utilize fermentation to obtain the energy they require. Pasteur observed this behaviour in experiments using a yeast which in the absence of air produced alcohol and carbon dioxide by fermentation. In the presence of air no alcohol was produced and carbon dioxide was the principal product of aerobic respiration. He also established that fermentation was a less efficient energy yielding process than aerobic respiration. In experiments in which a yeast inoculum was provided with the same quantity of sugar substrate under aerobic and anaerobic conditions less growth occurred in the anaerobic culture than in the aerobic culture. This discrepancy is due to the incomplete utilization of the substrate under anaerobic conditions which is evident in the presence of alcohol in the spent medium of the anaerobic culture in addition to carbon dioxide which is the only major product of the aerobic culture and indicative of complete oxidation and substrate utilization. The subsequent development of bacteriology into a fundamental science was achieved by Robert Koch who is generally accepted to be pre-eminent amongst pure bacteriologists. He established the basic methods for bacterial examination, many of which are little changed to this day.

1.2 History of sanitary microbiology

Sanitary or "public health" microbiology developed from medical bacteriology in the 1880s. With the advent of anaesthesia in the middle of the nineteenth century surgeons began to undertake longer and more complex operations which exacerbated the problems of surgical sepsis and frequently resulted in the death of the patient from infections which developed after the operation. A British surgeon, Joseph Lister, perceived the significance of Pasteur's earlier results and concluded that surgical sepsis probably developed as the result of microbial infection of the tissues exposed during the operation. The introduction of antiseptic surgery pioneered by Lister brought a dramatic reduction in post-operative deaths from surgical sepsis and provided powerful indirect evidence for the germ theory of disease.

Lister made a further important contribution to the development of bacteriology during 1878 when studying the lactic fermentation of milk. By diluting his bacterial samples he produced suspensions in which a single drop had the probability of containing a single cell on one occasion out of two. On the other occasion the drop was sterile and no lactic acid fermentation resulted. This experiment provided the basis for the isolation of pure cultures, in this case of the predominant organism since all others would have been diluted out. However, it was the development of solid culture media by Robert Koch, which is probably the single most important event in the history of bacteriology after the discovery of sterilization, that led to the routine isolation of pure cultures. The technique was based on Koch's observations of bacteria growing as discrete colonies on potato slices. He concluded from this observation that each colony had developed from a single bacterium which had settled upon the potato slice. This technique made it an easy matter to identify contaminants which appeared as distinct colonies and Koch was able to solidify a wide range of common media using gelatine. At a later date agar was substituted for gelatine since the latter could be liquefied by some organisms. The technique as it is recognized today was completed by one of Koch's assistants, Richard Petri, who invented his now famous plate to contain the solid medium.

The state of London's water supply in the nineteenth century made it the natural location for the development of public health microbiology. Although not proven, concern about the possible role of London's water supply in various epidemics had been expressed from the beginning of that century. The first slow sand filter, designed by James Simpson, was brought into operation in 1829 at the Grosvenor Road works of the Chelsea Water Company. From 1834 onwards Parliament insisted on some water companies including provision for sand filtration when seeking permission for the construction of new water works. The cholera epidemic of 1831–2 proved a stimulus in the development of a unified approach to London's sewage system resulting in the formation of the Metropolitan Commission of Sewers. A further severe epidemic in 1848–9 resulted in the Commission obtaining powers to compel all householders to connect their drains to public sewers and led to the practical elimination of cesspits within a few years. However, it was the research of Dr John Snow between 1849 and 1854 which established that the water drawn from a shallow well in Broad

Street, Gold Square, Soho, contaminated by materials drained from cesspits and possibly burial grounds, was the cause of an appalling outbreak of cholera that established beyond doubt the need for the public sewerage system.

Although the slow sand filter had been introduced early in the nineteenth century its importance in removing pathogenic bacteria from water was not perceived for more than fifty years. In 1885 Percy Frankland, then a lecturer in chemistry at the Royal School of Mines (now part of Imperial College, London) introduced Koch's gelatine-process of water examination to Great Britain and applied it to examining the effect of the slow sand filter on concentrations of viable bacteria in the treated and untreated water. Results of his studies were presented to the Royal Society in 1885 and the Institution of Civil Engineers in 1886. This research is best summarized in his own words, drawn from his book *Micro-organisms in Water* published in conjunction with his wife in 1894:

It is obvious that, with the information which the bacteriological investigation of water has furnished us, the subject of water-purification must be approached from an entirely novel point of view; that, whereas formerly the chemical standard was the only one which could be appealed to as a guarantee of the suitability or not of a water for domestic supply, we have now a far more delicate test as to the efficiency of the purification process employed in the biological examination to which a water can be submitted. Thus, perhaps, a concrete example will most clearly illustrate how the purification of water must now be regarded. Supposing that a water, derived from a course which is altogether unimpeachable as regards contamination with animal matters, is yet so highly impregnated with vegetable constituents as to be unpalatable, the question will arise how this water may be treated so as to free it from this blemish and render it suitable for drinking purposes. In a case of this kind it is obvious that chemical purification will be of paramount importance, whilst the removal of organic life from the water will be of less pressing consequence. On the other hand, if water which is known to have received sewage matters (and the *entire* exclusion of such from supplies drawn from rivers is practically impossible) is to be supplied for dietetic use, and if this water, as is so often the case, is not objectionable on account of the absolute quantity of organic matter, as revealed by chemical analysis, which it contains, but only because of the suspicious origin of a part of this organic matter, then it is evident that in the purification of such water the point to be taken primarily into consideration is how the organic life it contains can be reduced to a minimum.

In estimating the value of such processes of purification, it has hitherto been customary to assume that those processes which effect the greatest chemical improvement in water may also safely be considered to be biologically the most excellent; and, conversely, that those processes which effect little or no reduction in the proportion of organic impurity are not calculated to be of any service in removing organized matters.

The following chemical analyses of river-water before and after sand-filtration will sufficiently explain how, for example, the process of sand-filtration found comparatively little favour as long as the chemical analysis was the principal basis on which a judgment as to the hygienic value of water-filtration could be formed.

Chemical Analysis of Water of River Ouse before and after Sand-filtration

	River water	River water after sand-filtration
Total solid matters	28.40	26.20
Organic carbon	00.123	00.119
Organic nitrogen	00.025	00.022
Ammonia	00.0	00.0
Nitrogen as nitrates and nitrites	00.077	00.089
Total combined nitrogen	00.102	00.111
Chlorine	01.6	01.6
Hardness Temporary	11.5	10.9
Hardness Permanent	07.1	07.1
Hardness Total	18.6	18.0

Such slight chemical improvement as is here shown indicating that sand-filtration had comparatively little effect on the dissolved matters in water, led to a priori speculation that it would be unlikely to remove bacteria and other minute forms of life. During the early growth of the belief in the communication of zymotic diseases by micro-organisms the process filtration in general was viewed with great distrust, for it was not unnaturally assumed that the extremely small dimensions of these living creatures would admit of their passing through the comparatively large interstices of ordinary filtering media, with the same sort of facility that vehicles thread their way along a crowded thoroughfare.

The Franklands applied Koch's gelatine-process of water examination to the water supplies in London in 1885. They observed reductions in the bacterial content of water from the River Thames and the River Lea varying between 88.8 and 98.9% after sand filtration. At the request of the local Government Board they undertook regular bacteriological examinations of these rivers and the treated waters produced from them in the years 1886,

1887 and 1888. The conclusions of their study are best presented in their own words:

Thus, on the average, out of every 100 micro-organisms present in the untreated river-water, there were removed by the water companies before distribution in the case of the

	1886	1887	1888
Thames	97.6	96.7	98.4 micro-organisms
Lea (E. London Co.)	96.5	95.3	95.3 micro-organisms

With regard to the removal of this large percentage proportion of micro-organisms through the treatment adopted by the water companies, we may appropriately quote from a paper read by one of us at the York Congress of the Sanitary Institute in 1886.

"Although the organisms thus removed are probably in general perfectly harmless, it must not be supposed that their removal is of no importance, for it must be remembered that the micro-organisms which are known to produce disease, and which are termed pathogenic, do not in any way differ from the ordinary organisms in water so as to render it probable that they would behave differently in the process of filtration; but on the contrary, there cannot be any serious doubt that their behaviour under these circumstances would be precisely similar. Now such disease-organisms frequently do gain access to water, and it is obviously of the greatest importance to ascertain what sort of impediment this process of filtration, which is so largely practised, offers to their passing on to the consumer." By means of this bacteriological examination it is thus possible to obtain a far more satisfactory knowledge of the kind of filtration which water has undergone than by a mere appeal to the eye of the observer, and the vague terms "turgid", "slightly turgid", "clear" and the like, which have hitherto been employed to describe whether the filtration of water has been satisfactory or not, must now be replaced by this scientific and important standard which I have described. These investigations further brought to light some very interesting points in connection with sand-filtration, and have indeed placed that process on a sound basis by exhibiting what are the principal factors in determining its efficiency. Thus, amongst the London water companies there are 7 employing sand-filtration, and in the works of each the process has undergone to a great extent an independent evolution, so that in no two of them is the process at the present time carried on under precisely similar conditions. This, for experimental purposes, peculiarly fortunate circumstance enabled one of us to institute a comparison between the results achieved in the several modifications of the general process. In this connection we may quote from a paper read by one of us before the Institution of Civil Engineers in the year 1886, the conclusions which we then arrived at having been fully verified by further observations made both in this country and abroad since that date. The factors which, in my opinion, are more especially calculated to influence the number of micro-organisms present in the distributed water are the following: (a) storage capacity for unfiltered water; (b) thickness of fine sand through which filtration is carried on; (c) rate of filtration (d) renewal of filter-beds.

Biological wastewater treatment processes were developing in parallel with water treatment processes during the late 1800s, although the importance of the micro-organisms to these systems was not universally perceived. By 1860 the most prominent form of sewage treatment in urban areas which involved microbial activity was sedimentation followed by broad irrigation over agricultural land at sewage farms. It was an effective process but required large areas of land. In 1869 Sir Edward Frankland developed intermittent downward filtration in which the sewage was allowed to percolate periodically through specially prepared porous soil lying above underdrains. Periods of irrigation were separated by several hours to allow for aeration of the soil to facilitate oxidation of the organic matter. This technique and derivatives of it allowed for a five-to-tenfold reduction in the land area required for treatment over broad irrigation. However an area of 0.4 ha (1 acre) was still required to treat the sewage from approximately 1000 persons ($0.05 \, \mathrm{m^3 \, m^{-2} \, d^{-1}}$). Stimulated by the work of Sir Edward Frankland investigations were initiated into the use of local soils, sands and gravels for intermittent downward filtration at the Lawrence Experimental Station of the Massachusetts State Board of Health during 1889. During these studies it was noted that high rate of treatment ($0.5 \, \mathrm{m^3 \, m^{-2} \, d^{-1}}$) could be achieved when using coarse gravel. It was concluded that the treatment was not dependent upon the physical straining but by biochemical oxidation in the biological slimes covering the stones. From these observations came the impetus to construct a pilot scale filter, which fulfilled most expectations and provided for a further tenfold increase in the loading rate previously achieved in the most effective land treatments. Not only did this development permit a further reduction in the land area required for treatment but it removed the dependence of the treatment upon the suitability of the local soil, since sufficient coarse media could be imported. The development of percolating filters proceeded at a rapid pace in the UK between 1880 and 1910 due to sponsorship from the Royal Commission on Sewage Disposal. With the development of the travelling distributor it became a common form of treatment from 1890 onwards.

During the development of the percolating filter other treatment methods were being used or developed in parallel. These included contact beds, septic tanks, the Imhoff tank and physical-chemical treatment. None of these systems gave a high quality effluent and they usually embraced a period during which anaerobic conditions prevailed with a consequent odour problem. The liquefaction of part of the organic matter in sewage under anaerobic conditions was observed by Mouras in 1860. This formed the basis of the septic tank. In Massachusetts at the turn of the century H. W. Clark suggested that instead of subjecting the whole of the sewage to septic action the solids should be settled separately and "flushed" at intervals into a separate septic tank. It was found that the tank volumes required for effective treatment could be reduced by this expedient. The septic tank used for the treatment of the primary sludge was the forerunner of the anaerobic digestion process commonly used today. In Europe, Dibdin in about 1885 observed the "fermentation" of sludge allowed to become anaerobic, and noted the objections raised by his workers to the smell. In contrast, around the turn of the century slow anaerobic breakdown of sludge was found to be effective in reducing offensive odours and improved dewatering characteristics. Much of the early work on anaerobic digestion was done at the Saltley Works in Birmingham which remained the centre of expertise up to the 1930s. The first two-stage digester (1912), the first floating gas holders (1924) and experiments with cold digestion (1935) were notable milestones.

An appreciation of the microbiological basis of anaerobic processes was lacking for many years. Thus, they were at different times both the source of, and the solution to odour problems. In attempts to minimize the odour problems associated with putrefying sewage, workers studied the effect of blowing air into the waste waters. It is generally accepted that the first such study was conducted by Dr Angus Smith (1882). Other European workers included Dupré and Dibdin (1884) and Hartland and Kaye-Parry (1888) and in the USA Drown (1891), Mason and Hine (1891) and Waring (1892–4). These workers held the conviction that oxygen could achieve the oxidation of the waste waters, but this was not supported by their results. Aeration did delay the onset of putrefaction but achieved little improvement in effluent quality. At the Lawrence Experimental Station in the United States Waring and Lowcock artificially aerated filters with a consequential improvement in effluent quality. The pre-existing biomass retained in these fixed film systems was able to utilize effectively the oxygen provided.

As experiments which involved blowing air into sewage continued the importance of the suspended matter in enhancing the biological treatment began to emerge. In 1893 Mather and Platt Ltd reported that the presence of precipitated impurities at the base of the aeration tank resulted in a marked improvement in effluent quality. This belief was presented to the Royal Commission by Adeney in 1905, who clearly indicated that collected humus matters would accelerate treatment. In experiments on the aeration of waste water in 1897 Fowler also obtained a clear effluent lacking in odour in conjunction with rapidly settling deposits of particulate matter. However,

Fowler, working in Manchester, considered the production of this suspended matter a failure of the process. During the following ten years the values of aerating sewage in the presence of a biological humus or slime achieved a wide acceptance in both the USA and the UK. In New York Black and Phelps substituted closely spaced wooden laths for coarse rock media to increase the surface area for film growth under aerobic conditions. The system resembled the earlier Travis "Colloider" which had used wooden laths in an anaerobic chamber. In the UK Dibdin developed a similar approach using a slate bed reactor. In the USA slate bed reactors were also being studied by Clark and Gage at the Lawrence Experimental Station. During 1912 Fowler visited New York to examine the pollution of the harbour and subsequently visited the Lawrence Experimental Station and saw the experiments being performed there. Fowler found this visit thought-provoking and as a consequence reversed his original opinion on the value of aeration. He conceived an aeration process utilizing a suspended biomass to achieve treatment.

In 1913 Fowler and Mumford described their experiments with a bacterial strain designated M7 isolated from a pit water impregnated with iron. The process involved the preliminary removal of suspended solids from the sewage which could be achieved by sedimentation; the effluent from this stage passed into a reactor inoculated with M7 together with a small dose of ferric salt and the tank aerated by blowing air through the liquid. Fowler and Mumford reported that:

> A period of settlement is then allowed for precipitation of the coagulated matter, and eventually the clear liquid is run off, either for rapid final filtration or for direct discharge into the stream. The precise mode of action of the organism is not yet fully worked out, but it seems likely that simultaneous precipitation and solution take place, some of the organic matter being converted to amido derivatives, and some being coagulated and thrown down with the ferric hydroxide.

However, this system required continuous inoculation of the M7 strain. It fell to two of Fowler's students, Ardern and Lockett, to make the breakthrough that Fowler described as a "bombshell". In a paper presented in May 1914 they reported upon the desirability of recycling the humus solids and thus the activated sludge process was born. Preliminary experiments were undertaken by the aeration of sewage in bottles. Sewage required five weeks aeration to achieve complete nitrification after which a clear oxidized liquid was removed by decantation from the bottle. A further volume of raw sewage was incubated with the deposited solids and the process repeated several times. As the quantity of deposited solids increased the time for each oxidation cycle diminished until full oxidation was achieved within twenty-four hours. The authors stated that "for reference purposes and failing a better term the deposited solids resulting from the complete oxidation of sewage have been designated 'activated sludge' ". The conclusions drawn from these laboratory-scale fill and draw experiments at Davyhulme, Manchester, were highly perceptive and characterized the process very well even at this early stage. They were as follows:

1 That the resultant solid matter obtained by prolonged aeration of sewage, which has been termed activated sludge, has the property of enormously increasing the purification effected by simple aeration of sewage, or in other words it greatly intensifies the oxidation process.

2 The extent of the accelerating effect depends upon the intimate manner in which the activated sludge is brought into contact with, and upon its proportion to, the sewage treated.

3 That in order to maintain the sludge at its highest efficiency it is necessary that there should not be at any time an accumulation of unoxidized sewage solids. It is not necessary that the sewage should be kept in contact with the activated sludge until such conditions obtain, as its activity may be maintained by suitable aeration of the activated sludge alone.

4 That temperature exerts a considerable influence on the oxidation process. The purification effected is seriously diminished at temperatures constantly below 10°C. Up to 20–24°C no material difference in the clarification effect and general purification has been observed, although the nitrification change proceeds more rapidly as the temperature rises. At higher temperature the clarification effect is somewhat interfered with during the earlier period of aeration, with a consequent delay in the establishment of nitrification. Subsequently the rate of nitrification somewhat increases.

5 That under the conditions of experiment a well-oxidized effluent can be obtained by the aeration of average strength Manchester sewage in contact with activated sludge for a period of six to nine hours. The percentage purification effected, as measured by the usual tests, is at least equal to that obtained by the treatment of sewage on efficient bacterial filters. The period of aeration naturally depends upon the strength of the sewage treated and the degree of purification required.

6 That the activated sludge differs very considerably in character and composition from ordinary sewage sludge. It is in a well-oxidized condition and consequently entirely innocuous, can be readily drained on straining filters and possesses a high nitrogen content.

The laboratory-scale fill and draw experiments were repeated outdoors with equal success. Development of the process proceeded rapidly, the first continuous flow unit treating 250,000 gallons per day was brought into operation in September 1917 for the sum of £2,172 4s 8d at Withington,

Manchester. A large works followed at Davyhulme, Manchester – the site of early experiments – thus establishing biological sewage treatment as viable large-scale practice.

Despite the early success of Arden and Lockett which led to the widespread adoption of the activated sludge process, similar major advances in the understanding of the role of micro-organisms in wastewater treatment were not forthcoming. In 1865–6 the River Pollution Commissions had proclaimed that the treatment of sewage by filtration was not merely mechanical, but also involved chemical oxidation. The role of bacteria was not apparently acknowledged at that time. More surprisingly, more than sixty years later, and fifteen years after the completion of the Davyhulme activated sludge plant, there were still those, according to Wooldridge and Standfast, who believed that the aerobic breakdown of the organic matter in sewage was almost entirely due to chemical processes. Wooldridge and Standfast (1933) proved that the oxidation of sewage was due to bacterial action. In so doing they also made some interesting observations on the viability of the bacteria. Less than 1% of the bacteria present were viable, although their enzymes remained intact. The precise significance of this was not noted for another thirty years with the realization that not all substrate utilization is channelled into growth processes. This permitted the inclusion of the concept of cell maintenance and decay in the theory of continuous bacterial growth first put forward by Monod (1950), Novick and Szilard (1950) and subsequently extended to include systems with cell recycle. Thus, the successful adoption of the activated sludge process had preceded an understanding of the microbiological principles involved by several decades.

1.3 Principles of microbiology

Biology is the science of life; it has three major divisions:

> zoology – the study of animals,
> botany – the study of plants, and
> microbiology – the study of microbes.

These primary divisions may be divided further into specialities. For example, algology and mycology, the study of algae and fungi respectively, are subdivisions of botany. Protozoology, the study of unicellular animals, is a division of zoology whilst bacteriology and virology, the study of bacteria and viruses, are subdivisions of microbiology. Although knowledge about macroscopic animals and plants may be acquired through studies of anatomy and morphology, structure and form, our knowledge of the behaviour of micro-organisms is heavily dependent upon biochemistry.

There is no simple definition of life. It may be characterized by a list of properties which are shared by all living organisms, with the exception of the viruses, and which distinguish them from non-living matter:

Movement. It is characteristic of organisms that they, or some part of them, are capable of moving themselves. Even plants, which at first sight appear to be an exception, display movements within their cells.

Responsiveness. All organisms, including plants, react to stimulation. Such responses range from the growth of a plant towards light to the rapid withdrawal of one's hand from a hot object.

Growth. Organisms grow from within by a process which involves the intake of new materials from outside and their subsequent incorporation into the internal structure of the organism. This is called assimilation and it necessitates some kind of feeding process.

Feeding. Organisms constantly take in and assimilate materials for growth and maintenance. Animals generally feed on ready-made organic matter (heterotrophic nutrition) whereas plants feed on simple inorganic materials which they build up into complex organic molecules (autotrophic nutrition).

Reproduction. All organisms are able to reproduce themselves. Reproduction involves the replication of the organism's genetic "blueprint" which is encoded in a nucleic acid. Generally this is deoxyribonucleic acid (DNA) but in some viruses it may be ribonucleic acid (RNA).

Release of energy. To sustain life an organism must be able to release energy in a controlled and usable form. This is achieved by breaking down adenosine triphosphate (ATP). The energy to generate ATP is obtained by the breakdown of food by respiration. The occurrence of ATP in living cells appears to be universal.

Excretion. The chemical reactions that take place in organisms result in the formation of toxic waste products which must be either eliminated or stored in a harmless form.

Thus a living organism is a self-reproducing system capable of growing and maintaining its integrity by the expenditure of energy. All living organisms share the same basic unit structure termed the "cell". Each cell contains a nucleus which is surrounded by cytoplasm, these together constituting the protoplast which in turn is bounded by a cytoplasmic or cell membrane. Some cells, for example plant cells, also possess a cell wall, whilst animal cells do not. Cells may be organized in multicellular organisms such as the higher animals and plants or exist as unicellular entities, a mode which is common but not universal amongst the bacteria and protozoa. They may also occur in a multinucleate single cell form described as coenocytic, which is prevalent amongst the fungi. Further attention is given to these in Chapters 2, 3 and 4.

A prerequisite to the scientific study of the vast numbers of living organisms which exist is their classification and identification. This is the function of taxonomy. The primary division between the higher (vascular) plants and multicellular animals (metazoans) has been recognized since the beginning of biological study. Prior to the development of the microscope and the foundation of microbiology this bipartite division was a satisfactory basis for the identification and consequential naming of these organisms. A systematic approach to the naming of organisms was introduced in 1735 by the Swedish naturalist Carl Linnaeus. The binomial system that he developed is the basis of modern taxonomy which is applied to all organisms including microbes. This system identifies each organism by a generic and specific name. When printed the generic and specific names are always put in italics (when handwritten or typed they are underlined). The generic name beginning with a capital letter is only spelt in full on the first time of usage in any text and thereafter is abbreviated to the first letter only whilst the specific name beginning with a lower case letter is always written in full. Thus man has the name *Homo sapiens*. This name has been selected as a result of classification which is based upon evolutionary patterns and which categorizes organisms with similar characteristics into groups which in turn are subdivided into smaller groups with greater numbers of shared characteristics. Initially organisms are divided into two or three large groups called kingdoms, each of which is subdivided into a series of major subgroups called phyla (singular phylum). Each phylum is further divided into a series of successively smaller groups known as classes, orders, families, genera (singular genus) and finally the species (specific name). The functioning of this system can be examined in the classification of *H. sapiens*:

Kingdom	Animal	
Phylum	Chordata	Animals with notocords
Subphylum	Vertebrata	Types with vertebral columns
Superclass	Tetrapoda	Terrestrial, four limbs, bony skeleton
Class	Mammalia	Types with hair and milk glands
Subclass	Eutheria	Offspring develops in female parent, nourished by placenta
Order	Primates	Have fingers and flat nails
Family	Hominidae	Upright posture, flat face, stereoscopic vision, large brain, hands and feet
Genus	*Homo*	Double curved spine, long life span, long youth
Species	*Homo sapiens*	Well developed chin, high forehead, thin skullbones

There may be interpolation between the principal ranks by the use of the prefixes sub or super as in the examples of man above.

When the exploration of the microbial world commenced in the eighteenth century new groups of organisms were assigned to either the animal or plant kingdom. Thus fungi and bacteria were initially classified as plants, whilst microscopic motile forms were all classified as animals. As a result of the bipartite approach some very heterogeneous groups developed, which in turn led to dissatisfaction with the system. To overcome these problems E. Haekel proposed in 1886 the establishment of a third kingdom, the protists, to include all organisms with a simple biological organization, thus accommodating all of the new microbial groups. This concept was rapidly accepted and the new kingdom included the algae, protozoa, fungi and bacteria. This tripartite system survived until very recently. However, the development of the electron microscope from 1950 onwards permitted detailed examination of cell structure. These studies revealed a fundamental dichotomy in cell structure. It was evident that two basic cell types existed: the eucaryotic cell and the procaryotic cell. Whilst all multicellular animals and higher plants were composed of the eucaryotic cell, the protista was divided between the two types. The fungi, protozoa and most of the algae were eucaryotic whilst the bacteria and the so-called blue-green algae were procaryotic. As a consequence of these observations the term protist is now confined to eucaryotes only and the blue-green algae are now recognized as bacteria which form a separate group. These are summarized in Table 1.1. In the next three chapters each group is considered in detail.

Table 1.1 Distinctions between eucaryotic, procaryotic and acaryotic groups

Group	Cellular organization	Differentiation	Organisms
Eucaryotic	Multicellular	Extensive	Higher plants, (including ferns, mosses, liverworts) and vertebrate and invertebrate animals
	Unicellular, coenocytic or mycelial	Little or none	The protists (algae, fungi and protozoa)
Procaryotic	Unicellular	Little or none	The bacteria (including cyanobacteria)
Acaryotic	Not cellular	—	The viruses

References and further reading

Frankland, P. and Frankland, K. (1894) *Micro-organisms in Water*, Longmans, Green and Co., London.

Monod, J. (1950) La technique de culture continue: théorie et applications. *Ann. Inst. Pasteur, Lille*, **79**, 390–400.

Novick, A. and Szilard, L. (1950) Experiments with a chemostat on the spontaneous mutation of bacteria. *Proc. Nat. Acad. Sci.*, **36**, 708–719.

Stanbridge, H. H. (1976) *History of Sewage Treatment in Britain*, parts 1–6, Institute of Water Pollution Control, Maidstone.

Stanbridge, H. H. (1977) *History of Sewage Treatment in Britain*, parts 7–12, Institute of Water Pollution Control, Maidstone.

Wooldridge, W. R and Standfast, A. F. B. (1933) The biochemical oxidation of sewage. *Biochem. J.*, **27**, 183–192.

Chapter 2

Bacteria

2.1 Introduction

The bacteria and the blue-green bacteria, formerly known as blue-green algae, together constitute the kingdom Procaryotae. Of all organisms, this group is of most significance to the public health engineer, since biological wastewater treatment processes rely almost exclusively on the activity of bacteria. Most procaryotes are unicellular organisms with a single chromosome which is unbounded by a nuclear membrane. No internal organelles or specialist structures, such as are found in the more complex eucaryotic cells, are evident. Although some procaryotic cells display a certain degree of structural differentiation, their major diversity lies in their metabolic characteristics, i.e., their food sources and methods of producing energy.

These characteristics, which are defined in terms of the carbon source, energy source and the electron donor utilized, are indicative of the enzymatic composition of different micro-organisms and can thus be used to distinguish specific nutritional categories, as shown in Table 2.1. Bacteria which obtain carbon from an inorganic source (i.e. CO_2) are termed autotrophs, while those which utilize organic carbon for cellular synthesis are called heterotrophs. Similarly, two sources of energy can be identified; phototrophic micro-organisms use a light source to produce energy from the reaction with photosynthetic pigments such as chlorophyll, while chemotrophic micro-organisms oxidize organic or inorganic chemical compounds. The electron donor involved in such reactions may be organic or

Table 2.1 Nutritional categories of bacteria

Carbon source	Inorganic	Autotrophs
	Organic	Heterotrophs
Energy source	Light	Phototroph
	Chemical	Chemotroph
Electron donor	Inorganic	Lithotrophs
	Organic	Organotrophs

inorganic, giving a further division of organotrophism and lithotrophism respectively. The above categorizations are frequently combined to give a full description of a bacterium's metabolic requirements. Thus a phototrophic organism with an inorganic electron donor such as sulphide would be a photolithotroph, whereas one with an organic electron donor such as acetate would be a photoorganotroph. Likewise, chemotrophic organisms obtaining energy from oxidation of an inorganic electron donor such as hydrogen are chemolithotrophs and those oxidizing reduced organic compounds such as glucose are chemoorganotrophs.

The distinction between such nutritional categories is not, however, always clearly defined. Some of the largest group, the chemoheterotrophs, can derive both carbon and energy from a single source compound, while other micro-organisms may exhibit sufficient nutritional versatility to be classified under several headings. For example, some photoautotrophs may convert to chemoheterotrophy under conditions of darkness. The capacity or incapacity of an organism to use alternative nutritional modes is indicated by the terms facultative and obligate, respectively. Obligate autotrophs are thus restricted to using CO_2 as their principal carbon source, whereas facultative autotrophs may also be able to use organic carbon sources. Increased knowledge of the nutritional status of micro-organisms has shown that this latter facility is more common than previously thought, therefore a further term, mixotroph, has been used to describe bacteria which can utilize alternative energy sources separately or sometimes simultaneously. This metabolic variability displayed by bacteria is much wider than that of eucaryotic cells and consequently allows them to grow successfully in many more and varied types of environments.

2.2 Structure of the bacterial cell

The fine structure of the bacterial cell, as elucidated by electron microscopy, is shown in Fig. 2.1. Its most obvious feature is the lack of unit membrane systems, with the cytoplasmic membrane being generally the only exception. The cell essentially consists of the nucleoplasm and cytoplasm, which together constitute the protoplast, bounded by the cell membrane and the cell wall. Surface features which may occur include a slime layer or capsule, fimbriae, pili or motile organs such as flagella.

2.2.1 Nuclear material

Unlike the higher eucaryotic cells, the nuclear material of bacterial cells is not retained within a nuclear membrane but forms an irregular, low density zone of fibrillar appearance. The genetic information of the procaryotic cell is carried on a large, single, circular chromosome, which may be from 0.25 to 3 mm in length. The double stranded DNA forming the chromosome is

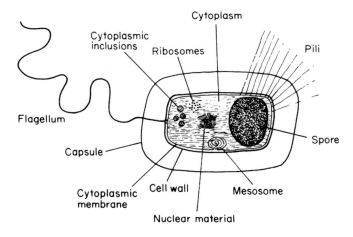

Fig. 2.1 The fine structure of the bacterial cell.

coiled or looped within the nuclear zone and is thought to have a more simple method of replication than the complex mechanism involved in eucaryotic mitosis. The chromosome becomes attached to the cytoplasmic membrane at certain points and as the membrane grows the chromosome divides lengthwise to produce two daughter chromosomes. Nuclear division frequently proceeds ahead of cellular division during rapid growth, resulting in multinucleoid cells. Having only a single chromosome, bacteria are haploid and thus reproduce by asexual means. Genetic transfer may occur, however, and often involves the transfer of plasmids, which are separate extrachromosomal elements, non-essential to the cell. Plasmids code for characteristics such as resistance to drugs and antibiotics; their artificial introduction into other bacterial species forms the basis for the genetic engineering of bacteria for specific purposes.

2.2.2 *Cytoplasm*

The cytoplasm appears under the electron microscope as a dark, granular area, containing a number of particles. These include ribosomes which are composed of RNA and protein and form the sites of protein synthesis for the cell. The type 70S ribosomes found in procaryotic cells are approximately 10 nm in diameter, and are composed of 30S and 50S subunits. The number of ribosomes present varies with growth rate; *Salmonella typhimurium* has >50,000 per cell when growing rapidly but <5,000 per cell when growing slowly. Other inclusions in the cytoplasm are storage granules which act as reserves of energy or scarce elements; their formation also depends on the growth conditions. Poly-β-hydroxybutyrate (PHB) and polyphosphate (volutin) are examples of such storage products. Elemental sulphur is stored in cells which use reduced sulphur compounds as an energy source, such as the purple sulphur bacteria, while in some species of *Bacillus* a single large protein crystal is formed, which acts as a potent insecticide. The only cytoplasmic inclusions which have a complex unit membrane structure similar to that of the cell membrane are the thylacoids, which are flattened membranous sacs containing the photosynthetic apparatus of the blue-green bacteria. Other inclusions which have simple, single-layered membranes of 2–3 nm thickness are the chlorobium vesicles which act as flotation aids in aquatic bacteria and carboxysomes which contain enzymes necessary for certain phototrophs and chemoautotrophs which use CO_2 as their carbon source.

2.2.3 Endospores

Certain bacteria, notably two genera of Gram-positive rods *Bacillus* and *Clostridium*, produce endospores, which are resting forms of the normal vegetative cells. Spore formation is a fairly complex process involving invagination of the cytoplasmic membrane and subsequent enclosure of cellular DNA and cytoplasm in a heavy multilayered spore coat. The coat is extremely resistant to adverse conditions such as heat, radiation and desiccation and thus permits the cell to survive for long periods under conditions which would render the vegetative cell inviable. This property makes heat sterilization of food problematical. Upon stimulation, usually by heating, the spore germinates and grows out into a single vegetative cell. Spores can be clearly seen in wet mounts as round or oval, highly refractive bodies occurring terminally or centrally within cells, sometimes forming irregular shapes such as "drumsticks" or "barrels" or distorting the normal cell shape.

2.2.4 Cytoplasmic membrane

Surrounding the cytoplasm is the cytoplasmic membrane (or plasmalemma) which is 7–8 nm thick, comprises approximately 10% of the cell dry weight and is composed mainly of protein (40–75%) and lipids (16–30%), predominantly phospholipids. The molecular structure of the membrane can be envisaged as a sea of lipids with globular proteins floating in it, the lipids oriented with their polar groups outward and their hydrophobic fatty acid residues inwards (Fig. 2.2). Under the electron microscope, the membrane appears as two dense lines between the cytoplasm and the cell wall. The lines delineate the three-layer structure with outer and inner electron dense layers of 2–4 nm thickness corresponding to the polar end groups of the lipid molecules and a low density layer of 3–5 nm thickness, formed by the fatty acid residues, between them. This structure, elucidated as the fluid

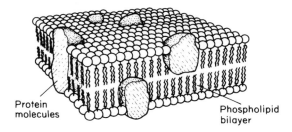

Protein
molecules

Phospholipid
bilayer

Fig 2.2 The molecular structure of the cytoplasmic membrane.

mosaic model, allows the active uptake of solutes required within the cell and also the passage of intracellular components outwards for, e.g. construction of extracellular materials.

In some procaryotes, the cytoplasmic membrane takes over the function of certain organelles such as mitochondria, which in eucaryotic cells function as the centres of energy production. Respiratory electron transport systems and their associated enzymes are incorporated into the cytoplasmic membrane of aerobic bacteria, and extensive infolding and invagination of the membrane greatly increases the surface area available for their siting. The deep invaginations of the membrane, which sometimes develop cross-septa, are known as mesosomes. Mesosomes in one group of photosynthetic procaryotes, the purple bacteria, contain the photosynthetic pigments and are thus analogous to the chloroplasts found in eucaryotic cells.

2.2.5 Cell wall

The cell wall is a rigid structure which confers shape on the bacterial cell; without the cell wall, the cellular contents assume a spherical shape. The main component of the cell wall is a structural polymer called peptidoglycan, also known as mucopeptide or murein. This polymer is found in nearly all procaryotic cell walls, including those of the blue-green bacteria, but does not occur in eucaryotic cells. The only procaryotic exceptions are the halophiles, the methanobacteria and the mycoplasmas. The latter have no cell walls whatsoever, but are bounded by a three-layered membrane stabilized with sterols; as pathogens or parasites the mycoplasmas' environment within a host reduced their need for a rigid cell wall.

The structure of the bacterial cell wall and its resultant response to a particular staining technique, the Gram stain (detailed in Chapter 5), allows a distinction to be made between two major groups of bacteria – those that react positively to the stain and become coloured blue and those which are negative and stain pink. This distinction is used as an important criterion in bacterial taxonomy. The difference in the cell walls of Gram-positive and Gram-negative bacteria is that the former contain as much as 95% peptidoglycan existing as a single layer. In addition, Gram-positive bacterial cell walls contain polysaccharides, techoic acids and techuronic acids. Only a limited number of amino acids are present in Gram-positive cell walls but those of Gram-negative bacteria contain protein (45–50%), lipopolysaccharide (25–30%) and phospholipid (25%). Electron micrographs show a clear difference in structure also between the two types of cell wall. That of Gram-positive bacteria appears amorphous, but the three-layered cell wall of Gram-negative organisms consists of the strengthening peptidoglycan layer (3–5 nm thick) which lies outside but remains in intimate contact with the cytoplasmic membrane, and a lower density region containing periplasmic spaces through which enzymes pass in transit to the outer, third layer, which is membranous in character and is 5–10 nm thick depending on the bacterial species.

2.2.6 External features of the bacterial cell

(a) Flagella

A number of cellular features occur externally to the cell wall, although originating from within the cell. One of the most common is the flagellum, which singly or in tufts provides the major means of motility for procaryotic organisms. Flagella occur either singly at one end of the cell, termed the monotrichate arrangement, or one at each end (amphitrichate), in tufts at one or, more rarely, both ends of the cell (lophotrichate) or distributed all over the cell surface (peritrichate). These organelles are not visible by light microscopy unless stained with a mordant that increases their diameter. Flagella are unbranched, helical filaments of uniform diameter, which may be sheathed or unsheathed depending on the species and strain of bacteria. Unsheathed flagella are 12–20 μm thick whereas sheathed forms may be up to 55 μm thickness, as in *Bdellovibrio bacteriovorus*. Three structural parts are recognized from electron microscope studies, the filament which is readily seen, a hook to which it is attached and a basal structure (Fig. 2.3). The basal structure is composed of a rod and a pair of discs which clamp onto the cytoplasmic membrane. Gram-negative bacterial filaments have a further pair of discs which clamp onto the cell wall.

The filament is made from a protein called flagellin, which has subunits of molecular weight 30,000–60,000 that appear as spherical bodies of 4.5 nm diameter. The subunits are arranged in helices to form fibres within the filament. Movement of the cells is brought about by a helical twisting of the flagellum and can reach speeds of 200 μm s^{-1}. The "flagellar motor" can reverse the direction of the twisting so that cells can themselves reverse, often done in a tumbling motion. The motion conferred upon bacteria by flagella allows them to move to regions of high nutrient concentrations or away

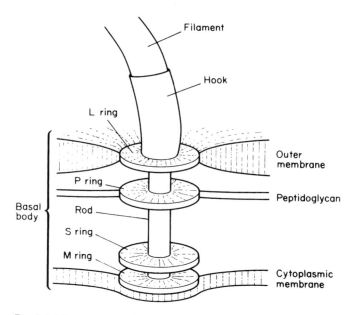

Fig. 2.3 The structure of a bacterial flagellum.

from toxicants (chemotaxis) or in the case of photosynthetic bacteria, away from areas of darkness (phototaxis).

In addition to flagella, two other mechanisms of motility exist. In the spirochaetes, axial fibres similar to flagella run the length of the helical organism but within its flexible outer envelope. These contract to cause flexion of the cell and thus induce propulsion in a twisting movement. The gliding bacteria also have flexible cell walls and when in contact with a solid surface exhibit gliding motility. This is characteristic of many blue-green bacteria and occurs in some groups of photosynthetic bacteria.

(b) Pili and similar structures

Shorter surface appendages known as pili or fimbriae occur on some Gram-negative bacteria. They are organelles made of polypeptide subunits that are similar to, but shorter, thinner and straighter than flagella. Generally covering the entire surface of the cell, they are thought to be involved in adhesion to surfaces, or in the case of pathogens, e.g. *Neisseria gonorrhoeae*, to the host cell. Cells in liquid culture frequently form a thick pellicle of surface growth, apparently held together by an interweaving of the pili. Both the effective increase in individual cell surface area and the flotation effect near the liquid/air interface probably aid in oxygen uptake by the cell. Certain special pili are involved in the transfer of DNA between donor and recipient cell during conjugation and are thus termed sex pili. Some bacteria, e.g. *Caulobacter vibrioides*, produce stalks which are used for attachment; bacteria having such stalks of filaments are prosthecate organisms. Some additionally have holdfasts at the outer tip, regions of adhesive secretions which allow them to attach firmly to a basal substrate.

(c) Extracellular polymers

Many bacteria produce a non-living layer of mucoid gel around the cell, generally termed a microcapsule (<200 μm thick), a capsule (>200 μm thick) or a slime layer (loosely associated material). Capsules are visible under the light microscope with negative staining with Indian ink. Fixed and stained capsules of bacteria such as *Streptococcus pneumoniae* have been shown by electron microscopy to be composed of fine fibrils, which may help to retain the capsule's structural integrity. Extracellular polymers are most commonly polysaccharide in nature, either forming homopolysaccharides containing only one sugar, e.g. dextrans and levans, or heteropolysaccharides containing a mixture of monomers. A minority of bacteria produce capsules of protein, e.g. *Pasturella pestis*, or polypeptides, e.g. *Bacillus anthracis*.

Some species of *Streptococcus*, *Leuconostoc* and *Bacillus* can form polymers externally, without substrate entering the cell, by the secretion of enzymes. However, most capsular and slime material is formed from within the cell by transfer of appropriate monomers by a carrier lipid to the cytoplasmic membrane, thought to be the site of synthesis. The degree of polymer synthesis varies with cultural conditions, with that of polysaccharide capsules increasing in media with a high C/N ratio and conversely that of polypeptide capsules increasing in a low C/N ratio medium. Extracellular polymers appear to be non-essential to the bacteria which produce them, since non-capsulated mutants of the same species remain viable. However, they are thought to aid in protection against bacteriophage attachment and desiccation and in nutrient uptake. Pathogenic organisms are frequently capsulated and as a result exhibit increased resistance to phagocytosis by white blood cells.

One of the most important functions of extracellular polymers with regard to public health engineering is their involvement in the process of flocculation, by which activated sludge bacteria are aggregated into flocs. The polymeric floc structure may also entrap non-capsulated bacteria, and protects the flocculated cells from predation by protozoa.

2.3 Identification of bacteria

Bacteria are identified on the basis of a number of different criteria, including their morphological, physiological and, more recently, their genetic characteristics.

2.3.1 Cell morphology

Some of the morphological features by which bacteria are recognized have already been mentioned, e.g. mode of motility, number, shape and location of motile organs, shape and location of spores, capsule formation and staining properties. Other aspects to be taken into consideration are the size, shape and arrangement of the individual bacterial cells. In general, bacteria are intermediate in size between fungi and viruses, with spherical organisms ranging from 0.2 to 1.5 μm in diameter. The ratio of diameter to length in non-spherical cells varies considerably between species. Typical examples are *Escherichia coli* which is between 0.4 and 0.7 μm in diameter and between 1 and 1.3 μm long, and *Bacillus anthracis* which is 1–1.3 μm in diameter and 3–10 μm long. The volumetric sizes of bacteria range from 5–50 μm^3 for photosynthetic bacteria to 0.1–2 μm^3 for spirochaetes and down to 0.01–0.1 μm^3 for the mycoplasmas. In considering the size of bacteria, it is important to remember that the cell dimensions are likely to vary with age and condition of the culture and also that the various techniques listed to stain and fix the cells prior to measurement under the microscope are likely to cause some degree of distortion. Shrinkage of up to 30% may occur in some electron microscopy preparations. Thus, quoted measurements of cellular size should be regarded as approximate rather than actual values.

The shape and arrangement of bacterial cells can be seen under the light microscope following staining or by phase contrast. There are three basic shapes: spheres, rods and spirals. Variations in both size and shape occur with age and culture conditions, a phenomenon known as pleomorphism. An example is *Acinetobacter* which is rod-shaped when young but forms pairs of spheres when mature. Spherical-shaped bacteria are called cocci and are identified by their tendency to divide irregularly in several planes. The disposition of the daughter cells gives rise to single cells (cocci), pairs of cells (sarcinae), clusters (staphylococci) and chains (streptococci). Rod-shaped or cylindrical bacteria may be arranged singly, in pairs end to end, in bundles, clusters, chains or in Chinese letter form where the individual rods lie perpendicularly to each other. Rod-shaped bacilli are also identified by the shape of the rod end, which may be square, rounded or sharply pointed. Spiral-shaped bacteria have cells which bend or twist to form curves or helices. Curved bacteria include the vibrios, which form a comma shape. These may be arranged singly, in S forms, semi-circles or wavy chains made up of S forms. The spirilla are distinguished by multiple curves or twists and form long helices, of up to 500 μm in the case of *Spirochaeta*. Examples of these shapes are given in Fig. 2.4.

2.3.2 Colony morphology

In addition to cellular morphology, species and even strains of bacteria can be recognized by their colonial morphology, typically demonstrated by

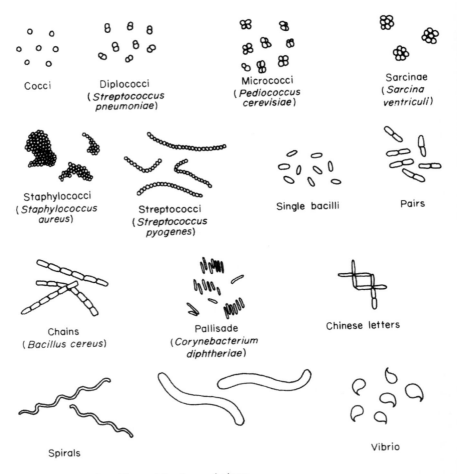

Fig. 2.4 Examples of bacterial cell morphology.

growth on solid agar media and/or the nature of their growth in liquid media such as nutrient broth. Observations are usually made every twenty-four hours and periodically thereafter. The appearance of colonies of the same bacterium may differ depending on a variety of factors such as oxygen and carbon dioxide availability, agar consistency and nutrient diffusion rates, therefore identification has to be made cautiously. Descriptions used for colonies growing on a solid medium include those of the shape, size and various aspects of the structure and appearance. Some examples of the terminology used are given in Table 2.2. Growth in liquid medium can be

Table 2.2 Some terminology used in describing colony morphology

Characteristic	Description
Shape	Circular, irregular, rhizoid (root-like)
Size	Colony diameter in mm after stated period of incubation
Vertical section	Raised, convex, dome-shaped, flat
Surface	Smooth, contoured, granular, ringed, rough, dull, glistening
Edge	Entire, undulate, lobate, crenated, fimbriate, effuse
Colour	Fluorescent, irridescent, opalescent
Opacity	Transparent, translucent, opaque
Consistency	Butyrous, viscid, friable, cohesive, membranous
Emulsifiability	Easy or difficult
Differentiation	Whether differentiated into a central and peripheral portion

described less specifically, but is important for the observation of characteristics such as a flocculent growth habit, which would not be evident from plate cultures. Identification may be facilitated in some cases by the production of coloured pigments. Bacterial pigments are of a range of colours and may be retained within the cell or diffuse into the surrounding medium.

2.3.3 Physiology

(a) Relation to free oxygen

Classification of different bacteria on a physiological basis involves the study of conditions necessary for growth, such as the partial pressure of oxygen and carbon dioxide, the optimum and limiting temperatures and the specific nutrient requirements. In terms of oxygen requirements, four classes of bacteria are recognized. Strict or obligate aerobes grow only in the presence of oxygen, while strict or obligate anaerobes grow only in its absence. The third class, facultative anaerobes, can grow both with and without oxygen although they usually grow less well in its absence. Certain aerobic bacteria can obtain oxygen from nitrates, thus tests for strict anaerobes should be made in largely nitrate-free media. The last class, the microaerophiles, require subatmospheric pressures of oxygen only, and grow well at elevated partial pressures of carbon dioxide. An example is *Sphaerotilus natans*, which proliferates in polluted streams where the oxygen concentration is low.

(b) Response to temperature

Bacteria are similarly divided according to their preferred temperature range. Most bacteria are mesophiles, which grow within the range 10–45°C with an optimum temperature for maximum growth occurring at 30–40°C. An example of a mesophilic bacterium is *E. coli*, which displays cardinal temperatures (i.e. minimum, optimum and maximum) of 10°C, 37°C and 45°C respectively. Thermophilic bacteria grow only poorly at temperatures of 50–60°C. An example is *Bacillus stearothermophilus* (30°C, 55°C, 75°C). Psychrophiles grow in cooler environments, below 20°C, and frequently have optima of 15°C or less. Some mesophilic bacteria with wide temperature tolerance can grow at 0°C; these are termed psychrotrophs. Many marine bacteria and those involved in biological wastewater treatment processes in cold climates are psychrophiles and psychrotrophs.

(c) Metabolism and genetic characteristics

In addition to the general metabolic classifications given in the introduction, some groups of bacteria are distinguished by their requirements for certain specific nutrients, e.g. amino acids, vitamins or trace elements. Tests for the ability of certain Gram-negative organisms to grow in mineral salts medium with ammonium ions and a range of single organic compounds are used for identification while other techniques assess the growth of a particular species on nutrient medium supplemented with, e.g., blood, serum or glucose.

Other physiological properties used in the identification of bacteria are their resistance to extremes of pH, salt concentration and heat, and their response to a range of biochemical tests. These commonly involve determination of the methods of degradation of sugars and other related compounds and the nature of the products formed, together with an evaluation of respiratory functions by means of catalase and oxidase tests and the ability to effect nitrate reduction. Serological tests are employed for bacterial groups such as the salmonellae which are classified on the basis of their antigenic character.

More advanced techniques include analysing the cellular material for its chemical composition. Further refinements in analytical techniques now also permit the determination of the genetic character of specific bacteria. The ratio of the base components of DNA, expressed as the moles per cent of guanine plus cytosine, can be used to show that a pair of apparently similar organisms are not closely related.

2.4 Classification of bacteria

Bacterial classification or taxonomy involves both identification, to distinguish between different genera and species, and subsequently a grouping of similar or related units to give a workable cataloguing system. The classical text of bacterial taxonomy is *Bergey's Manual of Determinative Bacteriology* (9th edn), which divides the bacteria into nineteen parts. The diversity of bacteria, however, precludes a definitive hierarchical taxonomic structure and thus their classification is constantly being amended. Bergey's text is currently being reorganized and published as a *Manual of Systematic Bacteriology*, containing a number of rationalizations and regroupings. As yet this is incomplete, therefore the following is based upon the nineteen-part classification, with emphasis on those groups of particular importance to the public health engineer.

2.4.1 Phototrophic bacteria

Phototrophic bacteria are predominantly aquatic organisms containing coloured pigments for photosynthesis. They include the blue-green bacteria, purple sulphur (Chromatiacae) and non-sulphur (Rhodospirilleae) bacteria and green sulphur (Chlorobiaceae) bacteria. Blue-green bacteria are ubiquitous, fix atmospheric nitrogen and can survive heat and desiccation. They are found in most environments including hot springs and soil and form "blooms" in nutrient-rich lakes, often rendering the water unpotable through toxin production. The purple and green sulphur bacteria are anaerobic photoautotrophs, which use H_2S as an electron donor and store elemental sulphur within or outside the cell. They are found in water or mud where light penetrates the anaerobic zone, usually at depths greater than the blue-green bacteria.

2.4.2 Gliding bacteria

These are aerobic, Gram-negative bacteria characterized by an outer slime layer and the ability to move by gliding. Found in soil and water, many are capable of degrading and assimilating other micro-organisms and some can digest cellulose. One of the two orders is characterized by production of fruiting bodies.

2.4.3 Sheathed bacteria

The sheathed bacteria are rod-shaped organisms which grow in chains surrounded by a gelatinous sheath. The sheaths may attach to solid surfaces by holdfasts or accumulate surface encrustations of iron and manganese oxides. Examples are *Leptothrix*, *Sphaerotilus* and *Streptothrix*, which occur in both clean and polluted water. *Streptothrix* and particularly *Sphaerotilus* filaments are common in activated sludge and are thought to be responsible for sludge bulking and poor settling.

2.4.4 Budding and/or appendaged bacteria

Bacteria which are morphologically identified by appendages such as stalks (e.g. *Caulobacter*) or hyphae (e.g. *Hyphomicrobium*) are included in this group. They are widely distributed in soil and water containing iron (II) salts and are thus occasionally problematic at water treatment works.

2.4.5 Spirochaetes

These are Gram-negative chemotrophs characterized by their flexible helically coiled shape. The group includes free-living, parasitic and pathogenic aerobes and facultative anaerobes. The largest in size, *Spirochaeta*, are free-living aquatic organisms found in fresh or polluted water, sewage or mud, often in anaerobic conditions. The smaller *Treponema* and *Leptospira* include pathogenic species causing syphilis (*T. pallidum*) and leptospirosis or Weil's disease (*L. icterohaemorrhagiae*), the latter being of particular concern to those working with sewage.

2.4.6 Spiral and curved bacteria

Distinguished by their rigidity and the possession of flagella rather than axial filaments, this group contains two Gram-negative genera, *Spirillum* and *Campylobacter*. These are either free-living aquatic aerobic or microaerophilic, e.g. *Sp. anulus*, or anaerobic or microaerophilic parasitic or pathogenic organisms, e.g. *Sp. minor*. *Campylobacter foetus* causes abortion in sheep and cattle. Related genera include *Bdellovibrio* which is parasitic on other bacteria.

2.4.7 Gram-negative aerobic rods and cocci

This group comprises five chemoheterotrophic families distinguished by their metabolic characteristics. The family Pseudomonadaceae includes the genera *Pseudomonas, Xanthomonas, Zoogloea* and *Gluconobacter*. *Pseudomonas* are flagellated rods which are extremely versatile; certain species can use over a hundred different compounds as their carbon and energy source. They are widespread in soil and water and are important in the degradation of organic materials. *Xanthomonas* are strict aerobes found in association with plants, which are noted for their ability to produce copious quantities of extracellular polymer. The genus *Zoogloea* also produce extracellular material but in the form of a gelatinous matrix in which the cells are embedded. *Zoogloea* are commonly found in activated sludge and are one of the primary floc-forming organisms.

The family Azotobacteriaceae includes genera such as *Azotobacter* and *Azomonas* which are free-living aerobes important for their ability to fix atmospheric nitrogen in soils. The family Rhizobiaceae also contains a genus which fixes atmospheric nitrogen in plant root nodules, *Rhizobium*. The two remaining families are the Methylomonadaceae, which utilize only

methane or methanol, and the Halobacteriaceae which grow only in habitats
with an extremely high salt content (>12% NaCl).

2.4.8 Gram-negative facultative rods

This group is one of the most important in terms of public health. It contains
many common organisms found either in water or in the intestinal tract and
includes some important human pathogens which are transmitted through
faecal contamination of water. The enteric bacteria, as the group is known,
are distinguished from other Gram-negative bacteria by their ability to
obtain energy either aerobically or by anaerobic fermentation or respiration.
The group contains two families, one of which, the Enterobacteriaceae,
includes the genera *Escherichia, Edwardsiella, Citrobacter, Salmonella* and
Shigella. The first three are intestinal bacteria and are normally harmless,
but the latter two are pathogenic and in areas where sanitation is poor cause
a range of infections. *Salmonella* species cause gastroenteritis (food poisoning) and enteric fevers, but the most dangerous is *S. typhi* which causes
typhoid. This severe, sometimes fatal epidemic disease of humans occurs
mainly in the Third World countries when untreated or insufficiently treated
water supplies are used. *Shigella* causes human bacillar dysentery.

Other pathogenic bacteria in the family Enterobacteriaceae are included
in the genus *Klebsiella*, species of which cause pneumonia and bronchial
infections, and in the genus *Yersinia*. *Yersinia pestis* causes bubonic plague
(Black Death), which is spread by rat fleas, while *Y. enterocolitica* causes
gastrointestinal disease. The second family, the Fibrionaceae, contains the
species *Vibrio cholerae* which causes another fatal epidemic disease,
cholera, through faecal contamination of water. Vaccines are largely
ineffective against such diseases, therefore they have to be prevented by the
provision of adequate sanitation.

2.4.9 Gram-negative anaerobes

These are strictly anaerobic, non-sporing rods which are found in large
numbers in the human intestine. Genera include *Bacteroides, Fuso-
bacterium* and *Leptotrichia*. One chemoheterotroph within this group,
Desulphovibrio, is found in anaerobic mud under mesophilic or psychrophilic conditions. This organism reduces sulphate to hydrogen sulphide
to obtain energy and can fix atmospheric nitrogen. Sulphate reducing bacteria are involved in the process causing crown corrosion of sewerage pipes.

2.4.10 Gram-negative cocci

The family of Gram-negative cocci, Neisseriaceae, includes four genera:
Neisseria (aerobic or facultatively anaerobic) and *Branhamella, Moraxella*

and *Acinetobacter*. The first three are pathogens or parasites found in the human body; *Neisseria* has two species which are the agents of meningitis and gonorrhea respectively. *Acinetobacter* are nutritionally versatile chemoheterotrophs found in soil and water and have been implicated in biological phosphorus removal in the activated sludge process. Two associated genera in this group are *Paracoccus* and *Lampropedia*, both of which are denitrifiers, reducing nitrate to nitrogen gas to obtain energy.

2.4.11 Gram-negative anaerobic cocci

The family Veillonellaceae includes three genera of strictly anaerobic bacteria. These are parasitic organisms inhabiting the intestinal tracts of humans and animals.

2.4.12 Gram-negative chemolithotrophic bacteria

This group contains bacteria important in the nitrogen and sulphur cycles in soil and water. The nitrifying bacteria, family Nitrobacteraceae, comprises two groups, one of which oxidizes ammonia to nitrite (genus *Nitrosomonas*) and one which oxidizes nitrite to nitrate (genus *Nitrobacter*). Nitrification of ammonia to nitrate followed by denitrification of nitrate to nitrogen gas forms an important aspect of biological wastewater treatment where standards are set for the levels of inorganic nitrogen compounds in effluents.

The sulphur bacteria are aerobes which oxidize sulphur to obtain energy. The group contains genera such as *Thiobacillus*, *Beggiatoa* and *Sulpholobus*. *Beggiatoa* is a gliding filamentous organism sometimes found in bulking sludge. *Sulpholobus* inhabits hot springs of low pH value (2–3). *Thiobacillus* can oxidize hydrogen sulphide or thiosulphate to elemental sulphur, which it accumulates around the cell, then oxidizes elemental sulphur and other sulphur compounds to sulphate. In cultures of *T. thiooxidans*, the sulphate forms sulphuric acid, responsible for concrete sewerage corrosion where hydrogen sulphide is provided by anaerobes.

The iron bacteria, family Siderocapsaceae, are found in water containing iron or manganese; they deposit metallic oxides which characteristically encrust the cell. The metabolic significance of this is as yet unclear.

2.4.13 Methane producing bacteria

These bacteria are also chemolithotrophs, using hydrogen as their energy source and electron donor, but are obligate anaerobes. The family Methanobacteriaceae contains three genera, *Methanobacterium*, *Methanosarcina* and *Methanococcus*. They are found in anaerobic environments such as the bottom mud in lakes and swamps, the intestinal tract of animals and in anaerobic sewage sludge digesters. Their metabolism is discussed further in Chapter 15.

2.4.14 Gram-positive cocci

The three families of Gram-positive cocci are Micrococcaceae, Streptococcaceae and Peptococcaceae. Micrococcaceae include the genera *Micrococcus*, found in soil and fresh water and which employ aerobic respiration; *Staphylococcus*, parasites or pathogens which are facultative anaerobes using respiration or fermentation; and *Planococcus*, marine aerobic organisms. Streptococcaceae metabolize by fermentation either with or without oxygen. The genus *Streptococcus* includes many significant human and animal pathogens, e.g. *S. pneumoniae*, as well as faecal organisms such as *S. faecalis*, which are used as indicators of faecal contamination of water. Also in this family are the lactic acid bacteria, used commercially in the dairy industry. Examples are *S. lactis* and members of the genus *Leuconostoc*.

2.4.15 Endospore-forming rods and cocci

This group contains bacteria in five genera, all of which produce spores and can thus survive adverse conditions for different periods. The two most important genera are *Bacillus* and *Clostridium*, found variously in soil, bottom muds and the intestinal tract. *Bacillus* are aerobes or facultative anaerobes, some of which can fix atmospheric nitrogen and one of which effects denitrification (*B. icheniformis*). *Bacillus anthracis* is the agent of anthrax, fatal to animals and man. *Clostridium* are obligate anaerobes; the genus contains many pathogens. *Clostridium botulinum* contaminates food and produces toxins causing botulism, which can kill. Infection of deep wounds by *C. tetani* causes tetanus and by *C. perfringens* causes gas gangrene.

2.4.16 Gram-positive non-sporing rods

The majority of the group form one family, the Lactobacillaceae. These are included in the association of lactic acid bacteria since they occur largely in milk and milk products. Other genera include the pathogens *Listeria* and *Erysipelothrix* which cause diseases in pigs and cattle.

2.4.17 Actinomycetes and related organisms

The order Actinomycetales contains bacteria which resemble fungi in that they produce a branching, mycelial growth. Most are found in soils, some are pathogens of animals while others, e.g. streptomycetes, are used commercially in antibiotic production. The genus *Nocardia* is associated with the formation of stable foams ("chocolate mousse"), on the surface of activated sludge tanks. Actinomycetes have also been implicated in taste and odour problems in reservoir water. A subgroup of medical significance includes the mycobacteria such as *M. tuberculosis* and *M. leprae*. The other major group is the coryneform bacteria, characterized by irregularly shaped cells. The genus *Corynebacterium* includes human pathogens, e.g. the facultative anaerobe *C. diptheriae*, aerobic pathogens of plants and non-pathogenic species common in soil and water.

2.4.18 Rickettsias

These are obligate intracellular parasites of arthropods such as fleas, ticks and lice, which cause diseases when transmitted to a vertebrate host through biting. Human diseases include epidemic typhus, from louse-borne *Rickettsia prowazekii*.

2.4.19 Mycoplasmas

These organisms have no cell wall. They are chemoheterotrophs with complex nutritional requirements, a factor distinguishing the two principal genera, *Mycoplasma* and *Acholeplasma*. Many are parasites or pathogens of animals or plants. Free-living mycoplasmas have been found in hot springs.

Further reading

Collier, L., Balows, A. and Sussman, M. (eds) (1998) *Topley and Wilson's Microbiology and Microbial Infections*, 9th edn, Vol. 2, *Systematic Bacteriology* (eds A. Balows and B. I. Duerden), Edward Arnold, London.

Hawker, L. E. and Linton, A. H. (eds) (1979) *Micro-organisms – Function, Form and Environment*, 2nd edn, Edward Arnold, London.

Holt, J. G., Krieg, N. R., Sneath, P. H. A., Staley, J. T. and Williams, S. T. (eds) (1994) *Bergey's Manual of Determinative Bacteriology*, 9th edn, Williams and Wilkins, Baltimore.

Krieg, N. R. and Holt, J. G. (eds) (1984) *Bergey's Manual of Systematic Bacteriology*, Vol. 1, Williams and Wilkins, Baltimore.

Stanier, R. Y., Ingraham, J. L., Wheelis, M. L. and Painter, P. R. (1987) *General Microbiology*, 5th edn, Macmillan Education Ltd, London.

Chapter 3

The eucaryotic protists

3.1 Introduction

In Chapter 2 it was stated that the bacteria are the most important group from the public health or environmental engineering standpoint. The microbial world also includes many eucaryotic organisms, however, which fall into three main groups: the algae, fungi and protozoa. These organisms will all be considered together in this chapter. This is not because they share similar characteristics, since the group is very diverse, but because they constitute all of the true micro-organisms with a eucaryotic cell organization.

3.2 Algae

The organisms which constitute this group are essentially plant-like. It includes both macroscopic as well as microscopic forms. The name of the group is derived from the Latin word for sea-wrack, macroscopic algal forms frequently washed onto beaches. Most algae are aquatic organisms and they may inhabit fresh or saline waters. They are usually free-living; however, some aquatic forms have adopted symbiotic relationships with marine invertebrate animals (e.g. corals and sponges). The terrestrial species normally grow in soil or on the bark of trees, but some have established symbiotic relationships with fungi to form lichens. These two-membered natural associations form slowly growing colonies in many inhospitable environments and in particular rock surfaces.

Some 70% of the Earth's surface is covered by water and it is probable that the algae fix more carbon dioxide than all the land plants combined. This may not be so readily apparent if one thinks only of the macroscopic algae which are confined to the shore line or shallow water. However, the overwhelming majority of algae are microscopic, unicellular, floating forms which constitute the phytoplankton. Their density is low and they therefore rarely impart any colour to the oceans. However, the enormous volume of the oceans which they occupy makes them the most abundant of all photosynthetic organisms.

The classification of the algae is based on their cellular properties, the nature of the cell wall, photosynthetic pigments and the arrangement of the flagella in motile cells. On this basis the algae may be divided into seven phyla (see Table 3.1). The algae all possess chlorophyll in their chloroplast and frequently additional secondary pigments which are responsible for a characteristic coloration. There are, however, some algae which have irreversibly lost their chloroplast and these organisms are referred to as leucophytes by botanists and algologists. These organisms can be clearly recognized as non-pigmented algae by their cellular characteristics and vestigial non-pigment chloroplast or even the presence of a pigmented eyespot. It is evident that these organisms have evolved through the loss of their photosynthetic ability at some stage.

This phenomenon can be demonstrated experimentally with certain strains of *Euglena*, which if exposed to ultraviolet radiation or high temperatures produce stable colourless organisms indistinguishable from non-photosynthetic flagellates of the genus *Astasia*. The existence of the leucophytic organisms which occur in several algal groups including the Euglenophyta are indicative of the similarity of some unicellular algae and

Table 3.1 Major divisions of the algae

Phylum	Main characteristics	Examples
Chlorophyta (green algae)	Major pigment chlorophyll; seventeen orders; quite diverse	*Chlamydomonas*
Euglenophyta (euglenoid algae)	Unicellular green or colourless flagellates	*Euglena*
Chrysophyta (yellow-green algae)	Flagellates, major pigments carotenoids	Diatoms
Pyrrhophyta (dinoflagellates)	Major pigments carotenoids, xanthophylls. Unicellular with two flagella. Mainly marine	
Phaeophyta (brown algae)	Fucoxanthin pigments; filamentous	Seaweeds
Rhodophyta (red algae)	Major pigment phycoerythrin; immotile	
Cryptophyta (blue and red flagellates)	Xanthophylls or chlorophylls predominant; unicellular flagellates	*Cryptomonas*

protozoa and of a transitional group. In this text these organisms are treated as algae although they are included in the protozoa by protozoologists. Thus the algal phylum Euglenophyta is equivalent to the superclass Phytomastigophora in the protozoan world.

The overwhelming majority of algal species have the ability to photosynthesize. As a result they have very simple nutritional requirements and in the light may grow in a completely inorganic medium. However, some algae are unable to do this because of specific vitamin requirements, most commonly for vitamin B_{12}. In the natural environment the source of these materials is probably bacteria sharing the same habitat. The ability to undertake photosynthesis does not preclude the use of organic compounds as the principal source of carbon and energy and in many algae a mixed type of metabolism is found and referred to as mixotrophic. Some algae even when growing in the light cannot reduce CO_2 due to defects in the photosynthetic apparatus which have occurred during its evolution. The green alga *Chlamydobotrys* is dependent on the availability of acetate or some other suitable organic compound to fulfil its carbon requirements.

Many algae which in the light use CO_2 can grow well in the dark utilizing a variety of organic compounds. These organisms are able to switch from autotrophic to heterotrophic metabolism, a change which is controlled by the prolonged absence of light. Organisms which are completely enclosed by a cell wall are only able to utilize soluble organic substrates in the dark. However, organisms which are only partially enclosed or have no cell wall can ingest bacteria and other small organisms and thus utilize a phagotrophic form of nutrition which represents a further similarity to the protozoa.

3.2.1 Morphology

The algal cell wall is typically fairly rigid, although highly variable in structure. Cells may also possess sheaths that lie external to their wall. These two structures, either or both of which may be formed from more than one layer, are generally of different composition. The separation between wall and sheath is often difficult to distinguish microscopically; there is also only limited precise information about the chemical composition of these two layers. Sheaths are predominantly formed from complex polysaccharides. Materials which have been identified include cellulose, alginic acids, pectin, fucin, chitin and mucopeptide. Some of these materials are confined to a single group of algae whilst others occur in several groups.

Examples of the range of vegetative morphology are shown in Fig. 3.1. There are six major types of cellular morphology in the algae, and these are listed in Table 3.2, together with some examples. Those algae which are

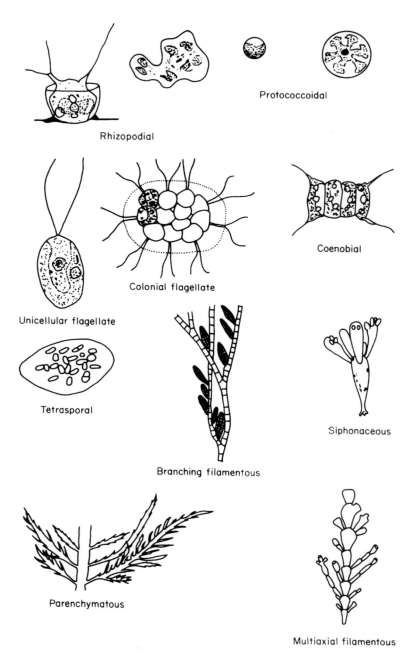

Rhizopodial

Protococcoidal

Unicellular flagellate

Colonial flagellate

Coenobial

Tetrasporal

Branching filamentous

Siphonaceous

Parenchymatous

Multiaxial filamentous

Fig. 3.1 Examples of the range of vegetative morphology in the algae.

Table 3.2 Cellular morphology in the algae

Type	Subdivisions	Examples
Unicellular	Non-motile	*Chlorella*
	Motile by flagella	*Euglena, Chlamydomonas*
	Motile by other means (e.g. mucilage)	Diatoms
Colonies of cells	Motile	*Volvox*
	Non-motile	*Pediastrum*
Filamentous	Unbranched	*Ulothrix*
	Branched	*Ectocarpus*
Complex filaments		The Rhodophyta
Parenchymous		*Fucus*
Coenocytic		*Cladophora*

important in environmental and public health engineering terms are principally the microscopic unicellular organisms (groups 1 and 2 in Table 3.2) and those which are motile by means of flagella regardless of the phyla to which they belong, collectively known as flagellates. Typical examples of this latter group would be *Euglena* and *Chlamydomonas*. Some multicellular algae are also included under this general heading, despite their immotility in the mature state. This is justified because during reproduction they frequently form motile cells or gametes which closely resemble unicellular flagellates and this demonstrates that these organisms are closely related. In addition several phyla also contain unicellular algae which are immotile or motile by means other than the use of flagella and which are of environmental significance. Examples of this group include the desmids which belong to the Chlorophyta and the diatoms which belong to the Chrysophyta and depend on the extrusion of mucilage for their motility. The cell wall of the diatom is reinforced with silica. These siliceous skeletons have accumulated in large fossil deposits called diatomaceous earth or Kieselguhr which is used in various filtration applications.

3.2.2 Reproduction

Asexual reproduction mechanisms exist in all the algae. These include fission, sporulation and the release of multinucleate fragments in multicellular forms. Fission is usually of the binary type and occurs within the cell wall if it is present. In some species the parental cell wall is discarded and a complete new wall is produced, whilst in others, notably the diatoms, the parental cell wall is shared, each of the progeny constructing the missing parts. Multiple fission may also occur as the result of sequential nuclear and cellular divisions producing four or eight or more progeny.

Most non-motile algae produce motile zoospores as the principal means of asexual reproduction. A vegetative protoplast may differentiate into a single zoospore by the development of flagella, or it may subdivide into smaller units, each of which develop a flagellum. The zoospores swim to a new location and germinate to produce a new vegetative organism. In many multicellular algae, vegetative reproduction may occur as the result of the accidental or deliberate release of a small number of cells from the multicellular unit.

Sexual reproduction also occurs widely in most algal groups resulting in progeny which are not genetically identical to their parents unlike the progeny of asexual reproduction mechanisms. The processes are very variable involving fusion of the entire organism in unicellular forms, whilst in multicellular forms, fission of indistinguishable gametes, dissimilar motile gametes or gametes which may be identified as male and female can occur.

The chromosome number of the algae exhibits three types of variation during the life cycle. These allow the process of meiosis (a reduction division) to prevent the genetic material increasing indefinitely by reducing it to the appropriate number of complete sets of chromosomes, i.e. one genome (N) in a haploid vegetative organism or two genomes (2N) in a diploid organism.

3.2.3 Classification

(a) Phylum Chlorophyta (green algae)

This is a large and diverse group of algae, some members of which are found in virtually every algal habitat. Some individual genera and species are ubiquitous. The majority of green algae contain only one chloroplast per cell. There are nine orders or major sub-divisions within the phylum which include the Volvocales, Tetrasporales, Chlorococcales, Chaetophorales, Desmidiales and Zygnemales, which occur in wastewater treatment processes and polluted waters. Unicellular and multicellular organisms are present in this group. A typical unicellular organism is *Chlamydomonas*, a member of the Volvocales. It is widely distributed in pools, ditches and oxidation ponds. The free-swimming vegetative cell is generally spherical or ovoid in shape. A cell wall composed of cellulose is always present. At the anterior end the cell bears two flagella of equal length and a red eyespot. Surrounding the nucleus is a small mass of colourless cytoplasm lying in the depression of a large cup-shaped chloroplast. Clearly visible in the chloroplast is the pyrenoid, a protein body which acts as the centre for the formation of starch which serves as a storage material.

Chlamydomonas reproduces asexually by multiple fission producing two, four or eight zoospores. The vegetative cell first becomes quiescent by retraction of its cilia and then the protoplast divides internally. The daughter cells undergo some enlargement to form a cell wall and a pair of cilia. The parent cell wall breaks down with the release of the daughter cells, which after further enlargement form new vegetative cells. If unfavourable conditions for vegetative growth occur then rather than the daughter cells formed as the result of asexual reproduction being released, a pallmelloid stage is retained in the parent cell wall which is covered by a thick layer of mucilage. Upon the return of suitable conditions for vegetative growth the daughter cells are released. *Chlamydomonas* also reproduces sexually by fusion of similar gametes which are produced by the division of a vegetative cell protoplast into sixteen or thirty-two subunits. The zygote undergoes a reductive nuclear division (meiosis) to produce four zoospores which are released and develop into vegetative cells.

(b) Phylum Euglenophyta

The cell of *Euglena* is also elongated, having a blunt anterior end and a pointed posterior. Its shape is maintained by a flexible pellicle which winds spirally around the body forming an outer membrane. The anterior end has a flask-shaped depression made up of a funnel-like mouth (the cytostome) and a short canal leading into a reservoir, which in turn leads to the contractile vacuole. A stigma or eye spot containing the light sensitive pigment carotene lies to one side of the gullet. *Euglena* has two flagella, one long and one short which arise from the base of the reservoir. The long flagellum extends through the cytostome and acts as a locomotory structure. The second flagellum does not extend beyond the cytostome. Movement is achieved through waves of contraction which spread from the base of the flagellum to its tip. The flagellum moves in a spiral manner, thus as *Euglena* draws itself forward through the water it traces out a helical path. At the same time the cell rotates about its own axis. This method of movement is rapid. However, the organism is also capable of a slower form of locomotion. The pellicle material is wound around the cell below the plasma membrane in strips. Articulation at the interface of these pellicle bands results in the alternating contraction and relaxation of the flexible pellicle. As a result of these changes in body shape slow creeping euglenoid movement occurs.

Normally *Euglena* swims towards the source of light with its long axis parallel to the rays of light. The photoreceptor at the base of the long flagellum working in conjunction with the stigma controls the phototactic response. The cytoplasm contains in addition the more usual organelles including chloroplasts and granules of a starch-like polysaccharide, paramylum. The shape and number of chloroplasts vary with the species. In *Euglena viridis* they take the form of astellate structures radiating from the centre of the cell. In the presence of light the chloroplasts are active and the organism functions as a photoautotroph. When light is absent for an extended period the photosynthetic pigments break down and if soluble organic substrates are available the cell becomes heterotrophic. If light once more becomes available the photosynthetic pigments return and the now green organisms then resume their previous form of metabolism.

(c) Phylum Chrysophyta (brown algae)

There are some two hundred species of golden brown algae which occur mainly in fresh water. Most members are unicellular flagellates lacking a cell wall. They may be free-swimming or attached. Some form colonies. Motile cells may have one, two or occasionally three flagella, of equal or unequal length, attached anteriorly. Asexual reproduction occurs mainly by longitudinal fission. In those species which are not motile zoospores may be produced. It is doubtful if sexual reproduction occurs in this group. However, a characteristic feature is the formation of cysts with a silicified cell wall and having a small plug at one end.

Another group included within this phylum are the diatoms. They constitute an isolated group whose relationships to the other algae are very uncertain. They include over five thousand species of unicellular plants occurring almost universally in fresh and salt water, as well as on damp soil. Diatoms may be either free-floating or attached. Frequently they form slimy brown coatings on mud at the bottom of shallow bodies of water, as well as on sticks, stones, shells, other aquatic plants and other debris. That they were more numerous in geologic times is shown by the great accumulation of diatomaceous earth found in various parts of the world. Although most diatoms are solitary, some form colonies of diverse types, the individuals being held together by a sheath of mucilage. Their colour, usually a golden brown, is due to the presence of chlorophyll in association with an excess of carotenoids, particularly carotene and several brown xanthophyll pigments. Diatoms are distinguished from other algae by their silicified cell wall. This consists of two valves, one overlapping the other like a lid and bottom of a pillbox. The place where the valves overlap is called the girdle. The cell wall is composed mainly of pectin impregnated with a large amount of silica. It is variously marked with numerous fine transverse lines that form regular and elaborate patterns. These make diatoms among the most striking and beautiful objects to be seen under the microscope. In most diatoms the nucleus is suspended in the centre of the cell by slender strands or by a broad transverse band of cytoplasm connected with a thin layer lying next to the cell wall. Embedded in the peripheral layer are one or more plastids that are usually brown, frequently yellow, or rarely green. Reproduction occurs chiefly by fission, the cell always dividing in the plane of valves. The two valves separate and each daughter protoplast forms a new wall on its naked side, the new wall fitting inside the old one. One of the daughter cells is always as large as the parent cell, but the other is smaller.

3.3 Protozoa

The name of this group is derived from the Greek words *protos* and *zoon* which translate to "first animal". There are approximately 32,000 living species in this group of eucaryotic organisms, of which 10,000 species are parasitic. All are unicellular, almost invariably microscopic. However, some species aggregate forming colonies of independent cells. They may be distinguished from other protists (algae and fungi) by their lack of a cell wall. This is not the only animal-like characteristic they share, because in addition all are motile at some stage.

The type of locomotory system utilized by the various groups of protozoa is of primary importance in their classification. Using this criterion four major groupings can be identified. The amoebae move by extending finger-like protrusions of their cells called pseudopodia. Ciliates move by means of the motion of tiny hairs called cilia which surround their cells. The flagellates propel themselves by means of flagella, usually located at the end of their cell. The sporozoa, which are invariably parasitic, are characterized by their lack of locomotory organelles. They move by gliding which is achieved by flexing their bodies.

The phylum Protozoa is divided into three subphyla. These are the Sarcomastigophora, the Ciliophora (ciliates) and the Sporozoa. The Sarcomastigophora is divided into two superclasses, the Mastigophora (flagellates) and the Sarcodina (amoebae).

Protozoa are readily distinguished from fungi by their marked tendency towards subcellular differentiation and the absence of a filamentous organization. The principal groups of protozoa have developed a unique characteristic which is the ingestion using specialized organelles of particulate foods. These organelles allow the protozoa to ingest particles which are too large to diffuse into the cell. Since some of the food of the protozoa are motile microbial cells, many protozoa are hunters and themselves motile. Non-motile protozoa usually retain their flagella, cilia or pseudopodia to collect particulate food. A further consequence of ingestion is the development of complex nutritional requirements, particularly for vitamins and amino acids, which are not widely available in the solution of the environment. This form of feeding determines that most of the protozoa are chemorganotrophs. However, some which can be classified as members of the Sarcomastigophora are able to photosynthesize. Protozoologists and algologists are unable to agree whether these organisms are protozoa or algae. For example *Euglena* spp. which occur in oxidation ponds are designated by protozoologists as members of the group Euglenida and by algologists as members of the Euglenophyta. They are nonetheless the same organism. Thus, whilst the protozoa are readily distinguishable from the fungi they are not so easily separated from microscopic unicellular algae.

The protozoa have effects in two important areas of interest for public health and environmental engineers. They are responsible for a number of water-borne diseases of man in areas lacking a safe water supply. In addition they play an important role in most aerobic biological wastewater treatment processes. More generally they fulfil an important role as a link in the food chains of many communities. For example, in the marine environment zoo-plankton, which are animal-like protozoa, feed on phytoplankton which include plant-like protozoa. The zooplankton in turn become the food for larger marine organisms. The food chain which is of importance in aerobic biological wastewater treatment systems predates such systems by many tens of thousands of years. In many aquatic environments including wet-lands and marshes, saprophytic fungi and bacteria decompose dead organisms and their excretary products and protozoa use the substances produced by the organisms involved in the final stages of the decomposition of this organic matter.

3.3.1 Morphology

The size and shape of the protozoa vary greatly. Some are spherical or oval, others are elongated and some are polymorphic (they have different forms during the different stages of their life cycles). Examples of the range of forms are shown in Fig. 3.2. Some protozoa are as small as 1 μm in diameter whilst some are as large as 2,000 μm (2 mm) and thus visible to the naked

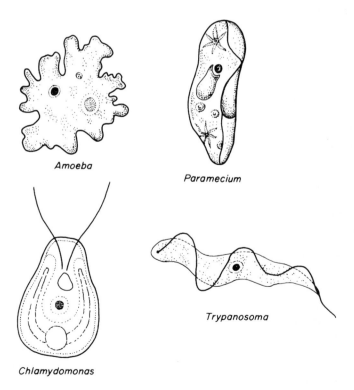

Fig. 3.2 Range of cell morphology of the protozoa.

eye. A typical protozoan cell is enclosed by a cytoplasmic membrane. Most have an outer layer of cytoplasm (the ectoplasm), which is distinguishable from the inner cytoplasm or endoplasm. These are not separated by any structure but contain different components and are of different consistencies. Most organelles are found within the endoplasm. Many protozoans have more than one nucleus throughout the greater part of the life cycle. In the ciliates, one large macronucleus controls the metabolic, growth and regeneration activities, while a small micronucleus controls the reproductive function. The cytoplasmic membrane is surrounded by a pellicle. In some species, e.g. amoebae, this is a thin and diffuse layer whilst in others, e.g. ciliates, it is rigid. The presence of a pellicle rather than a cell wall is one of the major distinguishing characteristics of this group of protists. Many protozoa form loose-fitting coverings, called exoskeletons, to their pellicles to give rigidity to their cells. These structures are called shells or tests and may be reinforced with inorganic substances such as calcium carbonate or silica. There are many other internal structures or organelles present in various members of the protozoa and these are outlined in Table 3.3.

A minimum number of each type of organelle is required by each new individual upon reproduction. There are three mechanisms by which this occurs:

1 replication of each organelle prior to reproduction;
2 resorption of the old organelle and reproduction of a new one for each daughter cell;
3 retention of an organelle by one member of the new generation and reconstruction of that organelle by the other member of that generation.

These mechanisms vary between species.

3.3.2 Reproduction

Asexual reproduction occurs in all the protozoa. The offspring are genetically identical to the parent cell. However, the mechanisms involved vary between species. Flagellates, some amoebae and ciliates commonly reproduce by subdivision of the parental cell in a process called fission. Binary fission gives rise to two progeny and multiple fission gives rise to several daughter cells. These offspring are physically identical to each other and their parent even when young. Suctoreans most commonly reproduce by budding, where one individual gives rise to a succession of smaller individuals.

Sexual reproduction also occurs within some species of protozoa, for example the ciliate *Paramecium*. The process of conjugation occurs between compatible mating types of the same species. As a result of meiosis and nuclear transfer, conjugation produces offspring with a variety of

Table 3.3 Organelles and structures which occur in protozoa (after Krueger *et al.*, 1973)

Internal structure	Description	Occurrence
Food vacuole	Transient, membrane-bounded vesicle formed at the cell surface, inner surface of cytostome, or inner end of cytopharynx to envelop food; digestion occurs as the vacuole migrates through the cytoplasm	In all protozoa that ingest particulate food
Cytopharynx	Passageway through which ingested particles pass from cytostome into a food vacuole; can be permanent or temporary	In most protozoa with cytostomes
Contractile vacuole	Membrane-bounded vesicle, usually fixed at a specific site within the cytoplasm; fluid from the cytoplasm accumulates within the vacuole and is expelled from it to the exterior	Usually amongst freshwater protozoa and of variable occurrence amongst marine and parasitic protozoa
Kinetosome	Also called a basal granule, body or structure, the intracytoplasmic basal portion of a flagellum or cilium	Occurs in the protozoa with flagella or cilia
Surface Muciferous body	Membrane-bounded organelle that lies just within the cell surface; it discharges its contents to the exterior; the contents may be used to construct cyst walls	Occurs in certain flagellates and ciliates
Trichites, trichocysts and haptocysts	Structures similar to and presumably evolved from muciferous bodies; when discharged their contents form a proteinaceous thread that aids in predation and possibly in defence against predators	Occur in certain flagellates and ciliates
Cytostome	A distinct, usually permanent area that can be opened for foods; generally circular when open but sometimes slit-like	Occurs in most ciliates and in some flagellates and amoebae
Oral groove	An indentation of the cell surface leading to the cytostome; aids in trapping food near the cytostome; a peristome is an elaborate oral groove	Occurs in a few flagellates; common in ciliates

Table 3.3 (continued)

Internal structure	Description	Occurrence
Undulating membrane	A flap of the cell cortex with which a flagellum is associated, and sometimes involved in locomotion	Only in certain flagellates
Tentacle	Slender cytoplasmic appendage with prehensile capacity; it is used for capturing prey	Typical of suctoreans; occurs in one group of ciliates
Cytopyge	A site on the cell surface that opens for elimination of undigested contents of food vacuoles; located within the oral groove or near the posterior end of the cell	Occurs in ciliates

genotypes. Self fertilization or autogamy may also occur, but this is obviously less important in bringing about genetic change.

3.3.3 Classification

(a) Subphylum Sarcomastigophora

This group contains the two superclasses Mastigophora and Sarcodina, distinguished largely on the basis of the possession of flagella by the former. The Mastigophora is one of the most diverse groups of protozoa and also probably the oldest group of these organisms. Protozoologists frequently include in this group those algae which are members of the Euglenophyta. In this classification scheme the Mastigophora is subdivided into two groups; the Zoomastigophora includes those organisms accepted as being exclusively protozoan and the Phytomastigophora includes the organisms of the Euglenophyta. This classification has been proposed because these organisms, for example *Euglena*, are able to metabolize both auto- and heterotrophically which clearly indicates that the division between the animal and plant kingdoms is not rigid at this level. In this text the Phytomastigophora are treated as algae and are included in Section 3.2. Colourless flagellates such as the *Trypanosoma* are exclusively heterotrophic and are typical of the Mastigophora. The most important trypanosomes are principally parasites of animals and man. Human diseases resulting from trypanosome infections include Chagas' disease, African sleeping sickness and kala-azar (leishmaniasis). Like many parasites the life cycle of *Trypanosoma* is very complex, the organism passing through four forms. Most trypanosomes are carried from the circulatory system of one host directly to another by blood-sucking arthropods, such as fleas, flies and bed bugs.

The superclass Sarcodina are the least specialized of the protozoa. They are divided into two classes, the Rhizopodea and the Actinopodea, and include probably the best known protozoans, the amoebae. Most of these organisms are free-living although some species such as *Entamoeba histolytica* cause amoebic dysentery in man. These organisms are common in fresh water, moist soil and seawater. Whilst some species of the free-living rhizopods have a shell or external siliceous skeleton the amoebae are almost naked protoplasts. Their cytoplasm is enclosed by a thin, highly elastic membrane called the plasmalemma. The cytoplasm has two layers: a thin transparent outer layer immediately beneath the plasmalemma, called the plasmasol, surrounds an inner granular layer called the plasmagel. Contained within the plasmasol are the various organelles including the nucleus and contractile vacuole.

The plasmagel and the plasmasol play a vital part in the organism's locomotion, which is achieved by protrusions of the cytoplasm called pseudopodia. When in motion the cell has a distinct posterior–anterior orientation. Pseudopodia are only formed at the anterior end of the cell. It is generally accepted that movement is achieved by changes in the viscosity of the cytoplasm which occur as plasmasol is converted to plasmagel. At the point where the pseudopodium is to be formed the plasmagel softens and plasmasol flows to this point of weakness to form the characteristic protrusion. At the tip of the pseudopodium the plasmasol is converted to plasmagel, whilst at the posterior end of the cell, plasmagel is changed into plasmasol and flows forward. This amoeboid movement can only be achieved when the cell is in contact with a solid such as a leaf or stone. When freely suspended in an aqueous medium amoebae exhibit no tendency to form pseudopodia and thus can only drift in the prevailing current.

Pseudopodia are not only organelles of locomotion but also of feeding. A wide variety of small micro-organisms are consumed, including small flagellated and ciliated protozoa. The proximity of a food particle stimulates the production of a pseudopodium which invaginates to form a cup-like depression which engulfs the prey. When finally enclosed the prey is taken into the cell, forming a phagosome. This process is described as phagocytosis. Enzymes are released into the phagosome and the prey digested and as the phagosome moves through the cytoplasm the soluble food products are absorbed. The undigested residues are passed to the cell surface and released. Amoebae also exhibit pinocytosis or "cell drinking". In addition to taking in water they may also obtain soluble nutrients by this means.

(b) Subphylum Ciliophora

The ciliates exhibit the highest degree of subcellular differentiation found in the protozoa. The cilia (small hair-like protrusions from the cell) are responsible for the rapid movement of these organisms through water. When located around the oral groove they also assist in feeding. Most ciliates are free-living. However, *Balantidium coli* is a parasite of humans causing bloody diarrhoea. The free-living forms are selective in their feeding, confining themselves to microbes of one genus or species. Generally, under a given set of conditions each species reaches a characteristic size which may vary from a few micrometres up to 2 mm. Size, shape and the extent of organelle development are influenced by environmental factors, especially by the sort of food available.

Paramecium, which is typical of the free-living members of the group, inhabits freshwater and sewage. It has an elongated cell body with a blunt anterior end and a tapered posterior. The cytoplasm is surrounded by a semi-rigid pellicle through which the cilia protrude in longitudinal rows. From the anterior end of the cell a shallow ciliated oral groove extends towards the posterior end and tapers to form a gullet at the end of which is the mouth or cytostome, which opens into the endoplasm. The organism is also equipped with a cytoproct (cell anus), for the excretion of solid wastes. The endoplasm also contains food vacuoles and two contractile vacuoles for osmoregulation. Each cell has two nucleii, a macronucleus and a micronucleus.

Whilst individual species of *Paramecium* are highly selective in their feeding, the genus as a whole will consume a wide range of micro-organisms.

Movement is achieved by the beating of the cilia which is co-ordinated in a regular manner called metachronal rhythm. The cell is able to move rapidly in a straight line whilst rotating about its longitudinal axis. If the path of the organism is obstructed or it encounters unfavourable environmental conditions, the beat of the cilia is reversed, the organism retreats a short distance, changes direction and moves forward again.

(c) Subphylum Sporozoa

All the sporozoa are parasitic in one or more animal species. The group is large and varied, having a simple cell organization. The adult forms have no obvious organelles for location. However, it is probable that all are motile by gliding at some stage in the complex life cycles characteristic of parasites. They are unable to engulf solid particles and feed on host cells or body fluids. All members of the group produce spores, which give the group its name. These spores are usually the infective stage. Each spore releases eight sporozoites upon germination and these are usually motile. Sporozoans occur as parasites of both vertebrates and invertebrates but their occurrence as endoparasites of man is of the utmost importance. A well known example is *Plasmodium*, the malarial parasite (see Chapter 19).

3.4 Fungi

Few members of the group are readily visible and when they are it is their fruiting body which is seen rather than the vegetative body of the organism. Perhaps the most familiar is the mushroom, whose Greek name *mykes* gives rise to the term applied to the scientific study of the fungi, mycology.

The fungi have a highly distinctive biological organization. Although some aquatic fungi and the yeasts are unicellular they are readily distinguishable from bacteria by their large cells and membrane-bound nuclei. Some aquatic fungi do show resemblances to flagellate protozoa. Fungi occupy a wide variety of habitats including the sea and fresh waters. However, the majority occupy moist habitats on land and are abundant in soil. At least 100,000 species are known. Although some produce macroscopic fruiting bodies (e.g. mushrooms) the overwhelming majority are microscopic.

The fungi are heterotrophic organisms requiring organic materials for their nutrition. Those which feed on dead organic materials are described as saprophytic. Saprophytes bring about the decomposition of plant and animal remains and in doing so release simpler chemical substances into the environment. In soil this is of vital significance in maintaining fertility by recycling essential plant nutrients. Fungi have been employed by man for many hundreds of years for the production of wine, beer, some cheeses and the leavening of dough for the making of bread. More recently fungi have found new industrial applications where large-scale fermentations permit the production of antibiotics (e.g. penicillin) and other biochemicals. They also play a significant role in biological wastewater treatment and the composting of municipal refuse. However, the effects of fungi are not all beneficial. They are responsible for the deterioration or rotting of many products made from natural materials. "Sewage fungus" is an unsightly indicator of pollution. There are also parasitic fungi of plants, animals and man.

The somatic structures, that is the vegetative parts of the organism as opposed to the reproductive part, exhibit little differentiation and no division of labour. Thus, with few exceptions the somatic parts of one fungus closely resemble those of another. Reproductive structures exhibit a variety of forms and provide the basis for classification. Very few fungi can be classified if the reproductive structures are not available. These organisms are termed Fungi Imperfecti and are placed for convenience in a single group called the Deuteromycotina.

The cytoplasm contains the organelles characteristic of eucaryotic organisms including mitochondria, ribosomes and an extensive endoplasmic reticulum. Vacuoles containing storage materials such as glycogen, lipids and volutin are also present. In a unicellular fungus such as the yeast

Saccharomyces spp. the protoplast is enclosed in a semi-permeable membrane, the plasma membrane, which is contained within a rigid cell wall. In filamentous species the protoplasm is concentrated in the tips of the young growing hyphae. The older hyphae are usually metabolically inactive and contain large vacuoles in their cytoplasm. The fungi all lack chlorophyll and are heterotrophic. A mycelium normally develops from the germination of a single reproductive cell or spore. Germination initially results in the production of a single long hypha which subsequently branches and ramifies to form a mass of hyphae which constitutes the mycelium.

3.4.1 Morphology

Although the yeasts provide an example of a unicellular fungus most fungi are filamentous. The body or thallus of a fungus contains two parts. These are the vegetative mycelium and the reproductive structure containing the spores. Examples of the structures of these are shown in Fig. 3.3. The mycelium is composed of a complex mass of filaments or hyphae. The hyphae are typically 5–10 μm wide but may vary from 0.5 μm to 1.00 mm according to the species. The hyphae have walls which are composed of cellulose or chitin or both. A common cytoplasm exists throughout the hyphae. As a result of this, fungi exhibit a unique form of cellular organization which may be divided into three types:

1 coenocytic, where the hypha contains a mass of multinucleate cytoplasm. This condition may also be described as aseptate;
2 septate with uninucleate protoplasts, where the hypha is divided by cross-walls or septa, each compartment containing a single nucleus;
3 septate with multinucleate protoplasts between the septa.

In septate species there is a central pore in the septum connecting the cytoplasm of neighbouring cells and permitting the migration of both cytoplasm and nuclei. Growth occurs by apical elongation of the hyphae but most parts of the organism are potentially capable of growth.

3.4.2 Reproduction

Fungi reproduce by a variety of means either asexually by fission (genetic division by mitosis) involving either budding or spore formation or sexually by fusion (genetic division by meiosis) of the nuclei of two parent cells. In fusion spores are formed by the division of a cell to produce two daughter cells, or in budding by an outgrowth from the cell which results in the

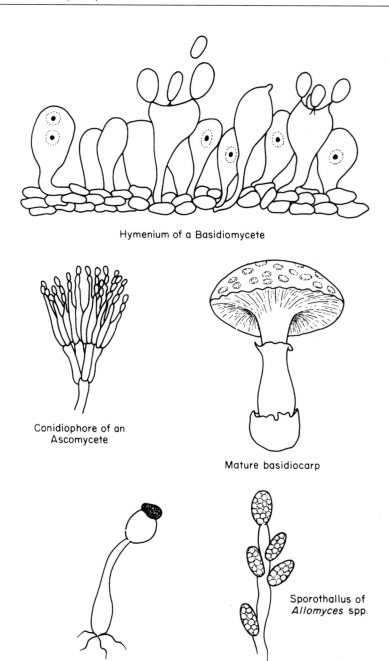

Hymenium of a Basidiomycete

Conidiophore of an
Ascomycete

Mature basidiocarp

Sporothallus of
Allomyces spp.

Fig. 3.3 Examples of fungal morphology.

formation of two cells. Asexual reproduction results in progeny (offspring) which are identical to their parent. They can be produced in large numbers and they achieve the dissemination of the species. There are several types of asexual spores:

1   Conidiospores or conidia, formed at the tip or side of a hypha. If the spores are single-celled they are said to be microconidia, if multi-celled they are termed macroconidia.
2   Sporangiospores, single-celled and formed inside a structure called a sporangium. Aplanospores are non-mobile sporangiospores. Zoospores possess flagella and are mobile sporangiospores.
3   Oidia or arthrospores, formed by the breakage of hyphae into single-cell portions.
4   Chlamydospores, thick-walled single-cell spores which are resistant to adverse conditions. They are formed from vegetative hyphae.
5   Blastospores, the buds formed by yeast cells.

Sexual spores are produced less frequently and in smaller numbers than asexual spores. They are formed only under certain conditions, most usually when the mycelium is in an inhospitable environment. The progeny produced by sexual spores are not identical to their parent. Sexual spores are of several types. They form the basis of classification of the fungi.

3.4.3 Classification

There are three major groups of fungi, the Phycomycetes, Ascomycetes and Basidiomycetes, together with a fourth group, the Fungi Imperfecti, which contains species whose correct classification is unknown, and which, therefore, is a provisional group only.

(a) Phycomycetes

This is a primitive group, containing both aquatic and terrestrial types. The most primitive aquatic types, the chytrids, consist of a sac, called a sporangium, anchored via several fine branched threads, called rhizoids, to a solid surface. During development, the sporangium divides internally to produce many zoospores which are eventually released as motile, flagellated cells which will settle elsewhere and develop into a new sporangium.

The aquatic Phycomycetes are quite varied in terms of their life cycles. In contrast to the chytrids, some aquatic Phycomycetes have alternative haploid and diploid generations. The diploid generation, called a sporophyte, produces haploid spores by a meiotic division. The spores develop into a haploid organism, called the gametophyte, which in turn produces motile male and female haploid spores (gametes). These fuse to produce a diploid zygote, which develops once again into a sporophyte. In addition to the haploid spores, a sporophyte also produces motile diploid spores which develop directly into new sporophytes.

Terrestrial Phycomycetes also reproduce both sexually and asexually, but unlike the aquatic organisms, the spores are not motile. The asexual spores are dispersed aerially, whereas sexual reproduction depends on actual contact between the hyphae of different organisms.

Despite the variations in life cycles of the Phycomycetes, they are distinguishable from all other fungi on the basis of two characteristics. Firstly, their asexual spores are always produced inside a sac-like structure. Secondly, the mycelium is always non-septate, i.e. it has no crosswalls.

(b) Ascomycetes and Basidiomycetes

These higher fungi produce haploid sexual spores borne in or on structures known as asci or basidia. After release, the spores of Ascomycetes germinate to produce mating structures which conjugate to form hyphae containing many haploid nuclei of two mating types. At the tip of the hypha two haploid nuclei fuse to form a diploid nucleus which undergoes a meiotic division followed by a mitotic division to produce the ascus containing eight haploid ascospores.

In the Basidiomycetes, sexual reproduction is similar, but instead of the formation of mating structures, binucleate cells are formed by the fusion of adjacent hyphae. These binucleate cells are ultimately borne on the characteristic fruiting body, the mushroom. Here, the nuclei fuse to form a diploid nucleus, which undergoes one meiotic division to produce a cell containing four haploid nuclei. Each of these develops into a basidiospore which is released from the fruiting body.

The Ascomycetes and Basidiomycetes also reproduce asexually by the formation of conidia, simple spores released from the tips of the haploid hyphae.

(c) Fungi Imperfecti

There are a number of higher fungi which cannot be placed in the Ascomycetes or Basidiomycetes because the sexual stages of their life cycles have never been observed. Some fungi may actually be incapable of sexual reproduction while others may not undergo sexual reproduction in isolation, requiring the presence of organisms of a different mating type. Organisms of this type are placed in the Fungi Imperfecti. However, if subsequently a sexual stage is discovered, the organism is reclassified as an Ascomycete or Basidiomycete, as appropriate.

References and further reading

Benson-Evans, K. and Williams, P. F. (1975) Algae and bryophytes, in *Ecological Aspects of Used Water Treatment* (eds. C. R. Curds and H. A. Hawkes), Vol. 1, Academic Press, London, pp. 153–202.

Curds, C. R. (1975) Protozoa, in *Ecological Aspects of Used Water Treatment* (eds C. R. Curds and H. A. Hawkes), Vol. 1, Academic Press, London, pp. 203–268.

Heritage, J., Evan, E. G. V. and Killington, R. A. (1996) *Introductory Microbiology*, Cambridge University Press, Cambridge.

Ingold, C. T. and Hudson, H. J. (1993) *The Biology of the Fungi*, 6th edn, Chapman & Hall, London.

Krueger, R. G., Gillham, N. W. and Coggin, J. H. (1973) *Introduction to Microbiology*, The Macmillan Company, New York.

Kumar, H. D. and Singh, H. N. (1979) *A Textbook on Algae*, The Macmillan Press Ltd, London.

Singleton, P. (1997) *Bacteria in Biology, Biotechnology and Medicine*, John Wiley & Sons, Chichester.

Tomlinson, T. G. and Williams, I. L. (1975) Fungi, in *Ecological Aspects of Used Water Treatment* (eds C. R. Curds and H. A. Hawkes), Vol. 1, Academic Press, London, pp. 93–152.

Chapter 4

Viruses and the metazoa

4.1 Introduction

The *Dictionary of Biology* of Thain and Hickman (1994) defines a micro-organism as a "microscopically small organism; unicellular plant, animal or bacterium". This definition would appear to exclude the metazoan (i.e. multicellular) animals and the viruses, which are neither plant, animal or bacterium. These two groups are clearly of interest to public health and environmental engineers, however, and for that reason are both considered in this chapter, despite the fact that they constitute the two extreme ends of the spectrum of life forms with which this book is concerned. The viruses are the smallest, simplest forms, so simple that many biologists do not consider them to be true organisms, while the metazoans are the most complex, including "all animals commonly recognised as animals, including man" (Thain and Hickman, 1994).

4.2 Viruses

The simplest possible form of a virus is merely an RNA molecule. Indeed, a number of plant diseases are caused by such entities, termed viroids because they are much smaller than the majority of viruses, which consist essentially of protein and nucleic acid in close association. A fundamental definition of a virus is that it is an entity which carries the information necessary for its replication, but does not possess the machinery for such replication. All viruses are therefore obligately parasitic, and rely upon the machinery of the infected host cell, which is "borrowed" and reprogrammed for the replication of viral nucleic acid, together with any structural proteins which may be associated with the mature extracellular form of the virus. Viruses are unable to multiply outside the host cell. As a result of their interference with the normal function of the cell, many viruses are responsible for disease.

Viruses are very much simpler and smaller than bacteria, and for this reason were unknown until the 1890s, when Ivanowski, a Russian scientist, found that tobacco mosaic disease was caused by an agent which could pass hrough a microporous filter designed for the retention of bacteria. These ɪgents were called "filterable viruses", later simplified to "viruses". Following Ivanowski's discovery, a variety of viruses responsible for dis-ɛases in plants and animals was discovered, and in 1915 the first virus para-ɟitic on a bacterium was found. Details of the morphology and biochemistry ɔf viruses were not elucidated until much later, because of the difficulties ɔosed by their extremely small size and their inability to be cultured except ɪn the host cell. Viruses typically range in size from 5 to 10 nm in diameter ɪnd up to 800 nm in length for some of the long thin ones.

4.2.1 Morphology

The protein component of most viruses is present as a coat which surrounds ɑ core of nucleic acid. The protein coat is usually constructed from many identical subunits, often several hundred of them, called capsomers. The entire structure is called a capsid and the fully assembled mature viral par-ticle is termed a virion. It shares few of the structural or functional charac-teristics of the surface layers of procaryotic or eucaryotic cells; for this reason, viruses are often termed acaryotic.

The nucleic acid may be DNA or RNA, single- or double-stranded, but in ɑny one virus is only of one type. In some of the smaller viruses, there is only sufficient nucleic acid to carry about ten genes (each gene carrying the information for the synthesis of one protein), although large viruses may contain the equivalent of several hundred genes. Clearly, in comparison with a bacterium, say, having a gene complement of 10,000 or more, viruses ɑre genetically quite simple.

There are three major types of virus which can be distinguished on the basis of the structural arrangement of the capsid. These are shown diagram-matically in Fig. 4.1.

(a) Helical viruses

Under the electron microscope, many helical viruses appear like hollow rods, about ten times longer than they are wide. Tobacco mosaic virus is an example of this type. The capsid consists of many identical protein subunits arranged regularly in a helix, with the nucleic acid embedded in a spiral groove. Some viruses are arranged in irregular helices, coiled up within a membranous envelope. The envelope is sometimes derived from the cell membrane of the host.

(b) Polyhedral viruses

In these viruses the capsid is a regular multi-faceted hollow structure. This is frequently an icosahedron, which has twenty equilaterally triangular

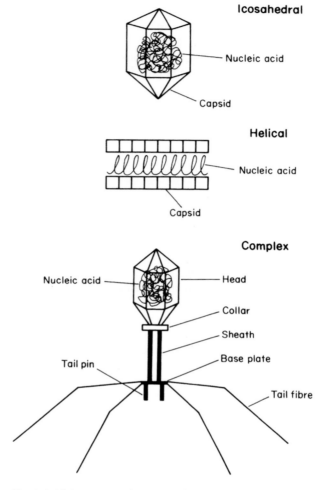

Fig. 4. I Major structural arrangements in viruses.

faces. Inside this the nucleic acid is packed in an unknown configuration. Polyhedral viruses can be enveloped (e.g. poliomyelitis) or naked (e.g. the papovaviruses, which produce tumours in animals).

(c) Complex viruses

Many of the bacterial viruses, called bacteriophages or just phages, consist of a polyhedral head connected to a helical tail, although the nucleic acid is only found in the head, the remainder of the particle being associated with the mechanism of entry into the host cell. The helical tail is contractile. Upon contact with the host cell by the tail plate which is held in position by the tail fibres, the tail sheath contracts and the cell wall and membrane are punctured by the inner tail core through which the nucleic acid is injected into the inner cell.

4.2.2 Reproduction

Since viruses can only reproduce inside the host cell, they exist in two fundamental states, as intracellular entities and as extracellular particles. They therefore have a life cycle which consists of entry to the cell, reproduction and release. This life cycle has two important variants, termed the lytic cycle, which is followed by virulent types often ultimately killing the host cell, and lysogeny, which is the state entered into by avirulent types.

(a) The lytic cycle

The first stage in the lytic cycle is penetration of the host cell. Animal cells, because they have no cell walls, are susceptible to virus penetration simply by adsorption and phagocytosis. In plants, the virus penetrates the cell wall either because it is mechanically damaged or via a vector, such as an insect. Bacteriophages with a complex morphology actually inject the nucleic acid into the cell, the capsid remaining outside.

Penetration is followed by the synthesis by the cell of proteins coded for by the viral nucleic acid together with replication of the nucleic acid itself. The products of these two processes are numerous molecules of viral nucleic acid and numerous protein capsomers. These are assembled together and released from the cell, often by extrusion through the membrane, and sometimes via lysis and hence complete destruction of the cell.

The lytic cycle of phages in coliform bacteria can be quite rapid. Following inoculation of a bacterial culture with the virus, a lag phase is observed, during which no viruses can be recovered from the culture. This represents the time during which the nucleic acid is replicated and viral proteins synthesized. The lag is known as the latent or eclipse phase and lasts for as little as five to ten minutes. It is followed by the rapid release of mature virions into the medium during the next ten minutes or so in what is called the rise period. Following this, a plateau is reached, termed the burst size, which corresponds to the number of virions released per cell. Each cell is capable of producing several hundred virions.

(b) Lysogeny

Many viruses are capable of an alternative interaction with the host in which the lytic cycle effectively proceeds to the point where the viral nucleic acid has entered the cell and goes no further. The viral nucleic acid retains its integrity, however, because it is replicated along with the host's nucleic acid and passed on to the next generation. The viral nucleic acid present in such a state is called provirus. Viruses which can form provirus are termed temperature viruses and the infected cells are said to be lysogenic. The lysogenic cells which carry the viral nucleic acid possess the potential to revert to the lytic cycle. Thus, the viral genes are not permanently inactive and will occasionally result in the formation and release of new virions from a previously lysogenic cell. Maintenance of the lysogenic state is due to the early production of a repressor molecule, coded for by the viral nucleic acid, which prevents further replication.

4.2.3 Classification

There is no single, consistent system for the classification of viruses. The most commonly used systems, unlike systems for the classification of other micro-organisms, do not attempt groupings on the basis of evolutionary relationships, but are based largely on structural and gross chemical characteristics. The properties involved are the type of nucleic acid (DNA or RNA), capsid geometry, capsid size and presence or absence of an envelope. An example of a grouping based entirely on virion structure is the picornaviruses, which include all small RNA viruses. Many other viruses, however, are classified simply on the basis of their observed effects on the host, either general (e.g. the pox virus) or specific (e.g. tobacco mosaic virus).

4.3 Metazoa

The metazoan animals of interest range from the Rotifera, through the round and segmented worms to the insects, spiders and mites. These organisms are generally of interest because they are found associated with percolating filters. It is obvious that the insects and higher organisms are not strictly micro-organisms, and therefore little attention will be given to these here. Generally, the contribution of these organisms to the efficiency of percolating filter treatment decreases as they increase in physical size.

4.3.1 Rotifera

This group constitutes one of the simplest of the multicellular invertebrate animals. They have been referred to colloquially as "wheel animals" because of the rotatory appearance of the ring of cilia called the corona which surrounds the mouth and which sweeps into it bacteria and other particulate organic matter.

These organisms are typically 50–250 μm in length. They were, at one time, confused with the protozoa and fulfil a similar role in the activated sludge process, consuming free-swimming bacteria and floc particles, thus contributing to the clarity of the effluent. Although some of the rotifers can swim, many are associated with a solid substratum and move by a creeping motion.

There are two orders of Rotifera containing genera associated with waste-water treatment. Despite their generally widespread occurrence and the fact that their presence is generally considered to be desirable, there is little quantitative information on their role in wastewater treatment.

4.3.2 Nematoda

The nematodes are small, unsegmented round worms, cylindrical in shape and typically 0.5–3 mm in length (i.e. about ten times the size of rotifers). They are the most abundant metazoan group in percolating filters. They are particularly suited to zones of moderate pollution (meso-saprobic zones).

Some nematodes feed on bacteria and others consume metazoans, including other nematodes. Their role as grazing animals in percolating filters is in the control of film accumulation. Although they are found in the activated sludge process their contribution to the efficiency of the process is minimal.

4.3.3 Annelida, Insecta and higher organisms

Members of the phylum Annelida (segmented worms) are almost unknown in the activated sludge process. They are frequently found in percolating filters, however, where they graze upon the film.

Insects are also found in percolating filters. Learner (1975) surveyed percolating filters at forty-eight sewage treatment works in the UK and found 186 different species. The main representatives were the Coleoptera (beetles) and especially Diptera (flies).

Diptera are frequently the cause of nuisance in the vicinity of filters. The larvae feed on the biological film and, after maturation, emerge from the filter as adult flies. *Sylvicola* and *Psychoda* spp. are common nuisance flies.

Although the insects are macroscopic, highly developed animals, they are of significance to the microbiology of percolating filters because of their effect on the biological films. Even larger organisms than the insects will occasionally be found in percolating filters, but these are mainly chance occurrences and have virtually no significance to the treatment process.

References and further reading

Doohan, M. (1975) Rotifera, in *Ecological Aspects of Used Water Treatment* (eds C. R. Curds and H. A. Hawkes), Vol. 1, Academic Press, London, pp. 289–304.

Fraenkel-Conrat, H. and Kimball, P. C. (1982) *Virology*, Prentice-Hall Inc., Englewood Cliffs, N.J.

Learner, M. A. (1975) Insecta, in *Ecological Aspects of Used Water Treatment* (eds C. R. Curds and H. A. Hawkes), Vol. 1, Academic Press, London, pp. 337–374.

Primrose, S. B. (1994) *Introduction to Modern Virology*, 4th edn, Blackwell Scientific Publications, Oxford.

Scheimer, F. (1975) Nematoda, in *Ecological Aspects of Used Water Treatment* (eds C. R. Curds and H. A. Hawkes), Vol. 1, Academic Press, London, pp. 269–288.

Solbe, J. F. de L. G. (1975) Annelida, in *Ecological Aspects of Used Water Treatment* (eds C. R. Curds and H. A. Hawkes), Vol. 1, Academic Press, London, pp. 305–335.

Thain, M. and Hickman, M. (1994) *The Penguin Dictionary of Biology*, Penguin Books Ltd, Harmondsworth.

Chapter 5

Laboratory techniques

5.1 Introduction

The detailed study of micro-organisms is complicated by the fact that their generally small size makes the manipulation of individual cells almost impossible in routine applications. Moreover, their ubiquity and the versatility of many groups mean that in many cases a given sample will contain very many different types, the differentiation of which is difficult without recourse to techniques for the selective isolation and cultivation of the species or groups of interest in the laboratory. The bacteria and other unicellular organisms are therefore generally studied at the population level, where large numbers of cells of the same species are generated in the form of pure cultures under defined conditions.

5.2 Safety precautions in microbiological laboratories

When working in a laboratory which handles cultures of micro-organisms or one which merely receives contaminated samples, all cultures, samples and materials with which contaminated samples or cultures have come into contact should be treated with caution due to the possible presence of pathogens. Steps should therefore be taken to avoid contamination of the skin and clothing, and to prevent ingestion or inhalation of contaminated materials. Inhalation of airborne organisms is a frequently underestimated risk, and can occur if aerosols (small suspended droplets) are dispersed from the breakage of glass culture vessels or by the excessive agitation of contaminated liquids in open containers.

Although the general incidence of pathogens in the working environment may be quite low it should be remembered that an integral part of many laboratory techniques is the intentional promotion of optimum growth of organisms which is rapid and prolific compared to their normal growth in the natural environment. The slightest contamination can therefore lead to rapid and unintentional proliferation of pathogens unless certain precautions are adhered to.

A mandatory precaution is the use of aseptic techniques at all times. This involves the use of apparatus and materials which are initially rendered free of living organisms and which, when in use, are manipulated in a manner which excludes the possibility of chance contamination from external sources.

5.3 Aseptic techniques

5.3.1 Sterilization

Sterilization is the process of removing or destroying all living organisms associated with a particular material. After treatment, provided that the material remains unexposed to external sources of contamination, sterilization should be effective permanently. Sterilization can be achieved by exposing the material in question to lethal physical or chemical agents or by the selective removal of micro-organisms from liquids by filtration.

(a) Heat

Heat is the most commonly used agent for sterilization. Laboratory glassware, pipettes, metal objects and other heat-resistant items can be sterilized in a temperature-controlled oven at 170°C for 2 h. As with all sterilization methods, ample provision for penetration of the lethal agent has to be made. If the oven is fairly well filled or if glassware is closely packed in metal containers, for example, a longer period, perhaps 3 h, should be used to ensure sufficient heating throughout. A hot air oven is obviously unsuitable for liquids and many heat-labile materials including plastics. Sterilization can be achieved by moist heat, which is effective at lower temperatures. Boiling of aqueous solutions will kill the majority of vegetative organisms present, but this is ineffective for bacterial endospores which can germinate and grow once the boiled medium has cooled. Heating to 100°C for 30 min on each of three successive days is necessary to kill bacterial spores which germinate in the periods between heating cycles.

The most efficient method of sterilizing liquid and solid culture media and other materials which are not very heat-labile is by autoclaving. In its simplest form an autoclave is similar in principle to a domestic pressure cooker. Heating under pressure in an atmosphere of saturated steam to a temperature of 121°C for at least 15 min (longer if relatively large volumes of liquid are to be treated) is sufficient to kill both vegetative forms and spores. This temperature is attained at a pressure of 103.5 kPa (15 lb in^{-2}) only if the autoclave is filled with saturated steam. The presence of air reduces both the temperature and the degree of heat penetration.

(b) Filtration

This method is effective for liquid culture media or solutions containing heat-labile substances. Filters can be made from earthenware or asbestos, but these have largely been replaced by modern membrane filters made from cellulose acetate or similar materials. Filters with a pore size of 0.45 μm have been used to retain bacterial cells, but a few smaller species of bacteria are only removed completely by 0.22 μm filters. Membrane filters of these sizes will not remove viruses from a suspension.

The filter is mounted on a porous support which forms the bottom half of the filtration unit. This is clamped to an upper reservoir and the assembly is connected to a receiving flask with a side arm plugged with non-absorbent cotton wool with the clamp only partially tightened to avoid breakage due to the differential expansion. The apparatus is wrapped in greaseproof paper, or similar material, and autoclaved. After cooling the clamp is tightened and the solution to be sterilized is drawn through the filter by suction applied to the side arm (see Fig. 5.1).

(c) Other methods

Irradiation with ultraviolet light or X-rays can be used for sterilization. The former is limited in its application because of lack of penetration; for

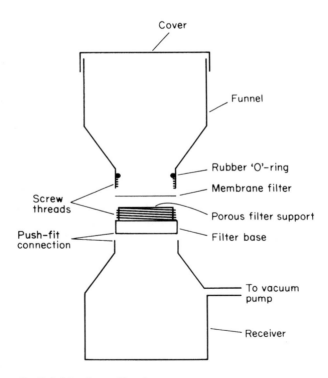

Fig. 5.1 Membrane filtration apparatus.

example it cannot pass through glass, and penetrates water to only a limited extent. Some commercially available items are sterilized by X-rays during manufacture but the method is rarely used routinely in the laboratory. Ethylene oxide, a gas which is readily soluble in water, can be used to sterilize glassware etc., but its highly toxic nature and the possibility of residuals remaining in the treated item limit its application.

5.3.2 Disinfection

A disinfectant is distinct from a sterilizing agent in that it is primarily used to inactivate potentially infectious micro-organisms. Disinfectant solutions in common use may be based on phenolic compounds or hypochlorite. They are not generally used for preparative sterilization. In the laboratory they are mainly used for swabbing benches and in containers into which used pipettes and microscope slides are discarded.

5.3.3 Inoculation and aseptic transfer

Since work in most laboratories is based on studies of pure cultures, care must be taken to avoid contamination during inoculation or transfer. Some laboratories are equipped with laminar flow hoods which provide an atmosphere of filtered air in the working area, but many are not. Contamination is avoided by working on a clean bench in an area which is draught-free and exposing cultures and sterile containers to the atmosphere for as short a time as possible. Operations are carried out in the vicinity of a Bunsen burner flame, which is used to decontaminate surfaces which are briefly exposed.

The transfer of bacterial growth from solid media is normally undertaken with a platinum or nichrome wire formed into a loop of approximately 5 mm in diameter at one end and inserted into a wire holder at the other. A wire loop can also be used to transfer small drops of liquid culture media. The loop is sterilized by holding in the flame of a Bunsen burner and heating to redness momentarily. The loop cools quite quickly. The loop should be sterilized both before and after use, when it should be introduced slowly into the flame to prevent spattering of any material adhering to it.

The transfer of larger measured volumes of liquid is undertaken with sterile pipettes plugged at the top with non-absorbent cotton wool. Pipettes are kept wrapped or in sterile canisters until required, and the barrels may be passed briefly through the Bunsen flame immediately prior to use.

When transferring material the cotton wool plug of the test tube or culture flask or the screw cap of the bottle is removed and the mouth of the container is flamed briefly before and after making the transfer. The internal surfaces of stoppers or caps may be flamed briefly before replacing them. These operations should be performed close to the Bunsen flame with the open vessels inclined towards it so that the updraught prevents airborne micro-organisms from settling inside.

5.4 Cultivation of micro-organisms

5.4.1 Culture media

Most bacteria can be grown or maintained on liquid or solid culture media. The basis of most solid media is an extract of seaweed called agar, which is available commercially in powdered form. This material dissolves in hot water and forms a firm gel when cooled to below 42°C. It only redissolves by heating to 100°C. It can be sterilized by autoclaving together with the appropriate nutrient components of the medium, and dispensed into Petri-dishes or test tubes prior to solidification. Agar is generally nutritionally inert, simply providing a solid support for growth. Liquid media are based on the same nutrients, but omitting the agar.

A culture medium should provide all of the nutrients and the energy source required by the organism it is intended to culture. The major nutritional requirements of all organisms, based on their relative abundance in living matter, are carbon, hydrogen, oxygen, nitrogen, phosphorus and sulphur. In addition, trace nutrients, including calcium, iron and other metallic ions are required. These together with phosphorus can generally be utilized as inorganic salts, but the utilization of the other elements depends on their chemical form.

The requirements for forms of carbon are probably the most complex aspect of microbial nutrition. Some organisms, called autotrophs, can use carbon dioxide as the major carbon source. Others, including the majority of the bacteria, require carbon in the form of organic compounds, and are termed heterotrophs. Some heterotrophs can use very few or only one organic compound while others are more versatile. For instance, *Pseudomonas multivorans* can use any one of more than ninety different organic compounds. The heterotrophs include organisms (phagotrophs) which ingest other organisms as well as those which utilize dissolved organic compounds.

Nitrogen and sulphur requirements vary widely in the microbial world. Some organisms can obtain these elements from nitrates and sulphates, while others need them in a reduced form, either as inorganic salts (NH_4^+ and S^{2-}) or in organic compounds such as amino acids or peptides (degradation products of proteins).

The requirement for free oxygen is an important characteristic in classifying micro-organisms. Oxygen, as the element, is ubiquitous as a component of water and organic compounds. Free oxygen is essential to the growth of many organisms which use it for aerobic respiration. Anaerobes do not require oxygen and many are killed or inhibited by it.

A major factor in microbial metabolism is the nature of the energy source. Phototrophs rely on light energy, while chemotrophs obtain it from chemical sources. The chemoautotrophs always use reduced inorganic compounds (e.g. NH_3, NO_2^-, Fe^{2+}, H_2S, S, $S_2O_3^{2-}$, H_2) while the chemoheterotrophs utilize organic compounds as sources of both carbon and energy.

A typical culture medium will contain water, a carbon source (autotrophs obtain carbon dioxide from the atmosphere which also provides oxygen for the growth of aerobes) either in the form of a specific organic compound or as a complex material (e.g. peptone, yeast extract, meat extract), inorganic or organic nitrogen compounds and trace nutrients. It would normally be buffered in order to prevent inhibition of growth due to excessive changes in pH. Within these constraints, the hundreds of different culture media range in composition from the very simple (inorganic salts plus an ammonium salt which will support the growth of some autotrophs) to the very complex, containing a mixture of nutrients of unknown composition (e.g. peptone, yeast extract), possibly supplemented with specific organic compounds and growth factors (vitamins, coenzymes) which fastidious organisms cannot synthesize themselves. Complex media are useful for the cultivation of a wide variety of organisms. Some media can be made selective. That is, they are formulated to favour the growth of specific organisms either by incorporating the minimal nutritional requirements of a particular organism, or by using agents which inhibit other groups, or both. Selective media are useful in the isolation of pure cultures and in testing for the presence of certain groups in a given sample.

5.4.2 Types of culture

Cultures on solid media can be grown in Petri-dishes or in test tubes. Petri-dishes are made of glass or plastic and consist of shallow dishes with over-lapping lids, typically 90 mm in diameter, into which 10–15 ml of molten agar is poured to form an "agar plate". Bacteria can be grown in the form of colonies both within and on the surface of the agar layer. Discrete colonies will only develop if the plates are allowed to dry to remove the surface film of water through which motile cells might migrate. Plates are inverted during incubation in order to prevent any condensed moisture from falling on the surface of the agar. The temperature of incubation varies according to the types of organisms present. Natural populations in aquatic samples can be incubated at 25°C, but pathogens for example will require incubation at 37°C.

Discrete colonies on an agar plate can be obtained from the inoculation of a suspension of cells by the pour plate or the spread plate technique. To make a pour plate, typically 0.1 or 1.0 ml of the cell suspension is pipetted into an empty Petri-dish, and molten agar poured over it. The two are mixed together by gentle rotation of the plate before the agar sets. A spread plate is prepared by adding 0.1 ml of suspension to a surface-dried agar plate. This is spread over the surface of the plate using a bent glass rod previously sterilized by dipping in ethanol which is burnt off in the Bunsen flame. The advantages of the latter method are that sensitive organisms are not exposed to the temperature of molten agar which may be inhibitory or lethal and that the colonies all develop on the surface of the agar where they are not limited by the diffusion of oxygen and are accessible for making transfers.

Agar slopes are made by dispensing molten agar into small screw-capped bottles or test tubes and then allowing them to set in an inclined position. Slopes are mainly used for storing isolates and for providing material in the form of surface growth for the inoculation of new cultures or for diagnostic tests.

Liquid cultures are normally grown in conical flasks plugged with cotton wool and are mainly used for studying the growth parameters of a particular species. Liquid culture media in test tubes or small screw-capped bottles are also used for certain diagnostic tests, including those where detection of gas production is required. The detection of gas is accompanied by the inclusion of a small inverted tube (Durham tube) in the bottle or test tube which fills with liquid when autoclaved. Gas production during incubation can be detected through the displacement of liquid from the Durham tube.

5.4.3 Isolation of pure cultures

In the case of organisms which form discrete colonies on agar media, the isolation of a single species from a population containing many different organisms may be obtained by the streak plate technique. A streak plate is made by inoculating a small area of the plate with a small amount of the sample using a wire loop. Material is spread across the plate from the origin in several stages using the loop, which is sterilized after each stage. One method of doing this is illustrated in Fig. 5.2. As the material is spread further from the origin the individual cells become more widely separated so

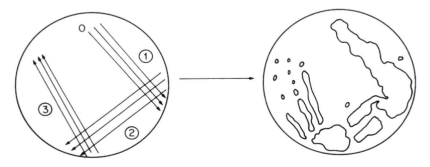

Fig. 5.2 Preparation of a streak plate.

that after incubation of the plate dense or confluent growth will be observed in the region of the initial inoculum and this will become more sparse with successive streaks until individual colonies can be distinguished which have arisen from the growth and division of a single cell. The existence of discrete colonies at this stage should not be taken to indicate a totally successful isolation. There is a possibility that a single colony may contain more than one species although this may not be macroscopically visible. Since no single culture medium can support the growth of all organisms in a sample, there is also the possibility that the colony may contain a dormant contaminant, perhaps only one or two cells, which could nevertheless grow if the colony were to be used to inoculate a different medium. It is normal, therefore, to take material from a single colony and make a second or even third streak plate to ensure that a pure culture is finally obtained.

Another useful technique involving liquid cultures is enrichment. If a selective medium is inoculated with a natural population and incubated under conditions favourable for the growth of the organisms of interest, the organisms for which the nutritional and incubation conditions are most suitable will outgrow the others and eventually become the predominant organisms in the culture. Depending on the degree of selectivity, this technique allows the selection of organisms of a known type from quite complex samples.

5.5 Microscopy

The optical microscope is a very useful tool in microbiology. Most microscopes used in microbiological studies are of the compound type, that is two lenses, the objective lens (the one nearest the specimen) and the ocular lens, or eyepiece. These are separated by a tube such that the eyepiece magnifies the image formed by the objective lens. In most cases the total magnification is that of the eyepiece multiplied by the magnification of the objective.

The essential components of a microscope are a source of illumination, the stage upon which the specimen (usually mounted on a slide) is placed, the lens system and a means of moving either the specimen or the lens system to bring the object into focus. Most modern microscopes have, in addition to these, a device for focusing the light source into the specimen, called the substage condenser, an iris diaphragm for controlling the amount of light entering the lens system and two or more objectives mounted on a rotating turret to provide various degrees of magnification. The most powerful objective is the oil-immersion objective, which has a magnification of nearly 100 times, giving a total magnification of almost 1,000 times. The oil immersion lens is so-called because it can only be used with a drop of special immersion oil between it and the slide, and touching both, which modifies its effective focal length. A schematic diagram of a microscope incorporating these features is shown in Fig. 5.3.

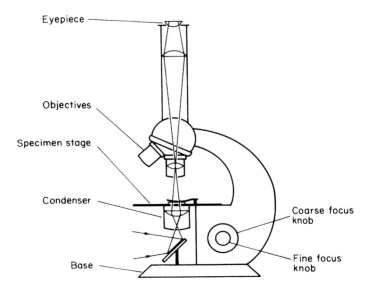

Fig. 5.3 Schematic of a compound microscope.

The limit of resolution of the optical microscope is approximately 0.2 μm. This is reached when two closely placed points can no longer be separated visually, or appear as one. Resolution cannot be improved by using higher magnifications since it is dependent on an inherent property of light itself. Thus, bacterial cells can be seen quite clearly under the microscope, but details of cellular structure are more difficult to discern. There are, however, techniques available which can enhance certain cellular structures to make them more easily visible. Some of these are described in Section 5.7.

5.6 Enumeration of micro-organisms

There are a number of ways in which the concentration of cells in a liquid sample can be measured. Direct determination of total cell numbers can be made visually, using a microscope, or electronically. The optical density or turbidity of a suspension can also be related to total cell numbers. In many applications a more useful parameter may be the number of viable cells (i.e. those capable of growth and cell division). In other circumstances it may be expedient to determine the concentration of a chemical component of the cells which can be related to their activity or numbers.

5.6.1 Total counts

A total count is a measure of the number of intact cells in a given volume, irrespective of whether they are alive or dead. Total counts are usually determined by direct microscopic examination of a known volume of the sample. Since dense liquid cultures can often contain hundreds of millions of cells per ml, the accurate measurement of a very small volume is required to make the task of differentiating and counting individuals manageable. Such a small volume may be obtained by the use of a specially constructed microscope slide, known as a counting chamber, the central part of which is lower than the two ends and is marked with a grid of known dimensions. A coverslip placed over a drop of the sample on the grid visible under the microscope corresponds to a known volume of liquid and from the number of cells counted in several squares their concentration in the original sample can be determined.

A total count can also be made with a Coulter Counter. This is a device in which the sample is passed through an orifice small enough to ensure that individual cells pass through one at a time. The instrument is equipped with an electronic detector which allows large numbers of cells to be counted in a relatively short time.

Total counts are limited in their applications. They are of little use in the examination for survivors in samples which have been sterilized or disinfected since the treatment may render the organisms permanently non-viable, while leaving superficial cell structure intact. In many samples a large quantity of debris of non-biological origin may be present, leading to difficulties in distinguishing the organisms both visually and electronically.

5.6.2 Viable counts

The basic assumption involved in viable counts is that when a small volume of a dilute and well-mixed suspension of organisms is spread in or on a solid medium, using the pour or spread plate technique, each separate colony which develops originates from a single cell. Viable counts are made by preparing a series of tenfold dilutions of the sample in sterile diluent such that an inoculum from the final dilution contains a number of cells equivalent to the optimum number of colonies which can be counted on an agar plate (usually between 30 and 300). For example, a sample containing 8,000,000 viable cells per ml would, after five successive dilutions and an inoculum of 1.0 ml, give rise to approximately 80 colonies on the plate.

The diluent is normally dispensed in 9 ml quantities in screw-capped bottles to facilitate making dilutions by successive transfers of 1.0 ml. Commonly used diluents include mixtures of inorganic salts, phosphate buffer and tap water or the plating medium itself, omitting the agar. For the study of organisms from a particular habitat (e.g. river water, sea water) the sterilized matrix itself or a synthetic version can be used.

5.6.3 Turbidity

Turbidity can be used for monitoring the growth of cultures in the laboratory since it is simple and rapid, unlike viable counting which only provides results after a certain period of incubation. Turbidity is normally measured using a nephelometer which measures the amount of incident light scattered by a suspension of particles, although a spectrophotometer or colorimeter may be used as an alternative. If turbidity is to be related to total or viable cell numbers a calibration curve has to be prepared. The calibration is only applicable to one particular organism and culture medium; others will give different curves. For this reason turbidity is of limited value in enumerating organisms in certain laboratory cultures or natural mixed populations.

5.6.4 Enumeration in water analysis

One important routine use of a counting technique is in the analysis of waters for contamination by human pathogens, particularly those responsible for enteric diseases such as cholera and typhoid fever. It is, however, fairly difficult to isolate or enumerate pathogens from many water samples since they are normally present in low numbers except in cases where contamination is highly localized or particularly severe. Instead of attempting to demonstrate the presence of pathogens the water sample is examined for indicator organisms. Enteric diseases are mainly transferred by the pollution of water with faeces, so in order to demonstrate faecal contamination and therefore the possibility of the presence of pathogens samples are examined for bacteria which are known to be present in normal faeces, even though they may not themselves be the agents of disease.

The main indicator organisms are the faecal streptococci and *Escherichia coli*, which are normally present in human faeces at levels of 10^6–10^8 per g dry weight. *E. coli* and other members of the coliform group can ferment lactose and are detected by incubation at 35°C for two days in tubes of lactose broth containing a Durham tube for the detection of gas production. Acid and gas production constitutes a presumptive test for coliforms. However, some coliforms are not of human or animal origin, being common inhabitants of soil and water. Confirmation of a faecal origin is obtained by incubating an inoculum from a positive tube at a higher temperature (44.5°C) or by using a medium selective for exclusively enteric organisms.

Enumeration of faecal coliforms cannot normally be carried out by viable counting techniques as described in Section 5.6.2. Acceptable levels of coliforms in potable water, for example, are usually <100 cells l^{-1}, which are obviously too dilute for viable counting. Instead a statistical treatment may be used to estimate the most probable number (MPN) of coliforms in a given sample based on observations of whether growth and acid production occur in tubes of lactose broth inoculated with different volumes of the sample.

Table 5.1 Dilution series for most probable number counting

Dilution of sample	No. of tubes	Volume of sample (ml)	Volume of broth (ml)
—	1	50	50*
—	5	10	10*
—	5	1	5
1/10	5	1	5

* Broth is prepared double strength to compensate for dilution by the sample.

For the purposes of MPN estimates it is assumed that a positive result is due to the addition of one or more cells in the original inoculum.

The way in which the inoculations are carried out is shown in Table 5.1. The numbers of tubes showing acid and gas production corresponding to each sample volume are recorded and the combination of results can be read in tables which give the corresponding MPN. Using the sample volumes shown in Table 5.1 MPN counts in the range 0–1,600 per 100 ml can be obtained. If all five tubes of each sample volume are positive an initial tenfold dilution of the sample can be made. It should be stressed that MPN values are based on statistical probability, and should therefore be regarded as approximate. They do not, nor would be expected to, agree with equivalent results obtained by viable counting. However, the MPN determination remains a useful tool in the analysis of the bacteriological quality of water.

Where direct viable counts are inappropriate an alternative to MPN determinations for fairly small numbers of bacteria in water samples is a membrane filtration technique. A sufficient volume of a water sample is passed aseptically through a membrane filter consistent with trapping a number of cells suitable for counting. The filter is then removed and placed on the surface of an appropriate solid medium in a Petri-dish, or alternatively on an absorbent pad soaked with the medium. Colonies develop on the nutrients which diffuse through the membrane and can be counted.

5.7 Staining methods

The amount of information which can be obtained from the microscopic examination of liquid films is limited. Only gross morphological characteristics of bacteria can be discerned, and the major structural components of the cells are generally indistinct or invisible, especially under bright illumination. Stains can be used to enhance certain structural characteristics of cells, such as flagella and capsules, and on the basis of their affinity for certain chemical components of the cells are useful for identifying unknown organisms.

5.7.1 Preparation of smears

Stains are normally applied to smears of bacterial suspensions spread onto clean microscope slides. The suspensions may be obtained directly from colonies on solid media by removing a small quantity of the growth and emulsifying it in a drop of water placed on the slide. The suspension is spread evenly and thinly over the slide to obtain a transparent film which is almost invisible when dry. The preparation is air dried and the bacteria are fixed to the slide by passing it two or three times through a Bunsen flame.

Smears are stained by placing the microscope slide on a rack over a sink and flooding it with the appropriate staining solution. The slide is left to stand for a specified period, which may be from a few seconds to several minutes. Excess stain is rinsed off gently but thoroughly by holding the slide under running tap water. The stained preparation may be gently blotted dry or left to dry on its own and then examined directly under a microscope. When stained with dyes such as crystal violet and methylene blue bacterial cells show up clearly against the unstained background.

5.7.2 Gram's stain

This is probably the most commonly used stain in the preliminary identification of bacterial isolates. The method involves a differential double staining technique. The smear is first treated with crystal violet which is fixed with iodine solution. The slide is then washed with ethanol, rinsed, and flooded with a second stain of a different colour known as a counterstain. Some bacterial genera retain the crystal violet when washed with ethanol and are unaffected by the counterstain, appearing blue or blue-purple on examination under the microscope. These organisms lose the crystal violet stain and assume the red or pink colour of the counterstain (usually carbol fuchsin or safranin). This differential staining depends on the chemical structure of the bacterial cell wall and its usefulness lies in the fact that the bacteria can be divided into two major groups on the basis of their Gram reaction, which permits the elimination of a large number of genera from consideration at the initial stage in the identification of an unknown isolate.

Some species are truly Gram-variable, while others may appear to be so if care is not taken to make smears from cultures which are between 1.5 and 2 days old.

5.7.3 Capsules

A simple method for demonstrating capsules, which may only appear as faint haloes surrounding the cells in preparations, is to mix a bacterial suspension with a drop of Indian ink placed on a slide, making a film by pressing a coverslip over the mixture. The capsule appears as a clear area around the cell, on a dark background, as shown in Fig. 5.4.

5.7.4 Flagella

Flagella cannot normally be seen by direct examination because of their small diameter. The application of a stain which forms a thick layer around the flagella allows them to be seen quite clearly, especially if a counterstain for the rest of the cell is used, and also allows their position on the cell to be

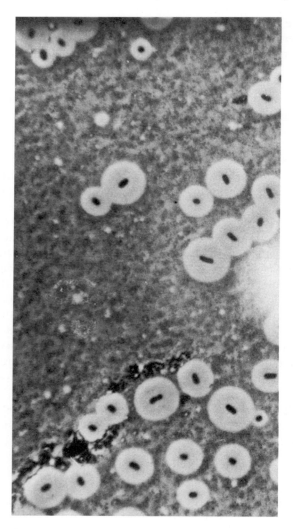

Fig. 5.4 Photomicrograph of *Klebsiella* ssp. showing capsules.

determined. The arrangement of flagella on the cell can be an important diagnostic property. The smear is made from material taken from an agar slope and is not heat-fixed since this disrupts the flagella. The smear is stained with basic fuchsin, and they appear red. If a methylene blue counterstain is used, the cells appear blue.

5.8 Microbial assays

There is considerable scope in the fields of pollution control and environmental engineering for the use of microbial assays. The determination of a single chemical compound of a given sample is in most cases most expediently undertaken by chemical or physical techniques. However, a biological assay, that is one which quantifies a biological response to a set of physical, chemical or environmental factors, can often yield more useful information than other analytical methods, especially where the potential toxicity or ecological impact of a given parameter is of interest. This is particularly true where the matrix of interest is very complex with respect to its physico-chemical properties.

This section describes only two microbial assays, included on the basis of their importance or general interest in the field of public health engineering.

5.8.1 Biochemical oxygen demand

This is a very widely used assay for measuring the concentration of biochemically oxidizable materials in a liquid sample. Biochemical oxygen demand (BOD) is defined as the quantity of molecular oxygen consumed by the micro-organisms present in a sample in a given time, expressed as mg of oxygen per l of sample. Normally, the BOD test is carried out at 20°C for five days (BOD_5) in the dark in a sealed container, and the BOD is taken to be the difference in dissolved oxygen concentration before and after incubation.

There are several chemical methods for measuring the organic strength of water, such as chemical oxygen demand (COD), which is the material oxidizable by potassium dichromate in boiling sulphuric acid and the concentration of total organic carbon (TOC). However, these determinations can frequently include organic compounds which cannot be oxidized by the microflora in the sample of interest. BOD, on the other hand, mirrors more closely the natural processes which occur when polluted effluents are discharged to natural watercourses and can give a more accurate indication of the pollution potential of an effluent or the biodegradability of a particular wastewater.

The requirements for conducting the BOD test are an active microflora to perform the oxidation and the maintenance of aerobic conditions by the provision of dissolved oxygen. Most sewage and sewage effluent samples already contain aerobic micro-organisms, but certain industrial wastes and other samples devoid of active bacteria will require a "seed" for the successful completion of the test. Such a seed can be obtained from a well-settled final effluent, but in certain circumstances a seed from the receiving water itself may be of more use in predicting the impact that a certain discharge may have.

The concentration of dissolved oxygen in water at 20°C at saturation is only 9.1 mg l^{-1}. For samples which have a higher BOD than this (all sewages, the majority of sewage effluents and many river waters) dilution of the sample is necessary to ensure that the oxygen supply is not depleted before oxidation is complete. Again, the receiving water for a particular effluent would be a logical choice for a diluent. However, the quality of such a diluent would be quite variable and it may itself exert an oxygen demand. Instead, well-aerated distilled water containing low concentrations of phosphate buffer is used. If a seed, such as sewage effluent, is required, 5 ml of this is added to 1 l of dilution water.

The test is performed by thoroughly mixing the sample and dilution water in the appropriate ratio while avoiding the entrainment of air bubbles. The diluted sample is then transferred to a 250 ml bottle with a tightly fitting stopper. It is important to fill the bottle completely before allowing it to stand for a while and then tapping gently to remove the last air bubbles prior to inserting the stopper. This avoids the possibility of atmospheric oxygen dissolving during the incubation period. Two bottles (or more if replicates are required) for each dilution are prepared, together with at least two bottles of dilution water only as controls. The dissolved oxygen concentration in one of each pair is determined immediately and in the others after incubation. The bottles are incubated in the dark in order to measure only the oxygen utilized for respiration. If incubated in the light, any algae present may liberate oxygen by photosynthesis, giving an erroneous result.

In normal circumstances, samples of river water will require dilutions of between 0 and 1/5, secondary sewage effluent up to 1/10, settled sewage from 1/15 to 1/100 and raw sewage 1/30 to 1/200. In most cases, where the nature of the sample is uncertain, more than one dilution will have to be made to ensure that there are some bottles which contain a residual dissolved oxygen concentration after incubation. It is desirable that this residual concentration is in excess of 2.5 mg l^{-1}, since at lower concentrations the rate of oxidation may be affected, giving rise to slightly erroneous results.

The BOD of a given sample may include that exerted by ammonia and organic nitrogen compounds which undergo nitrification. During five-day incubation periods the majority of the BOD will be due to reduced organic carbon. Nitrifying organisms grow slowly compared to heterotrophs and if nitrogenous BOD determinations are required, incubation should be extended to twenty days. The typical rate of oxygen consumption during a twenty-day incubation is shown in Fig. 5.5. In certain circumstances,

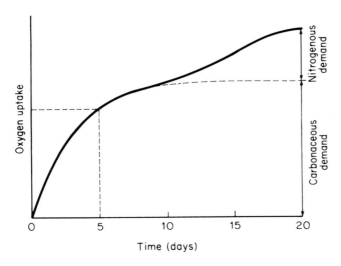

Fig. 5.5 Typical rate of oxygen consumption during a 20-day BOD test.

however, nitrification may commence earlier if it has already started in the sample before collection. Nitrification is also sporadic, and the organisms responsible are extremely sensitive to low levels of toxic agents, therefore attempts to determine nitrogenous BOD may yield misleading results. If nitrification is considered relatively unimportant in a particular analysis, it may be suppressed by adding allylthiourea (2%) to the sample.

5.8.2 Mutagenicity testing

A mutagen is an agent which induces chemical changes in DNA. These changes are normally restricted to individual genes, but may damage entire chromosomes. Such damage causes mutations, usually deleterious, which are inheritable, often resulting in gross changes in the growth and metabolic characteristics of a cell, tissue or organ. Many carcinogens are believed to act by mutagenic mechanisms.

Traditionally, examination of a substance for potential carcinogenic properties in man is a lengthy and expensive process. However, Ames and co-workers (1975) recently developed a rapid bacterial assay for mutagens using strains of *Salmonella typhimurium* which have a defective gene for the synthesis of histidine (an essential amino acid). The bacterial DNA in the region of this gene is particularly susceptible to mutagens, and the resulting mutations can lead to a reversion of the cell such that it can synthesize histidine.

The test is performed by inoculating a small volume of molten, dilute (0.6% w/v) agar containing the agent to be tested with an appropriate strain of *S. typhimurium*. This is then poured into a plate already containing minimal agar medium with just a trace of histidine. After incubation, a slight background growth should develop due to this trace of histidine and this is important for two reasons. First, its presence indicates that the test substrate is not toxic *per se*, and second it allows a number of cell divisions to occur, which is necessary for the detection of mutagens which act during replication of the DNA. If the test substrate is mutagenic, a number of distinct colonies will develop, superimposed on the background growth, which are due to some revertant whose ability to synthesize histidine has been restored. The number of colonies is compared to those arising on control plates due to spontaneous mutation, and confirmation of positive results may be obtained by establishing dose–response curves for the test chemical.

References and further reading

American Society for Microbiology (1981) *Manual of Methods for General Bacteriology*, American Society for Microbiology, Washington, D.C.

Ames, B. N., McCann, J. and Yamasaki, E. (1975) *Mutation Res.*, **31**, 347–364.

Collins, C. H., Lyne, P. M. and Grange, J. M. (1995) *Collin's and Lyne's Microbiological Methods*, Butterworth-Heinemann Ltd, Oxford.

Chapter 6

Introduction to chemistry

6.1 Introduction

Chemistry can be defined as the study and elucidation of the structure and properties of matter. All chemists are taught the fundamental principles of chemistry, a science which has concepts that are also important in environmental engineering. This chapter reviews basic chemical concepts that are related to the branch of inorganic chemistry.

6.2 The atom

Regarding the actual nature of matter, two hypotheses were put forward. One stated that matter is a structure that is infinitely divisible. The other stated that matter is comprised of discrete particles called atoms. This theory was supported by John Dalton, who believed that atoms were the solid building blocks of matter. Subsequent work by Rutherford and other scientists proved that the atom had electrical sub-particles and was in fact divisible.

The atom is considered to comprise the following sub-particles:

1   The *nucleus* which contains *protons* and *neutrons*. Protons have a charge of +1 and a mass of 1; neutrons have zero charge and a mass of 1. Particles present in the nucleus can also be termed collectively as *nucleons*.

2   Electrons which have a charge of −1 and a mass of 1/1836 of a proton. These negative particles can be considered as particles orbiting the nucleus on a given path, or as clouds of electrical charge surrounding the nucleus. The latter is thought to be more probable due to Heisenberg's uncertainty principle, which places a limit on the accuracy with which the position and motion of a particle may be calculated. This uncertainty is large for electrons.

6.2.1 Atomic number and mass

The atomic number (usually given the symbol Z) represents the number of protons present in the nucleus. In a neutral atom, this also corresponds to the number of electrons in the element.

The atomic mass number is the number of protons plus the number of neutrons in the nucleus, and is given by the symbol A. Thus, the number of neutrons can be given by $A - Z$. An element can therefore be defined as a substance consisting of atoms with the same atomic number. This does not apply to the atomic mass. Neutrons present in the nucleus do not affect the atomic number or the chemical reactivity of that element. However, with a mass of 1, the number of neutrons present in the nucleus will affect the mass of the atom. Chemically identical atoms that have different numbers of neutrons are termed *isotopes*. The simplest isotopes are those of hydrogen and are shown in Fig. 6.1.

As elements generally have more than one isotope, the atomic mass is known more accurately as the *relative atomic mass* (RAM), which takes into account the relative abundance of each isotope. The relative atomic mass can be defined as the mass of one atom of an element in comparison with $1/12$ of the mass of one atom of $^{12}_{6}C$. An example is chlorine (Cl) which has two main isotopes; one with an atomic mass of 35 ($\approx 75\%$ relative abundance), and the other with an atomic mass of 37 ($\approx 25\%$ relative abundance). Thus, the relative atomic mass of Cl is 35.46.

6.2.2 Electronic configuration

Electrons are seen to surround the nucleus in shells or orbitals, each one having a different energy level. The orbital with the lowest energy is filled first, a process known as the *Aufbau principle*. The first shell holds two electrons, the second eight electrons, the third holds 18 electrons and so on. This is illustrated in Fig. 6.2.

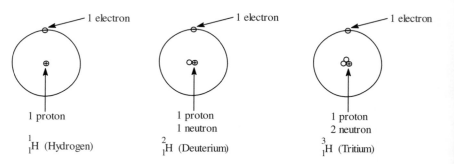

Fig. 6.1 Isotopes of hydrogen ($^{A}_{Z}H$).

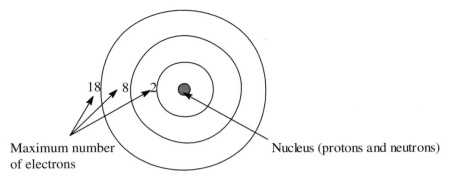

Maximum number of electrons

Nucleus (protons and neutrons)

Fig. 6.2 Filling of electron orbitals.

These shells (or orbitals) can be further subdivided, based on the angular momentum of electrons in the atom. Each subshell is defined by the shell number followed by the letter s, p, d or f. A statement of the number of electrons in each subshell for a particular element gives the electronic configuration for that element. Thus, hydrogen, which has 1 electron in the first shell, has the electronic configuration of $1s^1$; helium which has two electrons is $1s^2$ and so on. Table 6.1 shows the electronic configurations of several elements.

The subshells have differing energy levels and are not necessarily filled in the order one would expect. Figure 6.3 illustrates the energy levels for several subshells. For example, potassium (K) has the electronic configuration $1s^2 2s^2 2p^6 3s^2 3p^6 4s^1$. The 4s shell is filled first rather than the 3d shell as it has a lower energy.

Table 6.1 Electronic configuration of several elements

Element (symbol)	Electronic configuration
Lithium (Li)	$1s^2 2s^1$
Beryllium (Be)	$1s^2 2s^2$
Boron (B)	$1s^2 2s^2 2p^1$
Carbon (C)	$1s^2 2s^2 2p^2$
Neon (Ne)	$1s^2 2s^2 2p^6$
Sodium (Na)	$1s^2 2s^2 2p^6 3s^1$

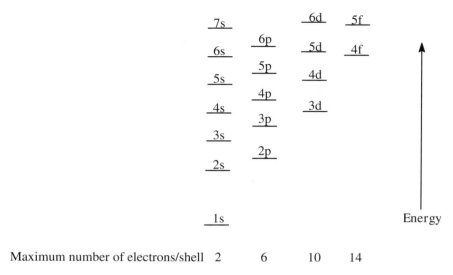

Maximum number of electrons/shell   2   6   10   14

Fig. 6.3 Energy levels of electron subshells.

6.3 Periodic Table and properties of elements

In 1869 a Russian chemist, Mendeleev, produced a table of chemical elements. He demonstrated that when the elements were arranged in order of atomic weight, they exhibited regularities and this allowed properties of several undiscovered elements to be postulated. The Periodic Table in use today arranges elements in order of increasing atomic number and shows that the properties of elements recur periodically. Figure 6.4 illustrates the Periodic Table. Horizontal rows of elements are termed periods; vertical columns of elements are called groups. The similarities within each group are governed by the electronic configurations of the atoms.

6.3.1 General properties

As you progress across the Periodic Table from left to right, the properties of elements move from metals to metalloids to non-metals. Two examples of this are group I (the alkali metals) and group VII (the halogens).

Group I elements all contain one electron in the outer shell and therefore have a tendency to loose this electron in order to achieve a full outer shell (i.e. a stable configuration). This results in a single positive ion. All alkali metals are known to react violently with water to form basic solutions. In terms of their metallic character, they are good conductors of heat and electricity.

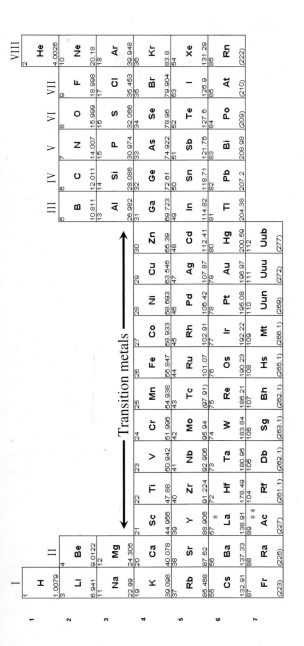

Fig. 6.4 The Periodic Table.

Group VII elements are classed as non-metals due to their poor conductance of heat and electricity. They all contain seven electrons in the outer shell and therefore wish to gain an electron in order to attain a stable configuration, resulting in a single negative ion.

At the extreme right of the Periodic Table are the so-called noble or inert gases (group VIII). The outer shells of these elements are full and this results in chemical stability.

6.3.2 Atomic radii

The atomic radii of elements can be seen to decrease on moving across a period. This is due to the increasing charge in the nucleus, which results in a greater attraction of the orbiting electrons towards the centre of the atom. Moving down groups in the Periodic Table, the atomic radii increase. This is because extra shells of electrons are added. Although there will also be an increase in nuclear charge this effect is outweighed by the addition of extra shells, as the pull of the nucleus on electrons decreases inversely with the square of the distance.

Positive ions of elements (e.g. Na^+) are smaller than their atoms due to:

1   the increase in pull of the nucleus on the electrons;
2   the loss of the outer shell electrons.

Negative ions of elements (e.g. Cl^-) will be larger than their atoms since the added electrons will reduce the effect of nuclear charge and therefore the electron cloud will expand.

6.3.3 Ionization potential

This is the energy required to remove one electron from a gaseous atom. Values for the ionization potential are always positive, as energy will always have to be provided in order to remove an electron from the attraction of its nucleus. Factors affecting the ionization potential include the size of the atom, the charge on the nucleus and the screening effect of the nucleus on the inner electrons.

An increase in atomic size will result in a lower ionization potential as the electron is further away from the nucleus and consequently the forces holding them together will be weaker. Increases in nuclear charge mean greater difficulty in removing an electron, resulting in higher ionization potentials. The so-called screening effect occurs when electrons in the inner shells "shield" outer electrons from the increased nuclear charge. This results in lower ionization potentials.

As the electron shell is filled going across a period, electrons become more tightly bound and therefore the ionization potential will increase. On moving down a group, the atom size and nuclear charge will increase. As explained above, atom size is the more important parameter and so the ionization potentials generally decrease down a group.

Successive ionization potentials (i.e. 2nd and 3rd) will rise as, due to the removal of the first electron, the effective nuclear charge increases and the electrons will be more strongly attracted towards the nucleus.

6.3.4 Electron affinity

The electron affinity can be defined as the energy released when an extra electron is added to a neutral gaseous atom. It is effectively the reverse process of ionization. Owing to the stability of the inert gas electronic configuration, the electron affinities of the halogens is high, as the addition of an electron will result in a full (stable) outer shell. Conversely, the inert gases have low electron affinities as the addition of an electron leads to the formation of a new electron shell.

6.3.5 Electronegativity

This can be defined as the tendency of an atom in a molecule to attract electrons to itself. On moving left to right across a period, the electronegativity is seen to increase as does the nuclear charge. Also, atom size is important and electronegativities decrease moving down a group due to increasing atomic radii. Atoms with nearly filled shells have higher electronegativities. One example is fluorine which has the highest electronegativity of any element in the Periodic Table. Electronegativity also influences the type of bonds formed between elements. Covalent bonds are generally formed from elements with similar electronegativities, whereas ionic bonds are usually formed from elements with a large electronegativity difference.

6.4 Chemical bonding

There are three main types of chemical bonding: ionic, covalent and metallic. All involve the interaction of electrons. The two theories which explain this phenomenon are the molecular orbital and valence bond theories. These will not be discussed here and the reader is referred to Harrison and deMora (1996) for an explanation of these theories.

Bonds are produced between chemical elements as a result of atoms trying to achieve a stable (inert gas) electronic configuration. This is essentially achieved in two ways: electron transfer and electron sharing.

6.4.1 Ionic bonding

An ionic bond is formed by the complete transfer of electrons from one atom

to another, resulting in the formation of ions. One atom will become a positively charged ion (a *cation*), the other a negatively charged ion (an *anion*).
Ionic compounds are usually made from a metal and a non-metal because of
their equal desire to lose and gain electrons. As these bonds are the result of
electrostatic attraction from oppositely charged ions, they are very strong
and once formed they are difficult to break.

For example, consider the production of sodium chloride from its parent
elements. Sodium (Na) has the electronic configuration $1s^2 2s^2 2p^6 3s^1$;
therefore to achieve an inert gas structure (full outer shell) it needs to lose
one electron. Conversely, chlorine (Cl) has an electronic configuration of
$1s^2 2s^2 2p^6 3s^2 3p^5$ (i.e. seven electrons in its outer shell) and wishes to gain

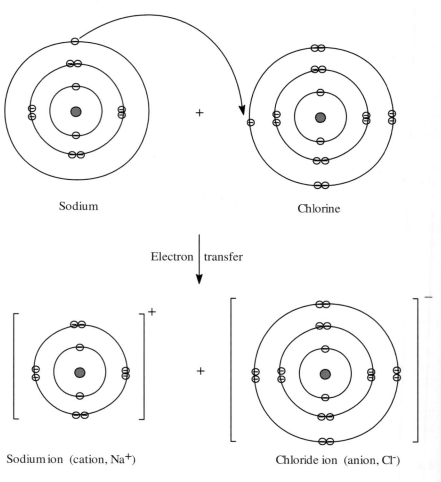

Fig. 6.5 Formation of sodium chloride (Na^+Cl^-).

an electron. Thus, these elements readily react to produce sodium chloride (NaCl), an ionic compound. This is shown in Fig. 6.5.

Characteristics of ionic compounds include strong electrical forces between ions, solids which have high melting points, are soluble in water, insoluble in organic solvents and are good electrical conductors.

6.4.2 Covalent bonding

The removal of an electron from an atom requires energy. This produces a cation and the removal of subsequent electrons will require more energy due to the increase in effective nuclear charge on the ion. Some elements near the middle of the Periodic Table need to lose (or gain) numerous electrons in order to achieve a stable electronic configuration. Rather than electron transfer which will require a large amount of energy, electrons are shared between the atoms. This results in the formation of a covalent bond.

An example of this is carbon which has the electronic configuration $1s^2 2s^2 2p^2$. Thus, to produce an ionic bond it would have to transfer four electrons which would require a large amount of energy. The sharing of electrons requires much less energy and a covalent bond is formed as a result of the sharing of a pair of electrons.

For example, consider carbon tetrachloride (CCl_4). Carbon ($1s^2 2s^2 2p^2$) needs to gain four electrons and chlorine ($1s^2 2s^2 2p^6 3s^2 3p^5$) needs to gain one electron. Therefore, these atoms share electrons from the outer shell to produce a stable electronic configuration for each atom. This is shown in Fig. 6.6.

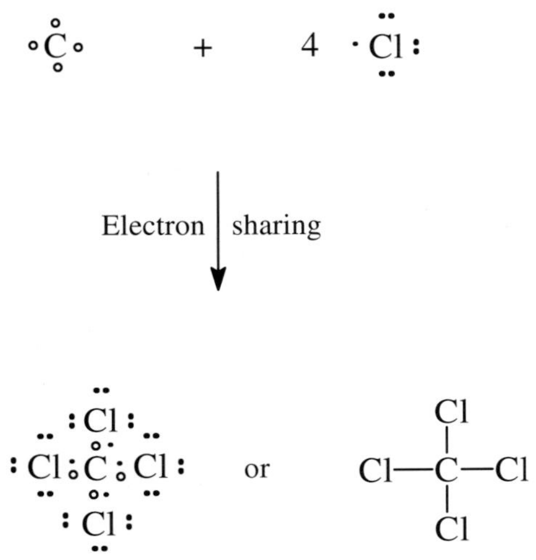

Fig. 6.6 Covalent bonding in carbon tetrachloride.

Atoms sometimes share more than one electron pair in order to achieve a stable configuration. This can be seen in carbon dioxide (CO_2) where two electron pairs are shared between the carbon and oxygen atoms:

$$O :: C :: O \qquad \text{or} \qquad O=C=O$$

This type of bond is a *double covalent bond*. Triple bonds are can also be produced, an example being nitrogen gas (N_2):

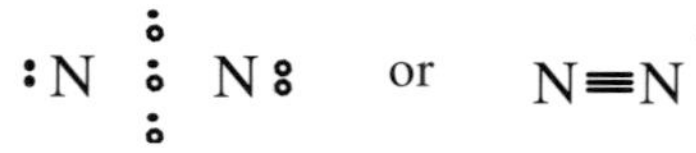

$$:N ⫶ N: \qquad \text{or} \qquad N\equiv N$$

It is also possible for atoms to be covalently bonded without all electrons being shared between atoms. For example, ammonia (NH_3), where nitrogen shares an electron pair with each hydrogen which results in a lone pair of electrons on the nitrogen atom.

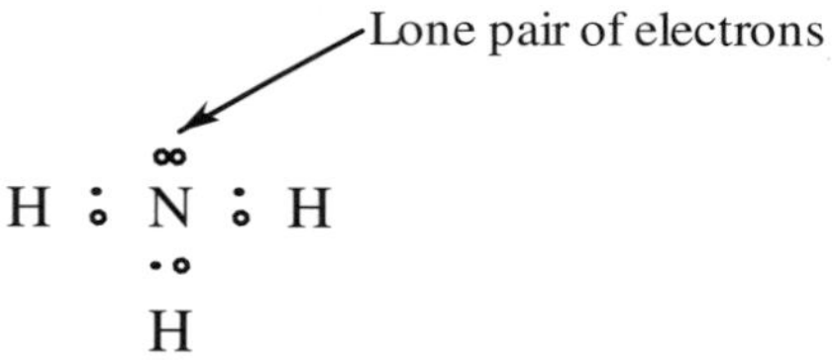

Lone pair of electrons

$$H : N : H$$
$$H$$

Lone pairs of electrons concentrate the negative charge in a certain area of the molecule. These molecules are said to be *polar*. Another example of a polar substance is water. This is discussed in further detail in Chapter 11.

In some cases, a polar molecule will react with a cation by contributing its lone pair of electrons to form a *dative covalent (coordinate) bond*. Atoms that donate a pair of electrons in this manner are usually highly electronegative (e.g. nitrogen or oxygen). Figure 6.7 shows the formation of an ammonium ion as an example of a dative covalent bond.

The forces between covalent molecules are relatively weak whereas the

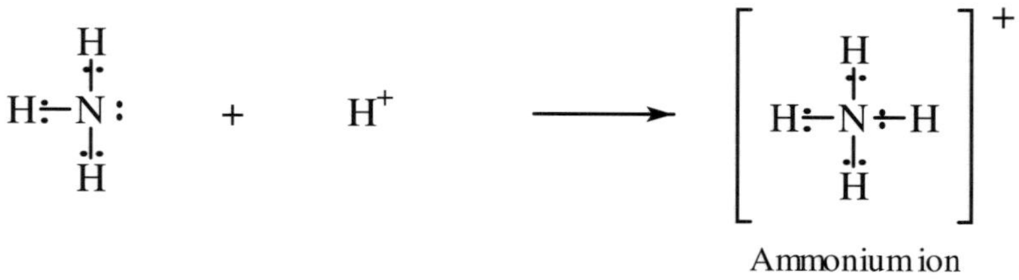

$$
\begin{array}{ccccc}
H & & & & \left[\; H\; \right]^{+} \\
H:N: & + & H^{+} & \longrightarrow & H:N:H \\
H & & & & H
\end{array}
$$

Ammonium ion

Fig. 6.7 Formation of a dative covalent bond.

actual covalent bond between the atoms is strong. Other covalent compound characteristics include low melting points, insolubility in water, solubility in organic solvents and non-conductability.

6.4.3 Metallic bonding

Metals can be thought of as cations in a 3-D crystal lattice structure. The outer electrons from each metal atom are shared between the ions and are said to be *delocalized* (i.e. spread out). The metal ions are present in a so-called "sea of electrons" (see Fig. 6.8). These electrons are mobile throughout the entire metal structure, which explains why metals are good electrical and thermal conductors.

6.4.4 Oxidation number/valency

The outermost electrons of an atom are sometimes referred to as the *valence electrons*. These are electrons that are to be utilized in bonding. Thus, carbon which has four electrons in its outer shell will have a valency of four if all the electrons are used in bonding. The oxidation number is the amount of charge on the atom in a compound after electrons have been transferred or shared in bonding. There are certain rules to be adhered to

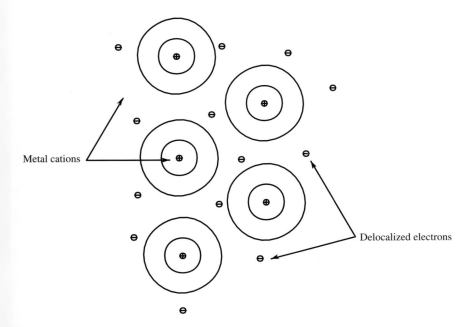

Fig. 6.8 Metallic structure.

when assigning oxidation numbers:

1   The oxidation number of hydrogen is +1, except in metallic hydrides (e.g. NaH) when it is −1, and in hydrogen gas when it is zero.
2   The oxidation number of oxygen is −2, except in peroxides (e.g. H_2O_2) when it is −1, and in oxygen gas when it is zero.
3   The oxidation number of an atom in an ionic compound is equal to the charge on that atom, whether it is positive or negative.
4   The oxidation number in a covalent compound is equal to the charge it would have in its most probable ionic formulation.
5   The oxidation number of fluorine is always −1, and generally all the halogens will have an oxidation number of −1 unless they are bound with more electronegative elements.

For example, in carbon tetrachloride (CCl_4), chlorine has an oxidation number of −1 and thus carbon must have an oxidation number of +4 in order to balance the charge on the molecule. As all of the carbon outer electrons (four in total) are used in bonding, carbon is said to have a valency of 4.

The oxidation number of sulphur in the sulphate ion, SO_4^{2-}, can be determined using the above rules. If oxygen has an oxidation number of −2, the total oxidation number given by the oxygen atoms is −8. As the anion has an overall charge of −2, then the sulphur atom must have an oxidation number of +6. This is simplified in the diagram below:

$$\begin{pmatrix} +6 & (-2 \times 4) \\ S & O_4 \end{pmatrix}^{2-}$$

Therefore in the sulphite ion, SO_3^{2-} the oxidation number of sulphur will be +4. Table 6.2 shows some common anions and their names.

Table 6.2 Nomenclature of common anions

Name	Anion
Sulphate	SO_4^{2-}
Sulphite	SO_3^{2-}
Carbonate	CO_3^{2-}
Nitrate	NO_3^{-}
Nitrite	NO_2^{-}
Phosphate	PO_4^{3-}
Hydroxide	OH^{-}
Oxide	O^{2-}
Chloride	Cl^{-}

6.5 Chemical quantities

6.5.1 Moles

The relative atomic masses (RAM) of any element will contain the same number of atoms. This is given by Avogadro's number which is 6.02×10^{23}. Therefore, 23 grammes of sodium (i.e. its RAM) will contain 6.02×10^{23} atoms.

The amount of substance which also contains 6.02×10^{23} particles can also be expressed in terms of *moles*. The particles may be atoms, molecules or ions. The number of moles of a particular element or compound can be calculated by the equation:

$$\text{moles} = \frac{\text{mass (g)}}{\text{relative atomic mass (g)}} \tag{6.1}$$

6.5.2 Molarity

A one molar solution (1M) contains one mole of solute dissolved in enough solvent to make one litre of solution (units: mol l^{-1}):

$$\text{molarity } (M) = \frac{\text{number of moles of solute}}{\text{number of litres of solution}} \tag{6.2}$$

It is useful to be able to relate this molarity concept to other units of concentration, e.g. calculation of the molarity of concentrated HCl.

The relative molecular mass of HCl is 36.45 g. The specific gravity for concentrated HCl is quoted as 1.16 g cm^{-3} and is $\approx 32\%$ solution. From the specific gravity it is determined that 1 litre weighs 1160 g, i.e. one litre contains

$$1160 \times \frac{32}{100} = 371.2 \text{ g HCl}$$

A 1M HCl solution will contain 36.45 g l^{-1}, so concentrated HCl containing 371.2 g l^{-1} is equivalent to $371.2/36.45 = 10.18\text{M}$.

Thus, having calculated the molarity of concentrated HCl, appropriate dilutions can be made to achieve the required molar concentration for use in the laboratory. Molar quantities should not be confused with molal quantities where a one molal solution is one mole of solute dissolved in one litre of solvent. The resulting volume will be slightly greater than one litre.

6.5.3 Normality

The concept of solution normality is very important in titrimetric analysis and ion exchange reactions which involve the exchange of equivalent quantities of reagents. It can be represented by equation (6.3):

$$\text{normality } (N) = \frac{\text{number of equivalents of solute}}{\text{number of litres of solution}} \tag{6.3}$$

A normal (N) solution is one which contains the gramme-equivalent weight of the substance in one litre of solution; i.e. the weight in grammes of the solute which is equivalent to 1.008 grammes of hydrogen.

6.5.4 Concentration units

The concentration of an element or compound can be expressed as its weight per unit volume (for solutions) or mass (for solids or samples containing mixed solid/liquid phases).

Solutions	Solid or solid/liquid samples
$mg\ l^{-1}$	$mg\ kg^{-1}$ (parts per million (ppm) equivalent)
$\mu g\ l^{-1}$	$\mu g\ kg^{-1}$ (parts per billion (ppb) equivalent)

6.6 Stoichiometry and chemical equilibria

6.6.1 Stoichiometry

In order to express a chemical reaction, chemical equations are used. These show the reactants (on the left side) and the products (the right side) and the amounts used in the reaction in terms of moles, for example

$$Na + Cl \rightarrow NaCl \tag{6.4}$$

$$Mg + 2Cl \rightarrow MgCl_2 \tag{6.5}$$

Thus, 1 mole of sodium reacts with 1 mole of chlorine to produce 1 mole of sodium chloride, and 2 moles of chlorine react with 1 mole of magnesium to produce 1 mole of magnesium chloride.

The law of conservation of matter states that matter is neither created nor destroyed in a chemical reaction. Thus, in a chemical equation the same number of atoms must appear on each side of the equation; i.e. it must be balanced. This provides a simple ratio of the amount of reactants needed for the reaction and is called the stoichiometry. For example, Na + Cl have a 1 : 1 stoichiometry and Mg + Cl have a 1 : 2 stoichiometry.

6.6.2 Chemical equilibria

Equilibrium deals with the extent to which reversible reactions proceed in a forward or backward direction. Reactions may be acid–base, solubilization–precipitation, complexation or oxidation–reduction (redox) reactions. A generalized equilibrium reaction can be written as:

$$aA + bB \leftrightarrow cC + dD \tag{6.6}$$

An expression for the equilibrium constant (K) can be written from equation (6.6):

$$K = \frac{[\text{products}]}{[\text{reactants}]} = \frac{[C]^c[D]^d}{[A]^a[B]^b} \tag{6.7}$$

where [ ] = concentration units (generally mol l^{-1} or activities).

Le Chatelier's principle states that "a system in equilibrium reacts to any change in its conditions in a manner that would tend to abolish this change". Therefore, an increase in D to a system in equilibrium will shift the above reaction to the left, consuming C and producing A and B until K is achieved once more. The driving force behind this adjustment is called the *mass action effect*. When the forward rate of reaction is exactly the same as the backward rate of reaction, the system is said to be in a state of *dynamic equilibrium*.

6.6.3 Acids and bases

Acids and bases have an important role in the determination of the composition of natural waters. There are three theories that can be used to interpret acid–base phenomena in solution. These are the Arrhenius, Lewis and Brönsted–Lowry theories. All three theories overlap, but the Brönsted–Lowry theory has the most value when dealing with aqueous solutions.

Brönsted and Lowry defined an acid as a substance which can donate protons. Therefore, a base is a substance that can accept protons. Depending on the strength of the acid in solution, it will undergo some degree of ionization according to the following equation:

$$HA \leftrightarrow H^+ + A^- \tag{6.8}$$

From equation (6.7), an expression for the equilibrium constant of an acid (K_a) can be obtained:

$$K_a = \frac{[H^+][A^-]}{HA} \tag{6.9}$$

For a strong acid, K_a will be large with the acid dissociating completely to give H^+ and A^-. The resulting base (A^-) produced from the corresponding acid (HA) is called the *conjugate base*. A strong acid will have a weak conjugate base and vice versa.

Similarly for bases, dissociation can be represented by the equation:

$$BOH \leftrightarrow B^+ + OH^- \qquad (6.10)$$

Therefore, the equilibrium (dissociation) constant (K_b) will be represented by:

$$K_b = \frac{[B^+][OH^-]}{[BOH]} \qquad (6.11)$$

The larger the K_b value, the stronger the base. Typical values of equilibrium constants for acids and bases range from 1×10^{-3} to 5×10^{-10}. This scale is quite cumbersome and so a more convenient logarithmic scale is used where:

$$pK_{a \text{ or } b} = -\log_{10} K_{a \text{ or } b} \qquad (6.12)$$

Thus, the smaller the pK value, the greater the strength of the corresponding acid or base. Table 6.3 shows some equilibrium constants for some acids and bases. It should be noted from Table 6.3 that for an acid and its conjugate base:

$$pK_a + pK_b = pK_w = 14 \qquad (6.13)$$

6.6.4 Dissociation of water

Water can be classed as *amphoteric*, i.e. it can behave as an acid or base.

Table 6.3 Equilibrium constants for acids and bases at 25°C

Acid	K_a	pK_a	Conjugate base	K_b	pK_b
HCl	1,000	−3	Cl^-	1×10^{-17}	17
H_2SO_4	1,000	−3	HSO_4^-	1×10^{-17}	17
HNO_3	10	−1	NO_3^-	1×10^{-15}	15
CH_3COOH	1.8×10^{-5}	4.74	CH_3COO^-	5.56×10^{-10}	9.26
H_2CO_3	4.3×10^{-7}	6.37	HCO_3^-	2.33×10^{-8}	7.63
H_2S	9.1×10^{-8}	7.04	HS^-	1.1×10^{-7}	6.96
NH_4^+	5.56×10^{-10}	9.26	NH_3	1.8×10^{-5}	4.74

Water can readily dissociate according to the following equation:

$$H_2O \leftrightarrow H^+ + OH^- \tag{6.14}$$

The expression for the equilibrium (dissociation) constant of water (K_w) can be described by the equation:

$$K_w = \frac{[products]}{[reactants]} = \frac{[H^+][OH^-]}{[H_2O]} \tag{6.15}$$

The value of $[H_2O]$ is taken to be a constant so that equation (6.15) now becomes

$$K_w = [H^+][OH^-] = 1.008 \times 10^{-14} \tag{6.16}$$

This value is generally referred to as the *ionic product of water*.

6.6.5 Buffers

Buffers can be defined as substances in solution which resist pH changes on the addition of an acid or a base. These solutions contain mixtures of weak acids and their conjugate bases, or weak bases and their conjugate acids.

Consider the equation for a weak acid in equilibrium:

$$K_a = \frac{[H^+][A^-]}{[HA]} \tag{6.17}$$

rearranging:

$$\frac{1}{[H^+]} = \frac{1}{K_a} \frac{[A^-]}{[HA]} \tag{6.18}$$

take $\log_{10}$:

$$pH = pK_a + \log_{10} \frac{[salt]}{[acid]} \tag{6.19}$$

From equation (6.19), it can be seen that the pH of a buffer solution depends on the ratio of the salt : acid concentration. One of the common buffer systems used is acetic acid (CH_3COOH)/sodium acetate (CH_3COONa). At concentrations of 0.1M acetic acid and 0.1M sodium acetate, the pH according to equation (6.19) will be:

$$pH = pK_a + \log_{10} \frac{[0.1]}{[0.1]}$$

$$= 4.74 + 0 = 4.74 \tag{6.20}$$

On addition of enough HCl to give a concentration of 0.05M, the following reaction takes place:

$$\underset{\text{salt}}{HCl + CH_3COO^-Na^+} \rightarrow \underset{\text{acid}}{CH_3COOH + Na^+Cl^-} \tag{6.21}$$

Thus, [salt] = 0.1 − 0.05 = 0.05M, and [acid] = 0.1 + 0.05 = 0.15M. Therefore the pH will now be:

$$pH = 4.74 + \log_{10} \frac{[0.05]}{[0.15]} = 4.26 \tag{6.22}$$

On addition of this quantity of HCl to an unbuffered solution, the pH would show a dramatic decrease. This would not occur in the buffered solution that was illustrated in the previous calculations.

6.7 Laws of chemistry

These laws describe the relationship between pressure, volume and temperature on a quantitative basis. *Boyle's law* states that the volume of a gas varies inversely with pressure at a constant temperature:

$$V \propto \frac{1}{P} \quad \text{or} \quad PV = constant \tag{6.23}$$

If the pressure remains constant, then the volume of gas is directly proportional to the absolute temperature (kelvin). This is known as *Charles' law*:

$$V \propto T \quad \text{or} \quad \frac{V}{T} = constant \tag{6.24}$$

If a constant volume is maintained, the gas pressure is also directly proportional to the temperature. This is referred to as the *Gay-Lussac law*:

$$P \propto T \quad \text{or} \quad \frac{P}{T} = constant \tag{6.25}$$

Equations (6.23)–(6.25) can be combined to produce the Ideal Gas Equation:

$$\frac{PV}{T} = constant \tag{6.26}$$

The following generally represents equation (6.26)

$$PV = nRT \tag{6.27}$$

where n is the number of moles of gas and R is the universal gas constant $(0.082 \, l \, atm \, K^{-1} \, mol^{-1})$. Using this constant, at standard temperature and pressure $(P = 1 \, atm; \, T = 273 \, K)$, 1 mole of any gas will occupy 22.41 litres. This volume will also apply to gas mixtures.

For gas mixtures, *Dalton's law of partial pressures* can be applied. The partial pressure of a gas can be defined as the pressure it would exert if it were the only gas present in the volume of the mixture. Dalton's law states that the sum of the partial pressures of several gases is equal to the total pressure of that gas mixture. This concept of partial pressures is utilized in *Henry's law*. This states that the amount of gas that can be dissolved in a liquid is directly proportional to the pressure of the gas, at a constant temperature. It is given by the equation:

$$[gas_{(aq)}] = K_H P_{gas} \tag{6.28}$$

where

$[gas_{(aq)}]$ = concentration of gas in solution $(mg \, l^{-1})$.
K_H = Henry's constant $(mg \, l^{-1} \, atm^{-1})$.
P_{gas} = partial pressure of gas (atm).

For example, K_H for oxygen at 20°C is $43.8 \, mg \, l^{-1} \, atm^{-1}$. If air contains 21% oxygen, then the partial pressure of oxygen in air when the pressure is 1 atm would be 0.21 atm according to Dalton's law. Thus, the concentration of oxygen in water under these conditions would be:

$$[gas_{(aq)}] = 43.8 \times 0.21 = 9.2 \, mg \, l^{-1} \tag{6.29}$$

The temperature can affect the solubility of gases. For example, gases generally become less soluble as temperature increases. This can lead to thermal pollution, e.g. reduced oxygen concentrations in rivers.

6.8 pH and activity

Sörensen proposed a definition for pH which is based on the hydrogen ion concentration and given by the formula:

$$pH = -\log_{10} [H^+] \quad \text{or} \quad pH = \log_{10} \frac{1}{[H]^+} \tag{6.30}$$

From equation (6.15), it can be seen that in water $[H^+]$ and $[OH^-]$ will be equal because water is neutral, producing a K_w of 10^{-14}. Therefore, $[H^+]$ and $[OH^-]$ will both have the value of 10^{-7}, and thus the pH of a "neutral" solution will be 7. Consequently if the $[H^+]$ is greater than $[OH^-]$ then the pH will be less than 7 (acidic) and vice versa when the solution will be termed alkaline.

It should be made clear that pH involves the measurement of the hydrogen ion activity and not the molar concentration.

The activity of a chemical species is a thermodynamic concept which denotes its effective concentration, i.e. how effectively it interacts with its surroundings, such as other solutes. According to Raoult's law, solvents only behave in an ideal manner in an indefinitely dilute solution (i.e. when there is no solute present and the solvent is pure). For solutes, a solution of unit activity (activity = 1) can be defined as a hypothetical $1 \, \text{mol} \, l^{-1}$ (M) solution of the solute in which it behaves as if it were actually present in infinitely dilute solution. Thus, the activity of a solute in solution can be equated with its effective molar concentration by means of an activity coefficient. This can be calculated by the following equation:

$$\text{activity coefficient } (\gamma) = \frac{\text{activity of solute } \{A\}}{\text{concentration of solute } [A]} \tag{6.31}$$

where { } are used to differentiate activity from concentration.

This represents the extent by which the solute departs from ideal behaviour (according to Henry's law) and shows that the activity coefficient approaches unity as the concentration of the solute approaches zero, i.e. concentrations $\rightarrow$ activities at low concentrations.

Ion activities in complex environmental solutions such as groundwaters and soil leachates may be difficult to determine, although their determination may be essential to perform accurate assessments of problems such as nuclear waste migration from underground repositories. Computer programs are generally used in this type of work.

6.9 Oxidation–reduction (redox) reactions

Redox reactions involve changes of oxidation states of reactants with transfer of electrons from one reactant to the other. Both processes occur simultaneously, i.e. one of the reactants is oxidized and the other one is reduced. The substance losing electrons is being oxidized (the reducing agent), and the substance receiving electrons is being reduced (oxidizing agent).

For example, soluble cadmium (Cd^{2+}) is removed from wastewater by reaction with metallic iron:

$$Cd^{2+} + Fe \rightarrow Cd + Fe^{2+} \tag{6.32}$$

This is the sum of two half reactions. Cd accepts two electrons and is reduced:

$$Cd^{2+} + 2e \rightarrow Cd \tag{6.33}$$

At the same time, iron is oxidized:

$$Fe \rightarrow Fe^{2+} + 2e \tag{6.34}$$

The addition of an oxidation half reaction and a reduction half reaction, each expressed for the same number of electrons so that the electrons cancel on both sides, results in a whole redox reaction. Overall redox reactions must always be the combination of two half reactions.

Important redox reactions in environmental systems include the following.

1   Reduction of insoluble Fe (III) to soluble Fe (II):

$$Fe(OH)_{3(s)} + 3H^+ + e \rightarrow Fe^{2+} + 3H_2O \tag{6.35}$$

2   Oxidation of ammonium to nitrate (nitrification):

$$NH_4^+ + 2O_2 \rightarrow NO_3^- + 2H^+ + H_2O \tag{6.36}$$

Many redox reactions in environmental media are controlled by microorganisms and can also be closely tied up with acid–base relationships.

Whereas the acidity of a solution is expressed as H^+ activity, the degree to which that solution is oxidizing or reducing can be expressed as electron activity.

By using a platinum electrode and a reference electrode, the redox potential of a system may be measured. The redox potential is the electromotive force (emf) in volts. This value can be used to determine the extent of oxidation and reduction in an environmental system which will affect the chemical speciation of elements. For example, positive redox potential values indicate that the system has oxidizing conditions and a negative potential indicates that the system is reducing. The larger the number the more highly oxidizing the system will be and vice versa. For example, surface waters which are generally oxidizing will contain species such as CO_2, NO_3^-, SO_4^{2-}, $Fe(OH)_3$ and MnO_2. On moving down the water column, conditions will become more reducing (i.e. less oxygen) resulting in the above species changing to CH_4, NH_4^+, H_2S, Fe^{2+} and Mn^{2+}. These reactions also have an effect on heavy metal mobility and speciation.

References and further reading

Harrison, R. M. and deMora, S. J. (1996) *Introductory Chemistry for the Environmental Sciences*, 2nd edn, Cambridge University Press, Cambridge, pp. 1–52.

Mackay, K. M. and Mackay, R. A. (1981) *Introduction to Modern Organic Chemistry*, 3rd edn, International Textbook Co. 349 pp.

Sawyer, C. N., McCarty, P. L. and Parkin, G. F. (1994) *Chemistry for Environmental Engineering*, 4th edn, McGraw-Hill, Singapore, pp. 1–186.

Chapter 7

Organic chemistry

7.1 Introduction

Organic chemistry is the chemistry of carbon and its compounds. Living systems form organic molecules and as such they are therefore essential to life. Proteins, nucleic acids, carbohydrates and fats are all organic compounds (i.e. the principal component is carbon).

Carbon-containing compounds were originally termed "organic" because it was thought that they could only be produced by living plants and animals. However, in 1828 Freidrich Wöhler converted lead cyanate (an inorganic compound) into urea (an organic compound) by treating it with aqueous ammonia.

$$\underset{\text{lead cyanate}}{Pb(OCN)_2} + 2H_2O + \underset{\text{ammonia}}{2NH_3} \rightarrow \underset{\text{urea}}{2H_2NCONH_2} + Pb(OH)_2 \qquad (7.1)$$

Since this discovery, more than six million organic compounds have been synthesized. These range from molecules such as penicillin to artificial sweeteners like saccharin.

Organic chemists are generally concerned with the various pathways involved in synthesizing organic compounds and improving the yield of the final product. Environmental chemists and engineers wish to know about the degradation of organic chemicals and how they will interact with all the environmental compartments (i.e. air, water and land). In order to achieve this, one must have a fundamental knowledge of organic chemistry. This chapter discusses these fundamental aspects and the various classes of organic compounds.

7.2 The carbon atom

Carbon can form a vast amount of compounds. This is partly due to the fact that a carbon atom contains four electrons in its outer shell and can thus form four single covalent bonds by sharing with four electrons from other atoms. Therefore, the carbon atom will attain a stable electronic configuration (i.e.

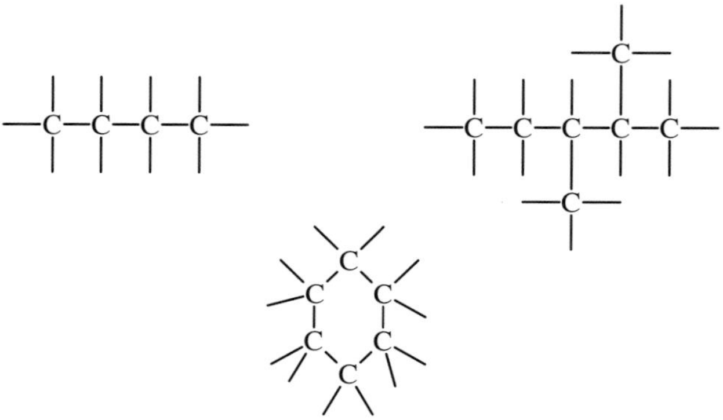

Fig. 7.1 Different arrangements of carbon chains.

eight electrons in the outer shell). Carbon can also form covalent bonds with other carbon atoms, a property which is essentially unique to carbon (although silicon also exhibits this). This can result in carbon chains of enormous length and variation. Some examples of this are shown in Fig. 7.1.

Carbon also has the ability to form multiple bonds with carbon and other elements (e.g. oxygen). Examples of this are given in Fig. 7.2. It follows that a triple covalent bond will be much harder to break than a single covalent bond as more energy will be required for the bond breaking process.

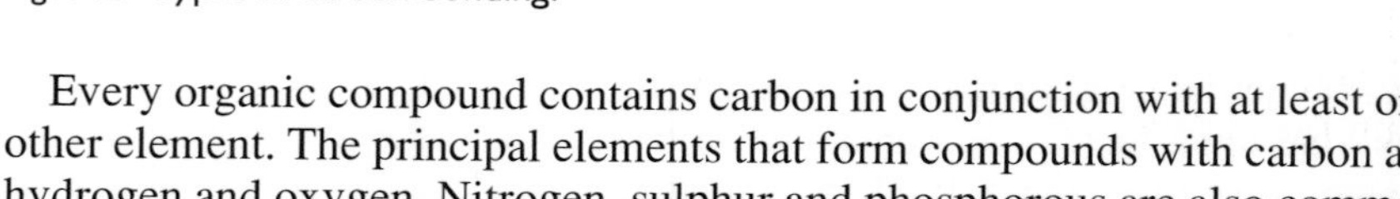

Ethene   Ethyne (acetylene)   Propanone (acetone)   Cyclopropane

Fig. 7.2 Types of carbon bonding.

Every organic compound contains carbon in conjunction with at least one other element. The principal elements that form compounds with carbon are hydrogen and oxygen. Nitrogen, sulphur and phosphorous are also common minor elements found in naturally occurring organic compounds. Several carbon compounds such as carbon dioxide and the carbonates are ionic and therefore are not classed under organic compounds.

7.2.1 General properties of organic compounds

Organic compounds compared with inorganic compounds have the following properties:

1. They are usually combustible.
2. They have lower melting and boiling points.
3. They are less soluble in water.
4. Several compounds exist for a given formula – called isomers.
5. Reactions are usually molecular rather than ionic.
6. The oxidation number (or valency) of carbon allows it to form multiple bonds.
7. They can serve as a food source for bacteria and other micro-organisms.

As the size of the organic molecule increases, the melting and boiling points will also increase. This is because more energy will be needed to break down the additional bonds in the carbon chain.

7.2.2 Formulae

There are several types of formula used in organic chemistry. *Empirical formulae* indicate the proportions/ratios in which the different types of atoms are present. For example, acetic acid will have the empirical formula CH_2O. This indicates that there are twice as many hydrogen atoms as there are carbon or oxygen. *Molecular formulae* give the exact number of different types of atoms in an organic molecule, i.e. the true composition of the molecule. Thus, acetic acid will have the molecular formula $C_2H_4O_2$. *Structural formulae* show the position of every atom and bond in a molecule and are applied to the molecular formula, e.g. acetic acid $C_2H_4O_2$:

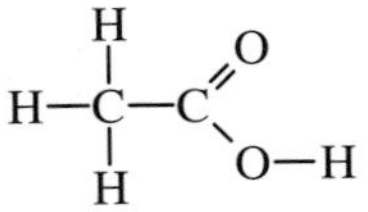

For the organic chemist, it would be a lengthy process to write out a complicated reaction sequence using structural formulae. Therefore, these are simplified to *condensed formulae*, for example

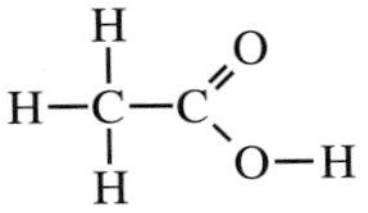

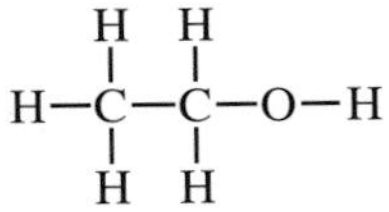

7.2.3 Isomerism

Isomerism can be defined as the existence of two or more chemical compounds with the same molecular formula, but having different physical and chemical properties due to the different arrangement of atoms within

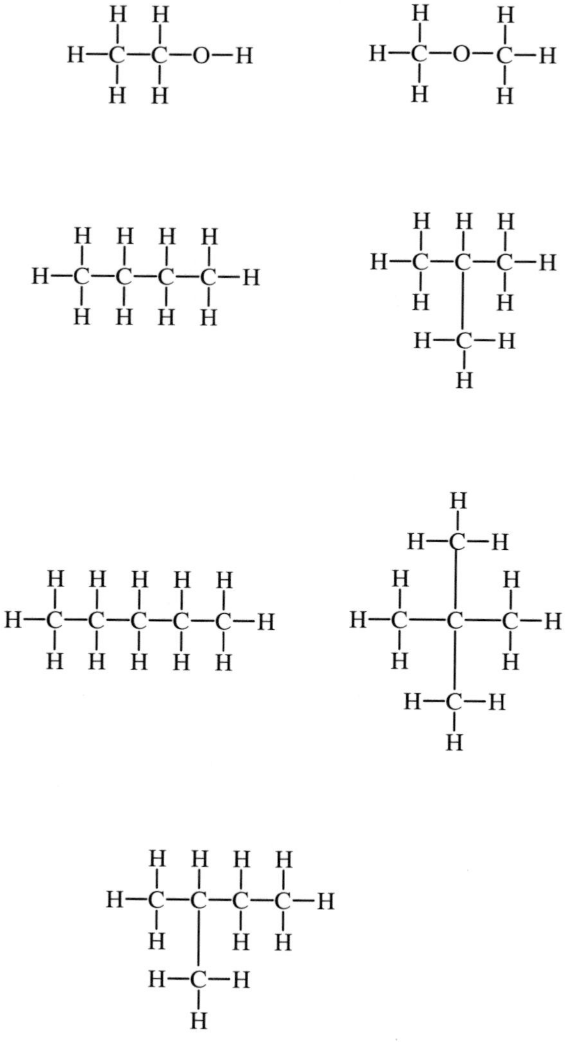

```
   H  H                          H       H
   |  |                          |       |
H—C—C—O—H                    H—C—O—C—H
   |  |                          |       |
   H  H                          H       H
```

```
   H  H  H  H                  H  H  H
   |  |  |  |                  |  |  |
H—C—C—C—C—H                H—C—C—C—H
   |  |  |  |                  |  |  |
   H  H  H  H                  H  |  H
                                  |
                               H—C—H
                                  |
                                  H
```

```
                                         H
                                         |
                                      H—C—H
   H  H  H  H  H                 H    |    H
   |  |  |  |  |                 |    |    |
H—C—C—C—C—C—H             H—C———C———C—H
   |  |  |  |  |                 |    |    |
   H  H  H  H  H                 H    |    H
                                   H—C—H
                                      |
                                      H
```

```
   H  H  H  H
   |  |  |  |
H—C—C—C—C—H
   |  |  |  |
   H  |  H  H
      |
   H—C—H
      |
      H
```

Fig. 7.3 Isomerism in alkanes.

the molecule. Some simple examples of this phenomenon are given in Fig. 7.3.

It can be seen from Fig. 7.3 that the more carbon atoms there are in the molecule, the greater the number of possible isomers. Therefore, it should become apparent why there is such a vast number of organic compounds.

Essentially there are three major types of organic compounds: aliphatic, aromatic and heterocyclic. Aliphatics contain only straight or branched carbon chains, aromatic compounds contain six membered carbon rings containing three double bonds: and heterocyclics possess a ring structure containing one element other than carbon.

7.3 Aliphatic compounds

7.3.1 Alkanes

Organic compounds containing only carbon and hydrogen atoms are termed hydrocarbons. Alkanes (sometimes called paraffins) are hydrocarbons that contain only carbon–carbon single bonds and carbon–hydrogen bonds. This type of compound is said to be *saturated*, i.e. they contain no double or triple carbon–carbon bonds which are capable of accepting subsequent hydrogens. Alkanes have the general formula $C_n H_{2n+2}$ where n is the number of carbon atoms in the compound. For example, if $n = 2$, $(2n + 2) = 6$ and the alkane would have the formula $C_2 H_6$ which is called ethane. Each successive member of the alkane series differs by the presence of a methylene group (CH_2). A series of compounds that are related in this manner produce a *homologous series* and can thus be given a general formula such as the one above.

Alkanes are classified into several types according to their structure. These are straight-chain alkanes, branched alkanes and cycloalkanes. Examples of these are shown in Fig. 7.4.

Branched alkanes are formed by the removal of a hydrogen atom from a methylene (CH_2) group and replacement by an *alkyl* group. Alkyl groups are produced when one hydrogen atom is removed from an alkane. Examples of this are methyl (CH_3), ethyl ($C_2 H_5$) and propyl ($C_3 H_7$). These species are

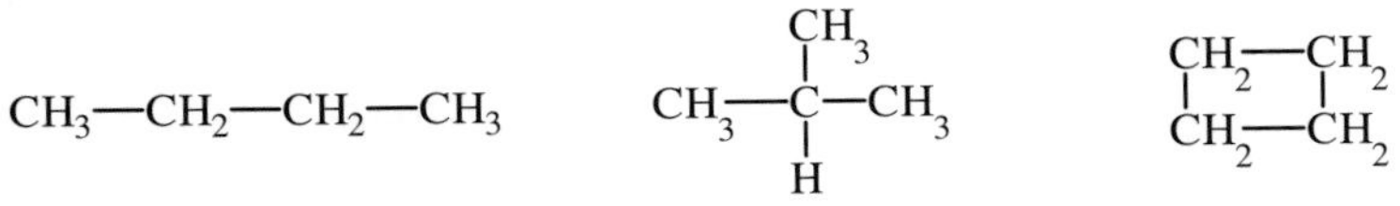

$$CH_3\!-\!CH_2\!-\!CH_2\!-\!CH_3 \qquad CH_3\!-\!\overset{\displaystyle CH_3}{\underset{\displaystyle H}{C}}\!-\!CH_3 \qquad \begin{array}{c} CH_2\!-\!CH_2 \\ | \quad\quad | \\ CH_2\!-\!CH_2 \end{array}$$

Fig. 7.4 Examples of different alkanes.

known as *radicals*. These are very reactive species due to the fact that they contain one unpaired electron and generally cannot be isolated.

(a) Nomenclature

The naming of organic compounds is very important. Systematic nomenclature has been adopted since 1892 and has been revised by the International Union of Pure and Applied Chemistry to produce the so-called IUPAC rules. The names of the first ten alkanes in the series are given in Table 7.1 along with some physical properties.

Alkanes are characterized by all the names ending in *ane*. The first four alkanes have special names that have been incorporated into the IUPAC system. After these, the names start with a Greek prefix which indicates the number of carbon atoms in present in the chain, e.g. pentane (five), hexane (six), heptane (seven) and so on.

The alkanes are considered to be parent compounds by chemists as they are used to produce a huge variety of organic chemicals. Moreover, the naming of different organic compounds is based on alkane nomenclature. The IUPAC rules for naming alkanes are as follows:

1. Choose the longest carbon chain and name this.
2. Name all groups (if any) which are attached to the longest chain as alkyl groups.
3. Number the longest carbon chain beginning at the end that is closest to an alkyl group.
4. Write the name of the alkane by naming all the substituents first in alphabetical order, followed by the alkane name for the longest carbon chain in the compound.

Table 7.1 Physical properties of alkanes

Name	Formula	Mpt (°C)	Bpt (°C)	Density at 20°C (g ml^{-1})
Methane	CH_4	−182.5	−161.7	0.555 (at 0°C)
Ethane	C_2H_6	−183.3	−88.6	0.509 (at −60°C)
Propane	C_3H_8	−187.7	−42.1	0.500
Butane	C_4H_{10}	−138.3	−0.5	0.579
Pentane	C_5H_{12}	−129.8	36.1	0.557
Hexane	C_6H_{14}	−95.3	68.7	0.660
Heptane	C_7H_{16}	−90.6	98.4	0.684
Octane	C_8H_{18}	−56.8	125.7	0.703
Nonane	C_9H_{20}	−53.5	150.8	0.718
Decane	$C_{10}H_{22}$	−29.7	174.0	0.730

These rules are illustrated in the following examples:

$$CH_3—CH_2—CH_2—CH_3$$

(I)

$$\underset{1}{CH_3}—\underset{2}{\overset{\displaystyle CH_3}{\overset{|}{CH}}}—\underset{3}{CH_2}—\underset{4}{CH_2}—\underset{5}{CH_3}$$

(II)

In the case of compound I, there are no substituents and the longest carbon chain contains four atoms. Therefore, the name of the compound is simply butane. For compound II, the longest carbon chain contains five atoms (i.e. pentane) and has one alkyl group situated on carbon number 2 (start numbering at the closest end to the substituent). Thus, application of rule (4) results in the name, 2-methylpentane.

If a molecule should contain more than one of the same alkyl groups, its name will be preceded by the prefix di, tri, tetra and so on. For example, the compound below would be called 2, 3-dimethylpentane.

$$\underset{1}{CH_3}—\underset{2}{\overset{\displaystyle CH_3}{\overset{|}{CH}}}—\underset{3}{\overset{\displaystyle CH_3}{\overset{|}{CH}}}—\underset{4}{CH_2}—\underset{5}{CH_3}$$

(b) Physical properties

The alkanes are colourless gases or liquids with higher alkanes (i.e. above 17 carbons in the chain) forming solids. They are generally insoluble in water but will readily dissolve in organic solvents. From the melting and boiling points given in Table 7.1, it can be seen that the values increase as the number of carbon atoms increases. This is because more energy will be needed to break down the additional bonds in the structure.

(c) Chemical reactions

Owing to their saturation and absence of functional groups, the alkanes are very unreactive. Despite this, there are a number of chemical reactions that alkanes will undergo.

Alkanes will burn in air (oxidation) to produce CO_2 and H_2O. This combustion process can be defined as a chemical reaction with oxygen (usually at elevated temperatures) in which an alkane (or other reactant) is converted into carbon dioxide and water (or other oxidized products). A simple example is shown in equation (7.2):

$$CH_{4(g)} + 2O_{2(g)} \quad \rightarrow \quad CO_{2(g)} + 2H_2O_{(g)} + heat \tag{7.2}$$

This reaction produces a large amount of heat energy, which is the reason these compounds are widely used as fuels.

Alkanes can be oxidized by the use of certain enzymes in microbial systems. The first step involves the addition of an oxygen to an alkane to produce an alcohol:

$$2CH_3CH_3 + O_2 \xrightarrow{\text{enzyme}} 2CH_3CH_2OH \qquad (7.3)$$

ethane   ethanol

An example of an enzyme capable of achieving this is *Methylococcus capsulatus*. Ultimately, further oxidation will take place and the hydrocarbon is eventually converted into CO_2 and H_2O. Because of this reaction, bacteria have been suggested as a possible means of cleaning up oil spills by oxidative degradation.

In alkanes, the C—C bonds are stronger than the C—H bonds. Therefore, it follows that the replacement of a hydrogen atom by another functional group will be a more energetically favourable process. This reaction is known as substitution and is illustrated in equation (7.4) by the chlorination of methane:

$$CH_{4(g)} + Cl_{2(g)} \rightarrow CH_3Cl + HCl \qquad (7.4)$$

chloromethane

This substitution reaction will not stop at chloromethane and subsequent substitutions will occur producing di, tri, and tetrachloromethane.

High molecular weight alkanes that are heat treated (pyrolysis) result in the C—C and C—H bonds being broken and smaller molecules being produced. This process is also known as *cracking*. Alkenes can also be formed and so the end products will be a complicated mixture of alkanes and alkenes. The reaction can be controlled by the use of special catalysts (zeolites) to increase the yields of lower molecular weight fractions which are used as petroleum or for synthetic chemical applications.

7.3.2 Alkenes

This class of compound is characterised by a carbon–carbon double bond. Owing to the presence of this double bond, these compounds are said to be *unsaturated*. Alkenes have the general formula C_nH_{2n} although the first compound in this homologous series will be ethene, C_2H_4. Higher alkenes in the series are formed by the addition of alkyl groups. Some simple alkenes are illustrated in Fig. 7.5.

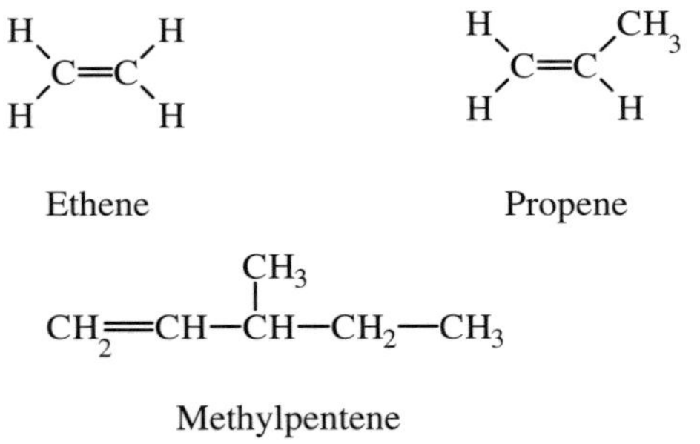

Ethene   Propene

Methylpentene

Fig. 7.5 Simple alkenes.

(a) Nomenclature

In IUPAC nomenclature, the suffix -*ene* is used to indicate the presence of a carbon–carbon double bond. This can be easily understood by observing the alkenes above. When the number of carbon atoms is $\geqslant 4$, then isomerism can occur and the position of the double bond in the carbon chain will be indicated by a number. The numbering of the carbon chain will start from the end that is closest to the double bond, and is shown in the following examples:

$$CH_2 = CH - CH_2 - CH_3 \qquad\qquad CH_3 - CH = CH - CH_3$$

1-butene   2-butene

Another type of isomerism can occur due to the fact that a double bond prohibits any free rotation of the carbon atoms about the bond, thus keeping the molecule fixed in one plane. This produces *stereoisomerism* and produces *cis* and *trans* isomers. These compounds have the same sequence of bonds but differ in their arrangement of atoms in space and as a result can exhibit different physical properties. An example of stereoisomerism is given in Fig. 7.6.

(b) Physical properties

Table 7.2 gives some names and formulae of selected members of the alkene series together with some physical data. The data show that the boiling

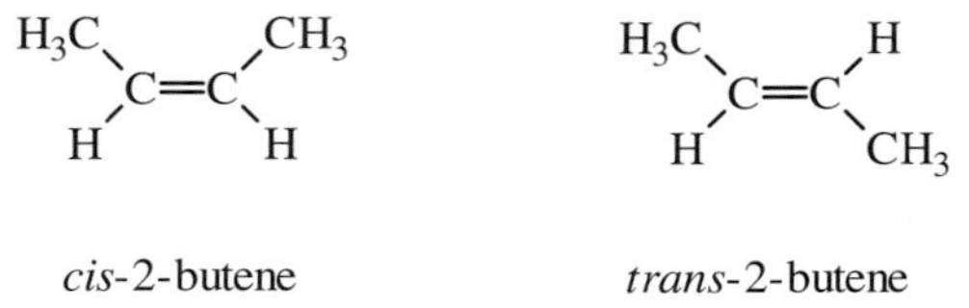

$$H_3C \diagdown \diagup CH_3$$
$$C = C$$
$$H \diagup \diagdown H$$

$$H_3C \diagdown \diagup H$$
$$C = C$$
$$H \diagup \diagdown CH_3$$

cis-2-butene trans-2-butene

Fig. 7.6 Stereoisomerism in an alkene.

Table 7.2 Physical properties of selected alkenes

Name	Formula	Mpt (°C)	Bpt (°C)	Density at 20°C (g ml⁻¹)
Ethene	$CH_2{=}CH_2$	−169.15	−103.71	0.384 (at −10°C)
Propene	$CH_2{=}CH{-}CH_3$	−185.25	−47.4	0.5193
1-butene	$CH_2{=}CH{-}CH_2{-}CH_3$	−185.35	−6.3	0.5951
2-butene	$CH_3{-}CH{=}CH{-}CH_3$	−105.55	1	0.6042
1-pentene	$CH_2{=}CH{-}(CH_2)_2{-}CH_3$	−138	29.96	0.6405
1-hexene	$CH_2{=}CH{-}(CH_2)_3{-}CH_3$	−139.82	63.35	0.6731
1-heptene	$CH_2{=}CH{-}(CH_2)_4{-}CH_3$	−119	93.64	0.6970
1-octene	$CH_2{=}CH{-}(CH_2)_5{-}CH_3$	−101.73	121.3	0.7149
1-nonene	$CH_2{=}CH{-}(CH_2)_6{-}CH_3$	−81	149.9	0.730
1-decene	$CH_2{=}CH{-}(CH_2)_7{-}CH_3$	−66.3 (freezes)	170.56	0.7408

points of alkenes are close to those of their corresponding alkanes. Ethene, propene and butenes are all gases at room temperature with higher alkenes in the series being liquids.

(c) Chemical reactions

Alkenes are more reactive than alkanes due to the presence of the carbon–carbon double bond. They will burn readily in air (oxidation) to produce carbon dioxide, water and heat, e.g. ethene:

$$C_2H_{4(g)} + 3O_{2(g)} \quad \rightarrow \quad 2CO_{2(g)} + 2H_2O_{(g)} + heat \tag{7.5}$$

Alkenes can undergo rapid reactions with halogens, hypochlorous acid (HOCl) and halogen acids (e.g. HCl) to eliminate the unsaturation in the molecule. This is shown in the following example:

$$\underset{\text{Ethene}}{\overset{H}{\underset{H}{>}}C=C\overset{H}{\underset{H}{<}}} \quad + \quad Br_2 \quad \longrightarrow \quad \underset{\text{1,2-dibromoethane}}{H-\overset{\overset{\displaystyle H}{|}}{\underset{\underset{\displaystyle Br}{|}}{C}}-\overset{\overset{\displaystyle H}{|}}{\underset{\underset{\displaystyle Br}{|}}{C}}-H} \qquad (7.6)$$

In cases where alkenes and unsaturated compounds in general are present in waste effluents, the chlorine demand will be high because of these reactions.

Hydrogen in the presence of a catalyst can also be added across the double bond to produce the corresponding alkane. This is known as *hydrogenation*.

$$\underset{\text{Ethene}}{\overset{H}{\underset{H}{>}}C=C\overset{H}{\underset{H}{<}}} \quad + \quad H_2 \quad \xrightarrow[\text{catalyst}]{\text{nickel}} \quad \underset{\text{Ethane}}{H-\overset{\overset{\displaystyle H}{|}}{\underset{\underset{\displaystyle H}{|}}{C}}-\overset{\overset{\displaystyle H}{|}}{\underset{\underset{\displaystyle H}{|}}{C}}-H} \qquad (7.7)$$

At high temperatures and pressures, and in the presence of a catalyst (e.g. free radical), alkenes can combine with each other to produce polymers, a process known as *polymerization*.

$$\underset{\text{Ethene}}{nCH_2=CH_2} \quad \xrightarrow[\substack{\text{1,000 atm} \\ >100°C}]{\text{ROOR}} \quad \underset{\text{Polyethene}}{-(CH_2-CH_2)_n-} \qquad (7.8)$$

Polymers have become an important part of our lives. Polystyrene, polyvinyl chloride (PVC), Teflon and many more polymers have desirable chemical and physical properties which make them useful for numerous applications. Table 7.3 gives some examples of common polymers.

Many of these polymers are not biodegradable, which can result in pollution problems. Indeed, wastes from the polymer industry will exhibit a high chlorine demand due to reactions with polymers. This may result in the production of very toxic organochlorine compounds.

7.3.3 Alkynes

Alkynes are hydrocarbons that contain a carbon–carbon triple bond. Hence, these compounds are also classed as *unsaturated*. They have the general formula C_nH_{2n-2}. Examples of some alkynes and their physical properties are given in Table 7.4.

Table 7.3 Common polymers

Monomer	Structure	Polymer
Ethene	$CH_2=CH_2$	Polyethene
Chloroethene (vinyl chloride)	$CH_2=CHCl$	Polyvinyl chloride (PVC)
Tetrafluoroethene	$CF_2=CF_2$	Teflon
Phenylethene (styrene)	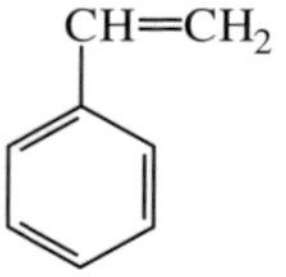	Polystyrene
Propenenitrile	$CH_2=C\overset{C\equiv N}{\underset{H}{\diagup\diagdown}}$	Orlon

Table 7.4 Physical properties of selected alkynes

Name	Formula	Mpt (°C)	Bpt (°C)	Density at 20°C (g ml^{-1})
Ethyne	$CH\equiv CH$	−80.8	−84.0(sub)	0.6208 (at 8°C)
Propyne	$CH_3-C\equiv CH$	−101.5	−23.2	0.7062 (at −50°C)
1-butyne	$CH_3-CH_2-C\equiv CH$	−125.72	8.1	0.6784 (at 0°C)
2-butyne	$CH_3-C\equiv C-CH_3$	−32.26	27	0.6910
1-pentyne	$CH_3-CH_2-CH_2-C\equiv CH$	−90.0	40.18	0.6901
2-pentyne	$CH_3-CH_2-C\equiv C-CH_3$	−101	56.07	0.7107

With alkynes, the suffix -*yne* is used to indicate the presence of a carbon–carbon triple bond. In cases where a hydrocarbon contains a double and
a triple bond, the resulting compound is termed an *alkenyne*. The following
example illustrates this:

$$\overset{6}{C}H_3-\overset{5}{C}H_2-\overset{4}{C}H=\overset{3}{C}H-\overset{2}{C}\equiv\overset{1}{C}H$$

3-hexen-1-yne

Alkyne compounds are known to polymerize very easily and can react very
violently. For example, ethyne (acetylene) can explode under pressure.

7.3.4 Alcohols

Alcohols may be regarded as hydrocarbons (alkanes) wherein a hydrogen atom has been replaced by a *hydroxy* (OH) group. In this case, the general formula for alcohols is given by R—OH, where R is the alkyl group (e.g. methyl, ethyl etc.).

(a) Nomenclature

Using the IUPAC rules, alcohols are considered to be derivatives of alkanes (i.e. they are the 1° oxidation product). The final -e on the alkane will be replaced by -*ol*. The position of the OH group may need to be given and this is achieved by numbering the carbon chain so that the position of the OH group has the smallest number. The examples below illustrate this:

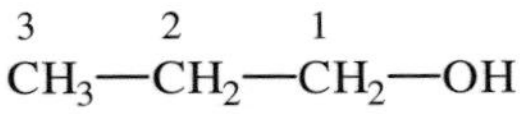

$$\overset{3}{C}H_3-\overset{2}{C}H_2-\overset{1}{C}H_2-OH$$

1-propanol

$$\overset{3}{C}H_3-\overset{2}{\underset{OH}{\overset{H}{C}}}-\overset{1}{C}H_3$$

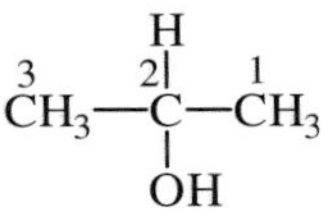

2-propanol

Alcohols can also be classified as primary, secondary and tertiary; this is dependent on where the OH group is attached. The general formulae for these three alcohol types are given in Fig. 7.7.

Any alkyl side groups present in the alcohol will be named according to the rules for hydrocarbons.

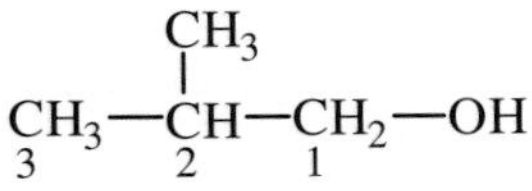

$$\overset{3}{C}H_3-\overset{2}{\underset{CH_3}{\overset{|}{C}H}}-\overset{1}{C}H_2-OH$$

2-methyl-1-propanol

The chemistry of these alcohols is different and thus one must be equipped with the knowledge to distinguish between them.

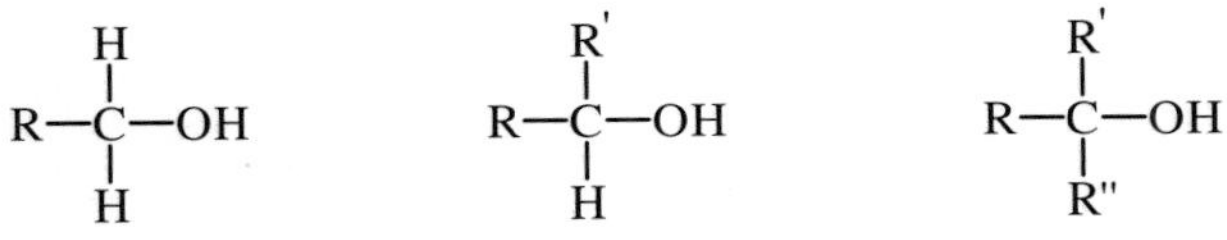

$$R-\overset{H}{\underset{H}{\overset{|}{C}}}-OH \qquad R-\overset{R'}{\underset{H}{\overset{|}{C}}}-OH \qquad R-\overset{R'}{\underset{R''}{\overset{|}{C}}}-OH$$

Primary   Secondary   Tertiary

Fig. 7.7 General formulae for 1°, 2° and 3° alcohols.

Table 7.5 Physical properties of selected alcohols

Name	Formula	Mpt (°C)	Bpt (°C)	Density at 20°C (g ml^{-1})
Methanol	CH_3OH	− 93.9	64.96	0.7914
Ethanol	CH_3CH_2OH	−117.3	78.5	0.7893
1-propanol	$CH_3CH_2CH_2OH$	−126.5	97.4	0.8035
2-propanol	$CH_3CHOHCH_3$	−89.5	82.4	0.7855
1-butanol	$CH_3(CH_2)_3OH$	−89.5	117.25	0.8098
1-pentanol	$CH_3(CH_2)_4OH$	−79	137.3	0.8144
1-hexanol	$CH_3(CH_2)_5OH$	−46.7	158	0.8136
1-heptanol	$CH_3(CH_2)_6OH$	−34.1	176	0.8219
1-octanol	$CH_3(CH_2)_7OH$	−16.7	194.45	0.8270
1-nonanol	$CH_3(CH_2)_8OH$	−5.5	213.5	0.8273
1-decanol	$CH_3(CH_2)_9OH$	7	229	0.8297

(b) Physical properties

Table 7.5 shows some physical properties of alcohols. Compared with the physical properties of alkanes (Table 7.1), alcohols have much higher boiling points. The reason for this lies in the ability of the alcohol molecules to form hydrogen bonds with each other as well as with water molecules. This is shown diagrammatically in Fig. 7.8. Although these bonds are relatively weak, so many are formed that to break them all requires a great deal of energy. Hence alcohols have elevated boiling points compared with the alkanes. Hydrogen bonding between alcohols and water can also take place. As a result, many alcohols are also water soluble.

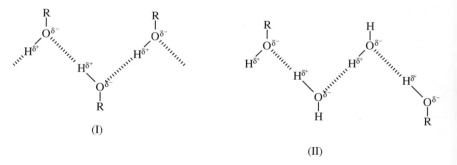

Fig. 7.8 Hydrogen bonding in alcohols.

(c) Chemical reactions

Primary alcohols can be oxidized to aldehydes according to the following equation:

$$R-CH_2-OH \;+\; \tfrac{1}{2}O_2 \;\xrightarrow{\text{oxidation}}\; H_2O \;+\; R-\overset{\displaystyle H}{\underset{}{C}}=O \qquad (7.9)$$

Primary alcohol   Aldehyde

The use of oxidation reagents such as chromium (VI) can be used to achieve this. There is a possibility that the reaction can go even further to produce a carboxylic acid (R—COOH). These compounds are discussed in Section 7.3.6. Secondary alcohols are oxidized to ketones:

$$R-\overset{\displaystyle H}{\underset{\displaystyle R'}{C}}-OH \;+\; \tfrac{1}{2}O_2 \;\xrightarrow{\text{oxidation}}\; H_2O \;+\; R-\overset{\displaystyle }{\underset{\displaystyle R'}{C}}=O \qquad (7.10)$$

Secondary alcohol   Ketone

Both reactions described are reversible. Overoxidation to carboxylic acids is more common with aldehydes and can be prevented by performing the reaction under anhydrous conditions. Ketones are more stable and are not readily oxidized.

Alcohols can react with carboxylic acids to produce esters:

$$CH_3-\overset{\displaystyle O}{\overset{\displaystyle \|}{C}}-OH \;+\; CH_3-CH_2-OH \;\xrightarrow{H^+}\; CH_3-\overset{\displaystyle O}{\overset{\displaystyle \|}{C}}-O-CH_2-CH_3 \;+\; H_2O$$

Ethanoic acid   Ethanol   Ethyl ethanoate   (7.11)

This reaction can also take place with inorganic acids. Esters are discussed in Section 7.3.7.

7.3.5 Aldehydes and ketones

Aldehyde and ketone chemistry is focused around the reactive *carbonyl* group, $>C=O$. In aldehydes, the carbonyl carbon atom is bonded to a carbon and a hydrogen. The general formula for aldehydes is R—CHO, the R group being any alkyl group. In ketones, the carbonyl group is attached to two carbon atoms. The general formula is:

$$R-\overset{\displaystyle O}{\overset{\displaystyle \|}{C}}-R'$$

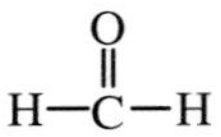

$$
\underset{\text{Methanal}}{H-\overset{\overset{\displaystyle O}{\parallel}}{C}-H}
\qquad
\underset{\text{Ethanal}}{CH_3-\overset{\overset{\displaystyle O}{\parallel}}{C}-H}
\qquad
\underset{\text{Propanal}}{CH_3-CH_2-\overset{\overset{\displaystyle O}{\parallel}}{C}-H}
$$

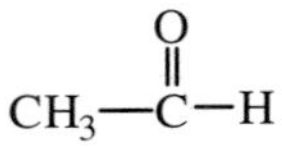

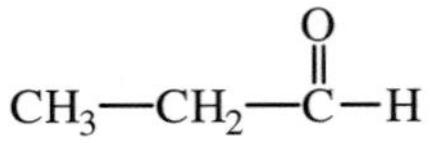

Fig. 7.9 Simple aldehydes.

(a) Nomenclature

Using the systematic (IUPAC) approach, aldehydes and ketones are treated as derivatives of alkanes. For aldehydes, the ending -e on the alkane is replaced by -al, thus turning an alkane into an alkanal. Examples of simple aldehydes are given in Fig. 7.9.

With aldehydes, the position of the carbonyl group is not stated. The carbonyl carbon is taken to be 1. Other substituents are then labelled according to their position in relation to the carbonyl group. Some examples can illustrate this:

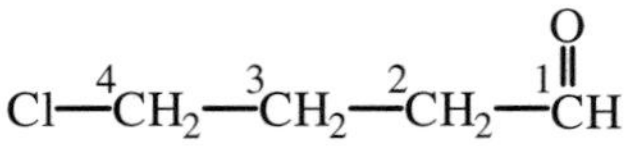

$$
Cl-\overset{4}{C}H_2-\overset{3}{C}H_2-\overset{2}{C}H_2-\overset{1}{C}\overset{\overset{\displaystyle O}{\parallel}}{H}
\qquad
CH_3-\overset{\overset{\displaystyle CH_3}{|}}{\underset{\underset{\displaystyle H}{|}}{C}}-\overset{\overset{\displaystyle O}{\parallel}}{C}-H
$$

$$
\text{4-chlorobutanal} \qquad\qquad \text{2-methyl propanal}
$$

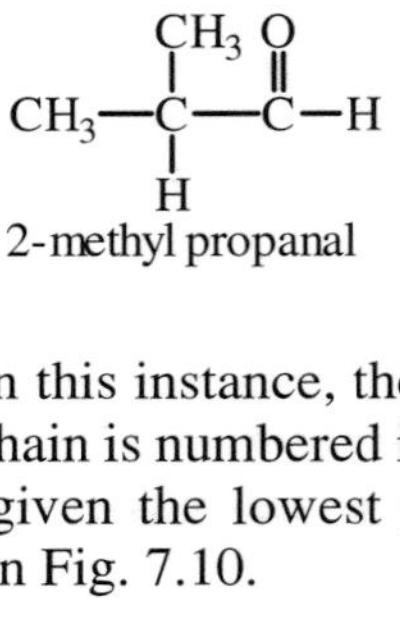

Ketones are also known as *alkanones*. In this instance, the -e on the end of alkane is replaced by -one. The carbon chain is numbered in such a way that the position of the carbonyl group is given the lowest possible number. Examples of simple ketones are shown in Fig. 7.10.

(b) Physical properties

Oxygen is more electronegative than carbon in the carbonyl group. As a result of this, the carbon–oxygen double bond will be polarized with a small positive charge on the carbon and a small negative charge on oxygen. Hence, a *dipole moment* is produced. This is shown in the following

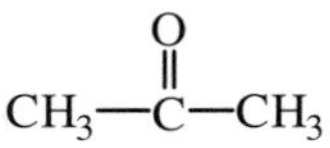

$$
\underset{\text{Propanone}}{CH_3-\overset{\overset{\displaystyle O}{\parallel}}{C}-CH_3}
\qquad
\underset{\text{Butanone}}{CH_3-\overset{\overset{\displaystyle O}{\parallel}}{C}-CH_2-CH_3}
\qquad
\underset{\text{3-Pentanone}}{CH_3-CH_2-\overset{\overset{\displaystyle O}{\parallel}}{C}-CH_2-CH_3}
$$

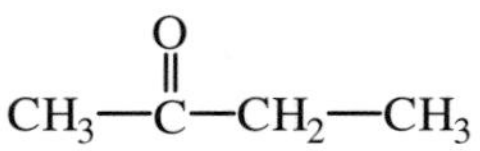

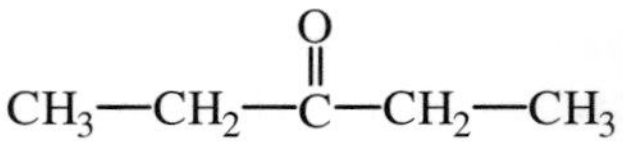

Fig. 7.10 Simple ketones.

diagram:

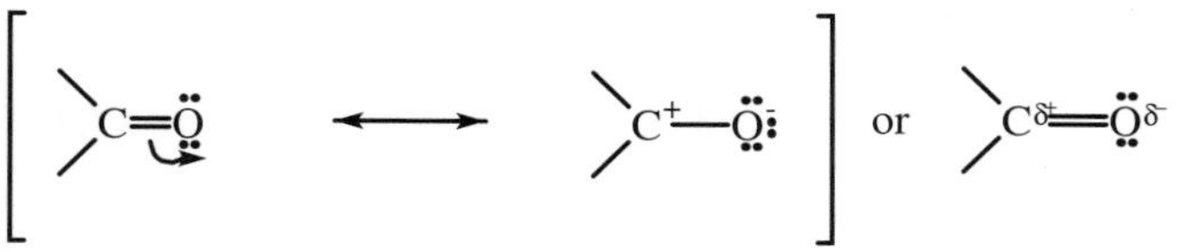

$$\left[\quad \underset{\diagup}{\overset{\diagdown}{C}}{=}\ddot{\underset{\cdot\cdot}{O}} \quad \longleftrightarrow \quad \underset{\diagup}{\overset{\diagdown}{C}}{}^{+}{-}\ddot{\underset{\cdot}{O}}{:} \quad \right] \quad \text{or} \quad \underset{\diagup}{\overset{\diagdown}{C}}{}^{\delta+}{=}\ddot{\underset{\cdot\cdot}{O}}{}^{\delta-}$$

Table 7.6 gives some physical constants for aldehydes and ketones. Owing to their dipole moments, lower carbonyl compounds (i.e. <6 carbon atoms) are soluble in water. As the carbon chain increases, so does the hydrophobicity and therefore the solubility of these compounds in water will decrease. In comparison with their corresponding alkanes, aldehydes and ketones have higher boiling points. This is due to the polarization of the carbonyl group.

(c) Chemical reactions

Aldehydes are more reactive than ketones due to the increased electrophilic character of the carbonyl carbon. Figure 7.11 illustrates the order of reactivity for carbonyl groups with various substituents attached.

Table 7.6 Physical properties of some aldehydes and ketones

Name	Formula	Mpt (°C)	Bpt (°C)	Density at 20°C (g ml^{-1})
Methanal	CH_2O	− 92	−21	0.815
Ethanal	CH_3CHO	−121	20.8	0.7834
Propanal	CH_3CH_2CHO	− 81	48.8	0.8058
Propanone	CH_3COCH_3	− 95.35	56.2	0.7899
Butanal	$CH_3(CH_2)_2CHO$	− 99	75.7	0.8170
Butanone	$CH_3CH_2COCH_3$	− 86.35	79.6	0.8054
Pentanal	$CH_3(CH_2)_3CHO$	− 91.5	103	0.8095
2-pentanone	$CH_3COCH_2CH_2CH_3$	− 77.8	102	0.8089
3-pentanone	$CH_3CH_2COCH_2CH_3$	− 39.8	101.7	0.8138

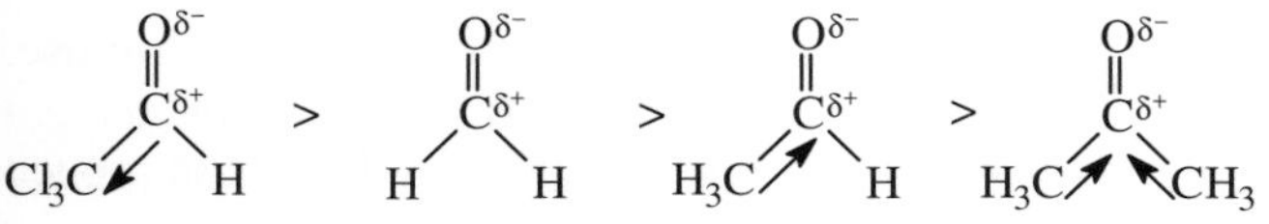

Fig. 7.11 Reactivity series for carbonyl groups with various substituents.

Thus, aldehydes are easily oxidized to produce corresponding carboxylic acids:

$$R-\underset{H}{\overset{}{C}}=O \quad + \quad \tfrac{1}{2}O_2 \quad \longrightarrow \quad R-\underset{OH}{\overset{}{C}}=O \qquad (7.12)$$

Ketones are more difficult to oxidize and usually result in the formation of two or more acids:

$$CH_3-\overset{O}{\overset{\|}{C}}-CH_3 \quad + \quad 2O_2 \quad \longrightarrow \quad H_2O \ + \ CO_2 \ + \ CH_3-\overset{O}{\overset{\|}{C}}-OH$$
$$(7.13)$$

These reactions can be carried out by micro-organisms with further reactions utilizing the organic acid products eventually to produce carbon dioxide and water.

Owing to the nature of the carbonyl group (i.e. polar), polar reagents such as water, alcohols and amines can undergo ionic addition. The general reaction is given in equation (7.14):

$$\underset{R'}{\overset{R}{\diagdown}}C^{\delta+}\!\!=\!\!O^{\delta-} \quad + \quad X^{\delta-}\!\!-\!\!Y^{\delta-} \quad \longrightarrow \quad R-\underset{R'}{\overset{OX}{\overset{|}{\underset{|}{C}}}}-Y \qquad (7.14)$$

Reactions of aldehydes and ketones with alcohols produce acetals ($R_2-C-(OR')_2$); reactions with amines produce imines ($R_2-C=N-R'$).

7.3.6 Carboxylic acids

If a hydroxyl group (OH) is attached to a carbonyl group (>C=O), then a *carboxyl group* is produced. This is generally represented by —COOH, and is characteristic of all organic acids. These organic acids can be utilized (oxidized) by micro-organisms as a food source, with carbon dioxide and water being the final products.

(a) Nomenclature

Some of the simple carboxylic acids still have their common names in use. For example, CH_3COOH is termed acetic acid (from the latin *acetum* meaning vinegar from which it was first derived). The IUPAC approach treats carboxylic acids as derivatives of alkanes and replaces the ending -e with -oic acid. Hence, the IUPAC name for acetic acid would be ethanoic acid, taken from the corresponding alkane, ethane. When numbering substituents, the carboxyl carbon is numbered 1 with the numbers increasing along the carbon chain. This is illustrated in the next diagram:

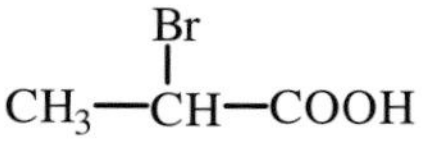

$$CH_3-\underset{\underset{\displaystyle Br}{|}}{CH}-COOH$$

2-bromopropanoic acid

$$\underset{5}{CH_3}-\underset{4}{CH_2}-\underset{3}{\underset{\underset{\displaystyle CH_3}{|}}{CH}}-\underset{2}{\underset{\underset{\displaystyle CH_3}{|}}{CH}}-\underset{1}{COOH}$$

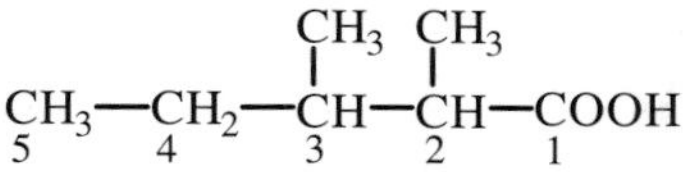

2,3-dimethylpentanoic acid

The position of a substituent can also be indicated by a Greek letter which shows its distance from the carboxyl group:

$$\underset{5}{\overset{\delta}{CH_3}}-\underset{4}{\overset{\gamma}{CH_2}}-\underset{3}{\overset{\beta}{CH_2}}-\underset{2}{\overset{\alpha}{CH_2}}-\underset{1}{COOH}$$

Dicarboxylic acids are termed *alkanedioic acids*. Several examples of this type of acid are given below:

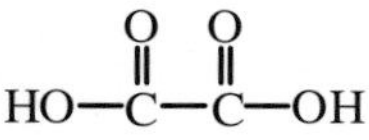

$$\underset{\text{Ethanedioic acid}}{HO-\overset{O}{\overset{||}{C}}-\overset{O}{\overset{||}{C}}-OH}$$

Ethanedioic acid
(Oxalic acid)

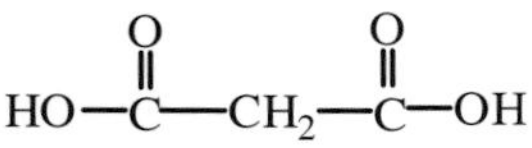

$$HO-\overset{O}{\overset{||}{C}}-CH_2-\overset{O}{\overset{||}{C}}-OH$$

Propanedioic acid
(Malonic acid)

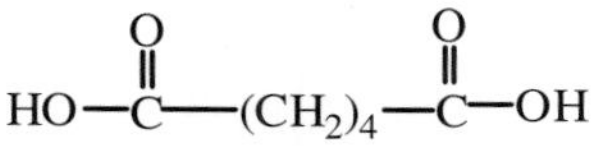

$$HO-\overset{O}{\overset{||}{C}}-(CH_2)_4-\overset{O}{\overset{||}{C}}-OH$$

Hexanedioic acid
(Adipic acid)

Adipic acid can be present in industrial wastes from the nylon industry as it is used in the manufacturing process of nylon fibres.

(b) Physical properties

Table 7.7 shows some physical data for carboxylic acids, together with the IUPAC nomenclature.

Table 7.7 Physical properties of carboxylic acids

Name	Formula	Mpt (°C)	Bpt (°C)	Density at 20°C (g ml⁻¹)
Methanoic acid	$HCOOH$	8.4	100.7	1.220
Ethanoic acid	CH_3COOH	16.6	117.9	1.0492
Propanoic acid	CH_3CH_2COOH	−20.8	140.99	0.9930
Butanoic acid	$CH_3(CH_2)_2COOH$	−4.26	163.53	0.9577
Pentanoic acid	$CH_3(CH_2)_3COOH$	−33.83	186.05	0.9391
Hexanoic acid	$CH_3(CH_2)_4COOH$	−2	205	0.9274
Heptanoic acid	$CH_3(CH_2)_5COOH$	−7.5	223	0.9200
Octanoic acid	$CH_3(CH_2)_6COOH$	16.5	239.3	0.9088
Nonanoic acid	$CH_3(CH_2)_7COOH$	12.24	255	0.9057
Decanoic acid	$CH_3(CH_2)_8COOH$	31.5	270	0.8858

The first nine carboxylic acids (C_1–C_9) are liquids at room temperature with the higher acids becoming greasy solids. As carboxylic acids can form hydrogen bonds, it should come as no surprise that the lower acids (up to butanoic acid) are soluble in water. In neat liquid form, hydrogen bonding will occur to produce dimers as shown below:

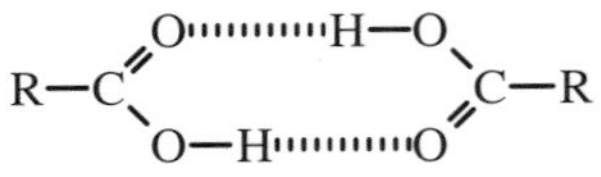

2 hydrogen bonds produced

Compared with their alkane derivatives they have relatively high melting and boiling points which is due to their hydrogen bonding ability. All carboxylic acids have characteristic odours with higher acids producing disgusting odours. For example, butanoic acid exhibits a rancid butter smell.

(c) Chemical reactions

As the name suggests, carboxylic acids are acidic in nature, more so than alcohols. They are partially dissociated in aqueous solution:

$$R\!-\!\overset{\displaystyle O}{\overset{\|}{C}}\!-\!OH \ + \ H_2O \ \rightleftharpoons \ R\!-\!\overset{\displaystyle O}{\overset{\|}{C}}\!-\!O^- \ + \ H_3O^+ \qquad (7.15)$$

In general, carboxylic acids have a pK_a of around 4 to 5 compared with HCl and HNO_3 with pK_a values of -2.2 and -1.3 respectively. Therefore, in comparison with inorganic acids, carboxylic acids are relatively weak acids. However, the pK_a of methanol is 15.5 compared with 4.74 for methanoic acid, which indicates that carboxylic acids are relatively strong acids with respect to other organic compounds.

Carboxylic acids can also form salts by treating the acid with a base such as NaOH:

$$CH_3\!-\!CH_2\!-\!COOH \ \xrightarrow[\text{NaOH}_{(aq)}]{} \ CH_3\!-\!CH_2\!-\!COO^-Na^+ \ + \ H_2O \qquad (7.16)$$

The ending -ic acid is replaced by -ate to give the name of the appropriate salt. For example, $CH_3CH_2COO^-$ Na^+ is called sodium propanoate. These salts are more soluble in water than their corresponding acids because the anionic group is readily solvated.

Esters are the most important carboxylic acid derivatives. These are formed by the reaction of a carboxylic acid with an alcohol:

$$R-\overset{\overset{\displaystyle O}{\|}}{C}-OH \quad + \quad R'-OH \quad \underset{}{\overset{H^+}{\rightleftharpoons}} \quad R-\overset{\overset{\displaystyle O}{\|}}{C}-OR' \quad + \quad H_2O$$

Carboxylic acid   Alcohol   Ester

$$(7.17)$$

The reaction is under equilibrium conditions and is readily reversible by having an excess of water present. The opposite of ester formation is called *ester hydrolysis*.

The sodium and potassium salts of long chain fatty acids have the ability to aggregate to produce clusters called *micelles*. In micelle formation, the hydrophobic carbon chains of the acids orientate themselves in such a way that they are exposed to as little water as possible. The hydrophilic so-called "head" groups interact with the water and create a wall to protect the hydrocarbon chains from the water. This is illustrated in Fig. 7.12.

This micelle formation can reduce the surface tension of aqueous solutions and creates a foaming that is characteristic of simple soaps. Soaps dissolve water-insoluble materials by incorporating them into the hydrophobic part of the micelle and, because the surface tension is reduced, the soap solution can permeate through clothing.

7.3.7 Esters

Esters are formed by the reversible reaction of alcohols with carboxylic acids. This reaction is analogous to the production of salts in inorganic chemistry by reactions of acids and bases. The following is the general

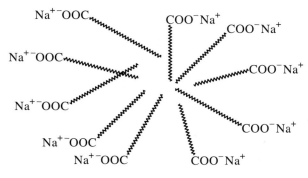

Fig. 7.12 Micelle formation.

formula of an ester:

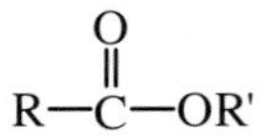

$$R-\overset{\overset{\displaystyle O}{\|}}{C}-OR'$$

Esters are termed *alkyl alkanoates*, the actual ester functional group, —COOR, is called *alkoxycarbonyl*. The structural formulae and IUPAC names of some esters are given in Fig. 7.13.

Esters have characteristic pleasant odours and are used in fragrances and for flavourings. Esters with lower carbon contents are used as solvents and this can often lead to considerable discharges in industrial wastes.

7.3.8 Ethers

Ethers contain the characteristic *alkoxy group*, —OR, where R represents the alkyl group. IUPAC nomenclature calls ethers *alkoxyalkanes*. They can be represented by the general formula R—O—R', with the smaller substituent being incorporated into the alkoxy group. This can be illustrated by the following examples:

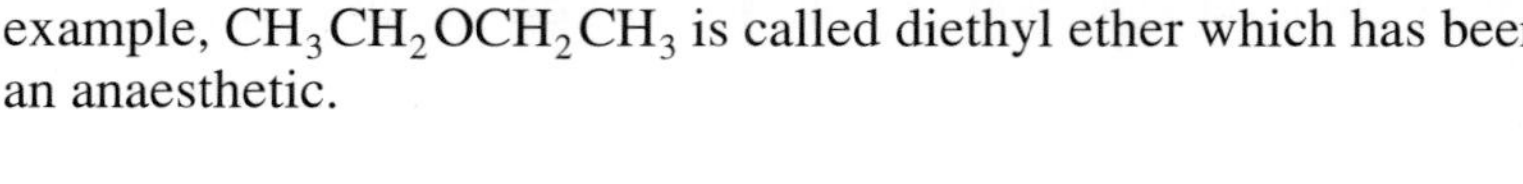

$$CH_3-O-CH_2-CH_3 \qquad CH_3-CH_2-O-CH_2-CH_3$$

Methoxyethane   Ethoxyethane   Oxacyclopentane

Some of the common names are still employed when naming ethers. For example, $CH_3CH_2OCH_2CH_3$ is called diethyl ether which has been used as an anaesthetic.

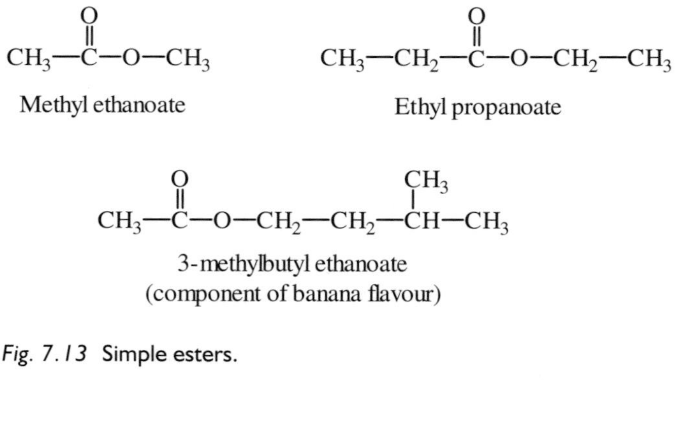

$$CH_3-\overset{\overset{\displaystyle O}{\|}}{C}-O-CH_3 \qquad\qquad CH_3-CH_2-\overset{\overset{\displaystyle O}{\|}}{C}-O-CH_2-CH_3$$

Methyl ethanoate   Ethyl propanoate

$$CH_3-\overset{\overset{\displaystyle O}{\|}}{C}-O-CH_2-CH_2-\overset{\overset{\displaystyle CH_3}{|}}{CH}-CH_3$$

3-methylbutyl ethanoate
(component of banana flavour)

Fig. 7.13 Simple esters.

Table 7.8 Boiling points of ethers and alcohols

Ether name	Formula	Bpt (°C)	Corresponding alcohol	Bpt (°C)
Methoxymethane	CH_3-O-CH_3	−23	Ethanol	78.5
Methoxyethane	$CH_3-O-CH_2-CH_3$	10.8	Propanol	97.4
Ethoxyethane	$CH_3-CH_2-O-CH_2-CH_3$	34.5	Butanol	117.25
1-butoxybutane	$(CH_3-CH_2-CH_2-CH_2)_2-O$	142	Octanol	194.45

The molecular formula of basic ethers is the same as the alcohol series, i.e. $C_nH_{2n+2}O$. Table 7.8 shows the boiling points of ethers compared with corresponding alcohols.

In ethers there is no hydrogen bonding and thus the boiling points of ethers are much lower than their alcohol counterparts. Ethers are also generally insoluble in water. These compounds are fairly unreactive and for this reason they are used widely as solvents.

7.4 Nitrogen-containing compounds

7.4.1 Amines

Amines are derivatives of ammonia and can be classified according to three types – primary, secondary and tertiary:

$$R-N\begin{smallmatrix}H\\H\end{smallmatrix} \qquad R-N\begin{smallmatrix}R'\\H\end{smallmatrix} \qquad R-N\begin{smallmatrix}R'\\R''\end{smallmatrix}$$

Primary amine   Secondary amine   Tertiary amine

In primary amines for example, one of the hydrogens in ammonia is replaced with an alkyl group, in secondary amines two hydrogens are replaced and so on. The amines and other nitrogen-containing compounds are some of the most abundant organic molecules as they are components of amino acids, peptides and proteins.

(a) Nomenclature

Amines are classed as *alkanamines*, i.e. the ending *-e* on the alkane is replaced by *-amine*. The position of the functional groups is indicated in a manner similar to that of the alcohols. Examples of some amines are given in Fig. 7.14. Alternatively, amines can be named as *alkyl amines*.

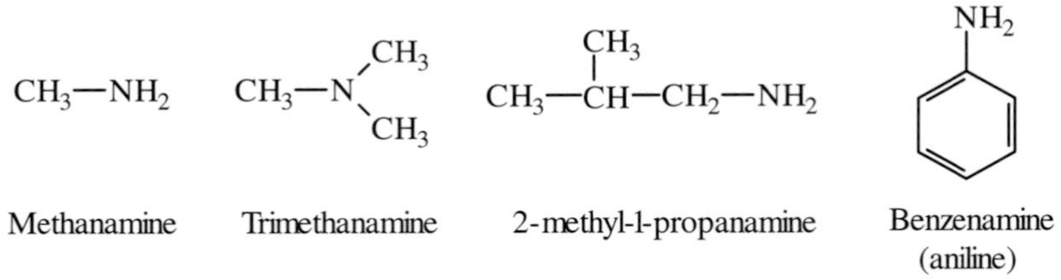

$$CH_3-NH_2 \qquad CH_3-N\underset{CH_3}{\overset{CH_3}{<}} \qquad CH_3-\underset{CH_3}{\overset{CH_3}{\underset{|}{CH}}}-CH_2-NH_2$$

Methanamine   Trimethanamine   2-methyl-l-propanamine   Benzenamine
(aniline)

Fig. 7.14 Amines.

Tertiary amines can react with alkyl halides to produce *quaternary ammonium salts*. This is shown in the following equation:

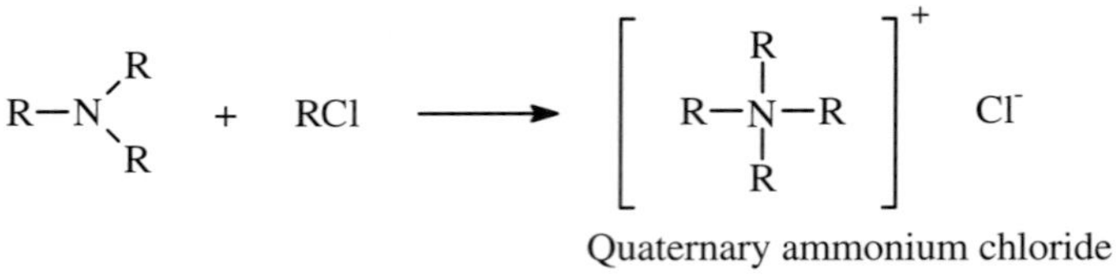

$$R-N\underset{R}{\overset{R}{<}} \quad + \quad RCl \quad \longrightarrow \quad \left[R-\underset{R}{\overset{R}{\underset{|}{\overset{|}{N}}}}-R \right]^{+} \quad Cl^{-} \qquad (7.18)$$

Quaternary ammonium chloride

These salts have bactericidal properties, which makes them useful disinfecting agents. All amines are generally basic, with basicity increasing from primary to tertiary. They have a characteristic "fishy" odour and can therefore be detected at low concentrations by smell.

7.4.2 Amides

Amides are carboxylic acid derivatives. They are called *alkanamides* using the same systematic approach utilized in the naming of amines. Like amines, there can be primary, secondary and tertiary amides.

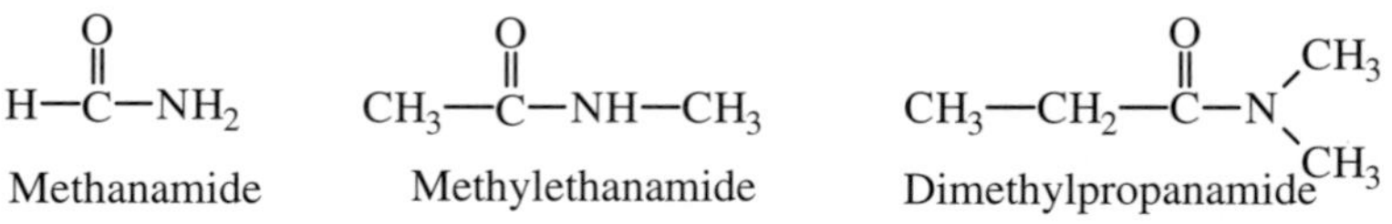

$$H-\overset{O}{\overset{||}{C}}-NH_2 \qquad CH_3-\overset{O}{\overset{||}{C}}-NH-CH_3 \qquad CH_3-CH_2-\overset{O}{\overset{||}{C}}-N\underset{CH_3}{\overset{CH_3}{<}}$$

Methanamide   Methylethanamide   Dimethylpropanamide

One important reaction of amides is the Hofmann rearrangement. In this halogenation reaction, the carbonyl group is removed from the molecule to produce an amine containing one less carbon atom in the chain. This is shown below:

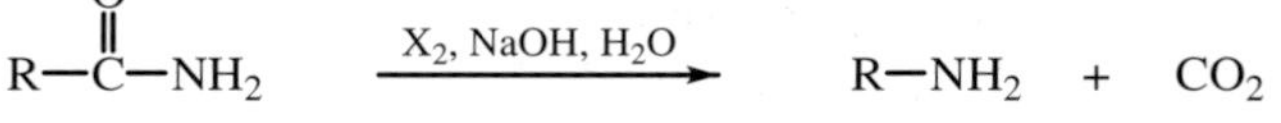

$$R-\overset{O}{\overset{||}{C}}-NH_2 \quad \xrightarrow{X_2,\ NaOH,\ H_2O} \quad R-NH_2 \quad + \quad CO_2 \qquad (7.19)$$

7.4.3 Halogenated organic compounds

These compounds (called *haloalkanes*) are used frequently in organic synthesis. They can pose a significant environmental threat as they are generally toxic to humans. Simple haloalkanes have the general formula $R—X$, R being the organic substituent. The common names for these compounds are *alkyl halides*.

One example of a simple haloalkane is chloroethane, C_2H_5Cl, its common name being ethyl chloride. This compound is used to produce tetraethyl lead $((C_2H_5)_4Pb)$, a compound used as an anti-knocking agent in petrol.

Many polyhalogenated compounds are utilized in industry as solvents, paint strippers, pesticides and so on. Many of these substances are highly toxic, with some being carcinogenic. Examples of polyhalogenated compounds are chloroform (trichloromethane, $CHCl_3$), carbon tetrachloride (tetrachloromethane, CCl_4) and chlorofluorocarbons (CFCs).

7.5 Aromatic compounds

Aromatic compounds are so named due to their strong aroma. They can be classed as those compounds which contain a benzene ring structure.

Benzene (C_6H_6) is an unusually unreactive compound and its electronic arrangement gives it this stability both thermodynamically and kinetically. The electrons in the double bonds (π electrons) are delocalized, i.e. spread around the entire ring system. Thus, benzene is often represented by structure III in Fig. 7.15 and is considered to be the "parent" aromatic molecule. Owing to its alternate double bonds in the ring, benzene is said to exhibit *conjugation*.

7.5.1 Nomenclature

Benzene can be written as any of the three structures given in Fig. 7.15. Simple benzene derivatives are named by adding the substituent in front of *benzene*.

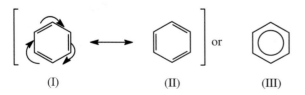

    (I)   (II)   (III)

Fig. 7.15 Resonance in benzene.

F

Fluorobenzene

CH₃

Methylbenzene
(toluene)

With disubstituted benzenes, there are three possible arrangements. These are given the prefixes 1,2 (*ortho*), 1,3 (*meta*) or 1,4 (*para*). Examples of this are illustrated in Fig. 7.16.

It can be seen that dimethylbenzene has three isomers and also exhibits isomerism with ethylbenzene as they have the same molecular formula. Substituted benzenes are termed *arenes*. When an aromatic compound is present as a substituent it is called *aryl*, and is given the symbol Ar. The parent aromatic aryl is C_6H_5 and is called phenyl.

7.5.2 Benzene derivatives

There are numerous derivatives of benzene, some of which have significant impact on the environment. Figure 7.17 illustrates some benzene derivatives giving the systematic and common names for these compounds.

7.5.3 Chlorinated benzene compounds

These types of compounds are widely used as pesticides and industrial solvents, and can be present in wastewater and groundwater. They are fairly volatile and moderately soluble. An example of a chlorinated benzene derivative is pentachlorophenol (PCP) which is illustrated on the next page.

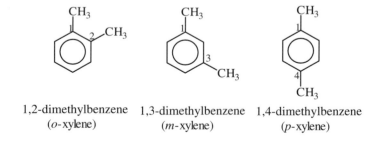

1,2-dimethylbenzene   1,3-dimethylbenzene   1,4-dimethylbenzene
(*o*-xylene)   (*m*-xylene)   (*p*-xylene)

Fig. 7.16 Possible arrangements of disubstituted benzenes.

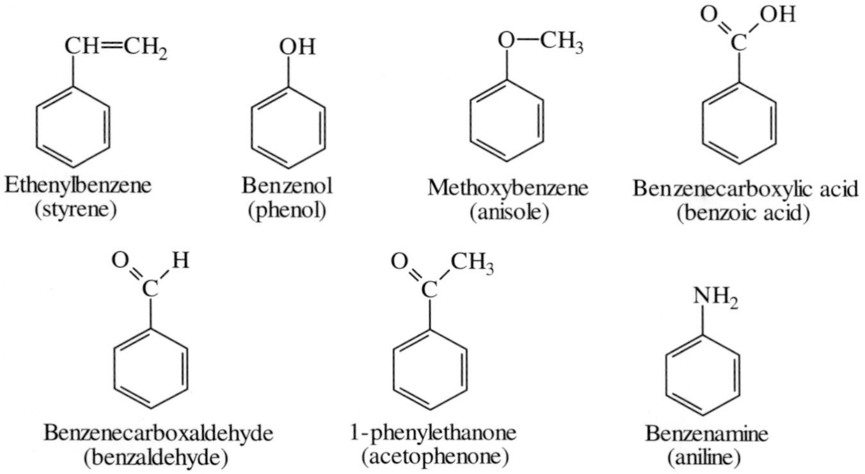
CH=CH$_2$
OH
O—CH$_3$
O OH C
Ethenylbenzene (styrene)
Benzenol (phenol)
Methoxybenzene (anisole)
Benzenecarboxylic acid (benzoic acid)
O H C
O CH$_3$ C
NH$_2$
Benzenecarboxaldehyde (benzaldehyde)
1-phenylethanone (acetophenone)
Benzenamine (aniline)

Fig. 7.17 Derivatives of benzene.

This compound is used as a pesticide and is very toxic due to its high number of chlorine atoms.

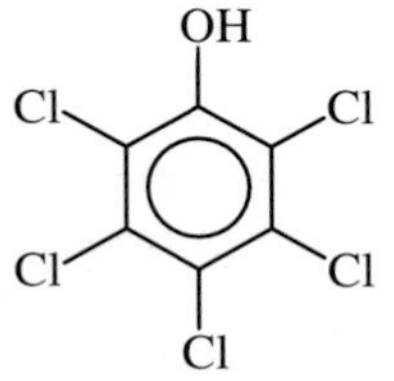
OH
Cl
Cl
Cl
Cl
Cl

Pentachlorophenol (PCP)

Polychlorinated biphenyls (PCBs) have the following structure:

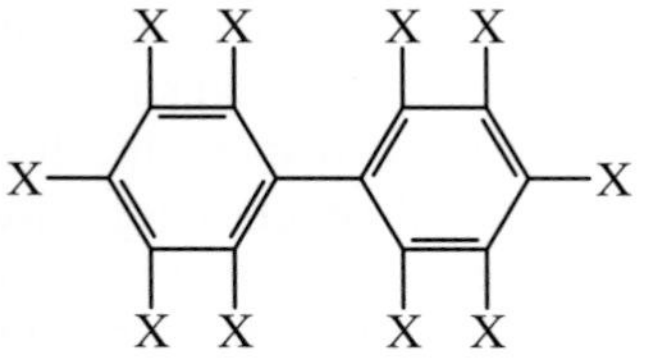
X X X X
X
X
X X X X

where X=Cl or H atom.

PCBs are stable compounds and are utilized in coolants, solvents and hydraulic fluids. They are environmentally important as they are subject to bioaccumulation in the food chain where significant concentrations have been found in fish and birds. Although PCB disposal is controlled, they

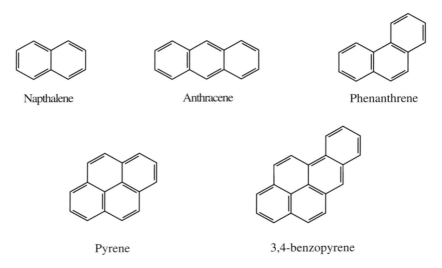

Napthalene   Anthracene   Phenanthrene

Pyrene   3,4-benzopyrene

Fig. 7.18 Polyaromatic hydrocarbons (PAHs).

continue to pose a potential environmental problem as they can resist bio-degradation by micro-organisms.

7.5.4 Polyaromatic hydrocarbons (PAHs)

These compounds are produced by fuel combustion, incineration, forest fires and cigarettes, and are considered to be important pollutants in the atmosphere and aquatic environments. PAHs represent a class of compounds wherein benzene rings are fused together. They are known by their common names as there is no other system of nomenclature for these compounds. Figure 7.18 gives some examples of PAHs.

The majority of PAHs are derived from coal tar which is produced on heating coal in the absence of air. One example of a coal-tar-derived PAH is napthalene, which is a moth repellent and insecticide and was used in the production of mothballs. It is also used in the production of dyestuffs.

Many PAHs are carcinogenic and are thought to be responsible for the first recognized human cancer which was observed in chimney sweeps around 1775. One strongly carcinogenic environmental pollutant is 3,4-ben-zopyrene. It has been estimated that the annual release into the atmosphere by the USA is around 1,300 tons.

7.6 Heterocyclic compounds

Heterocyclic compounds can be defined as having at least one ring carbon replaced by a heteroatom such as nitrogen, oxygen, sulphur, phosphorus, etc. The most common heterocycles usually contain either nitrogen or

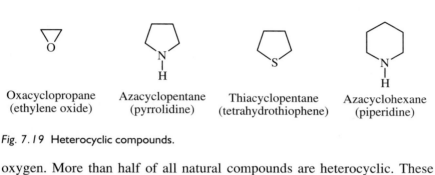

Oxacyclopropane (ethylene oxide)	Azacyclopentane (pyrrolidine)	Thiacyclopentane (tetrahydrothiophene)	Azacyclohexane (piperidine)

Fig. 7.19 Heterocyclic compounds.

oxygen. More than half of all natural compounds are heterocyclic. These compounds can be aliphatic or aromatic in nature.

7.6.1 Nomenclature

For saturated heterocycles, the presence of the heteroatom is given the following prefixes: *aza* – nitrogen, *oxa* – oxygen, *thia* – sulphur, *phospha* – phosphorus. Some examples are given in Fig. 7.19.

The epoxides (e.g. ethylene oxide) are of particular environmental concern. Ethylene oxide is a regulated air pollutant and has been used as a sterilizing agent and a pesticide. They are also very reactive and can act as intermediates in producing other toxic organic compounds.

With unsaturated heterocycles, the common names are used as they have been firmly established. Figure 7.20 shows some simple unsaturated heterocycles.

Of the compounds shown, by far the most important are purine and pyrimidine. Derivatives of purine are adenine and guanine; pyrimidine derivatives are cytosine, uracil and thymine. The compounds constitute the major components of nucleic acids, which are the fundamental building blocks of life.

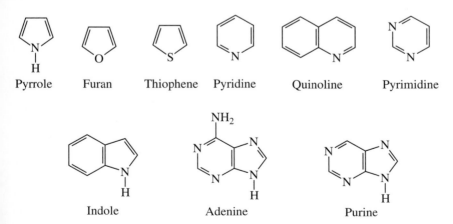

Fig. 7.20 Unsaturated heterocyclic compounds.

References and further reading

Harrison, R. M. and deMora, S. J. (1996) *Introductory Chemistry for the Environmental Sciences*, 2nd edn, Cambridge University Press, Cambridge, pp. 160–172.

Sawyer, C. N., McCarty, P. L. and Parkin, G. F. (1994) *Chemistry for Environmental Engineering*, 4th edn, McGraw-Hill, Singapore, pp. 187–287.

Vollhardt, K. P. C. (1987) *Organic Chemistry*, W. H. Freeman and Co., New York, 1275 pp.

Weast, R. C. (1978) *Handbook of Chemistry and Physics*, 59th edn, CRC Press, USA, pp. C80–C548.

Chapter 8

Biological molecules

8.1 Introduction

The existence of a cell depends upon its structural components, which maintain its physical integrity, and its functional components, which control its growth, metabolism and reproduction and which regulate these processes. In a great many instances structural and functional roles are closely integrated, even in the simplest cell.

The highly ordered, but extremely complex nature of living cells is due to the great variety of organic compounds which may be formed, inherent in the chemical properties of carbon itself. Its ability to form four covalent bonds with other atoms and itself, generating long carbon chains, gives rise to an almost limitless number of possible compounds, including macromolecules with molecular weights as high as 100 million. In all cases, these macromolecules are formed from the linkage, in a series, of smaller molecules, termed monomers, and hence are called polymers. Even if constructed from a pool containing a relatively small number of different types of monomer molecules the number of different possible sequences of these subunits is theoretically very large indeed. If only a small fraction of all of the possible combinations of monomers were to be associated with a distinct structural or functional characteristic of the polymer molecule, this number would be more than sufficient to account for the enormous diversity and complexity of living organisms. There are, however, a large number of characteristics common to all cells, irrespective of their origins, which is indicative of a moderately consistent selection of only a few of the possible combinations in each case, suggesting that cellular chemistry is organized and regulated in a *highly precise manner*.

There is, in addition to the macromolecules, a great variety of other organic compounds which are important by virtue of their role as nutrients and intermediary metabolites. However, this chapter focuses only on those which are unique and integral components of cells – the macromolecules.

In spite of the potential for infinite variety, there are only four major classes of biological polymers. These are proteins, polysaccharides (carbohydrates), nucleic acids and lipids. They each have closely defined, distinct functional and structural characteristics, but it should be noted that the chemical reactions involved in the assembly of their subunits are the same in each case. Polymer chains are constructed from condensation reactions (the removal of a molecule of water) between a hydroxyl group and hydrogen-containing groups on adjacent monomers.

8.2 Isomerism

Isomerism is defined as the existence of two or more chemical compounds with the same molecular formula but having different properties because of different arrangements of atoms or functional groups within the molecule. An example of this is the two structures of the compound having the formula $C_3H_6O_3$. Glyceraldehyde has the structure $CHO—C(H)OH—CH_2OH$ and dihydroxyacetone is $CH_2OH—CO—CH_2OH$. These two compounds are said to exhibit keto-aldo isomerism. Further examination of the three-dimensional structure of glyceraldehyde reveals that it has the potential to exist in two forms each having the same functional groups around the central carbon atom, but with different spatial arrangements in each case. A three-dimensional model of methane (CH_4) is shown in Fig. 8.1 which indicates quite clearly the spacing of the four covalent bonds. The four groups on the central carbon of glyceraldehyde are spaced similarly. The terms D and L are applied by convention to compounds which contain an asymmetric carbon atom (that is one surrounded by four different functional groups). It should be noted that dihydroxyacetone has no asymmetric carbon and therefore does not exhibit stereoisomerism.

Although D- and L-glyceraldehyde have identical structural formulae, their three-dimensional configurations are different. This can be appreciated if one imagines that an attempt is made to superimpose the D and L isomers on top of each other. No amount of rotating or inverting the two rigid

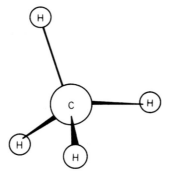

Fig. 8.1 A three-dimensional model of methane.

structures will result in the four functional groups lying in the same spatial configuration: the two isomers are mirror images of each other. The different shapes of stereoisomers have important implications regarding the specificity of biological systems.

8.3 Proteins

8.3.1 Structure

The monomers from which proteins are assembled are all amino acids, having the general formula:

$$
\begin{array}{c}
\text{H} \\
| \\
\text{H}_2\text{N}-\text{C}-\text{COOH} \\
| \\
\text{R}
\end{array}
$$

There are about twenty naturally occurring amino acids, each with a different side chain functional group (R). In the simplest, glycine, R is a hydrogen atom, but others have quite large groups, e.g. lysine ($-(CH_2)_4-NH_2$) and arginine ($-(CH_2)_3-NH-C(NH_2)=NH$). In the structure above, the central carbon atom is termed the α-carbon, to which are linked the α-carboxyl and α-amino groups. This serves to distinguish these from other amino and carboxyl groups which may be present on the R side chain, since only the α-groups take part in condensation reactions to form the protein chain. A typical protein will contain 100–1,000 amino acid residues in the chain. The sequence of amino acids in the chain is termed the primary structure of the protein. There are, however, higher levels of structural organization. Many proteins do not exist as chain-like or linear molecules, but "fold up" to greater or lesser degrees. The repeating C—C and N—C linkages along the peptide chain allow considerable flexibility of the molecule, permitting bending and rotation. The term secondary structure is generally used to describe regular or repeating three-dimensional forms arising from this. One secondary conformation is the α-helix, which is due to hydrogen bonding between the C$=$O group of one peptide bond and the N—H group of the third peptide bond along the chain from it.

Another regular conformation is the β-pleated sheet, which is formed by hydrogen bonding between the N—H and C$=$O groups of peptide bonds on two stretches of the primary chain running parallel to each other but in opposite directions. Few proteins, however, consist entirely, or even mainly, of α-helices or β-pleated sheets, since other interactions can occur between regions of the primary chain leading to a greater variation in the folding and bending, giving rise to a more complex three-dimensional form, termed the tertiary structure. One phenomenon which disrupts the α-helix is the presence of a proline residue in the polypeptide chain. Proline, in fact, is an α-imino acid, where the nitrogen atom is linked to two carbons, such that the formation of a peptide bond leaves it without a hydrogen atom with which to form a hydrogen bond. This introduces bends in the α-helix where proline residues are located. Other interactions which stabilize the tertiary structure of proteins occur between the side chain (R) functional groups. These interactions take the form of covalent, ionic and hydrogen bonds and hydrophobic and hydrophilic orientations. An important type of covalent bond is the linkage between two adjacent cysteine residues (R=CH$_2$—SH) in the folded protein chain to form a disulphide bridge.

$$-SH + HS- \longrightarrow -S-S-$$

Ionic interactions can occur between amino acids which normally carry positive (e.g. histidine, lysine and arginine) and negative (e.g. aspartic acid and glutamic acid) charges, although the extent to which they occur is limited since such interactions can occur over only short ranges and extensive folding of the chain is required to bring the amino acids within range.

Some of the side chain (R) functional groups are only sparingly soluble in water. These include the CH_3, $CH(CH_3)_2$, $CH_2CH(CH_3)_2$ and $CH(CH_3)CH_2CH_3$ groups of alanine, valine, leucine and isoleucine respectively. They are termed hydrophobic (i.e. water-hating) because they contain no ionizable functional groups. In an aqueous environment, they tend to cluster towards the centre of the protein molecule as far away as possible from the surrounding water matrix.

Finally, many proteins have a quaternary structure, which is the result of two or more complete protein molecules aggregating to form a functional unit with a number of molecular subunits. The tendency to aggregate in this way also appears to be due largely to the presence of hydrophobic groups, not buried in the interior of the protein structure but on the surface. Another example of the role of surface-hydrophobicity is given in Section 8.6.2.

8.3.2 Function

(a) Food storage and structural proteins

It will be seen that the synthesis of macromolecules from small molecular subunits requires the input of energy, and conversely, their breakdown liberates energy and releases the monomers into the central metabolic pool. In this respect, biological macromolecules may be considered as food or energy storage compounds. However, this is a very minor function of proteins in micro-organisms, although it is reasonably well developed in some higher organisms.

Probably because of their potential for existing in any shape or size, and with varying degrees of water solubility, many proteins have a structural role. Again, these functions are relatively well developed in higher organisms, but in micro-organisms are limited.

(b) Enzymes

Probably the most important role of proteins is as biological catalysts. A catalyst is an agent which accelerates the rate of a chemical reaction without itself being consumed in that reaction. An example of chemical catalysis is the action of manganese dioxide (MnO_2) on hydrogen peroxide (H_2O_2). The latter breaks down spontaneously to form oxygen and water but under normal conditions in the laboratory this occurs quite slowly and is generally unnoticeable. The addition of MnO_2 causes rapid effervescence and when the reaction is complete the MnO_2 remains completely unchanged. Hydrogen peroxide is fairly toxic and some bacteria possess a catalyst to break it down. The effects of adding H_2O_2 to a smear of bacteria on a slide are much the same as those observed if MnO_2 were to be present instead. Such biological catalysts are termed enzymes and all of those which have been characterized so far have been found to be proteins. Put simply, all enzymes are proteins, but not all proteins are enzymes.

The important characteristics of enzymes are their rapidity of catalysis, their specificity and their amenability to control. The powers of acceleration of reaction rates are immense. Some enzymes can make a chemical reaction occur 10^{12} (1 million million) times faster than it would in their absence. A single molecule of the enzyme catalase which breaks down H_2O_2 can thus deal with almost 100,000 molecules of H_2O_2 per second. It should be pointed out that enzymes can only accelerate a reaction which can already proceed without catalysis, albeit slowly. If it is chemically impossible for a reaction to occur, then no amount of enzyme will promote it. Why, then, does a cell need enzymes if the chemical reactions they catalyse will occur anyway? One reason is that quite simply, some reactions occur so slowly that the cell cannot derive any usefulness from them – it cannot wait for them to occur. Secondly, the intervention of an enzyme amounts to a selection of the particular metabolic pathway which is required. Certain compounds, known as intermediate metabolites, can undergo a large number of transformations. Let us assume that in a hypothetical cell all of these transformations occur synchronously at a slow rate. Ultimately as this situation persists throughout the entire metabolic network the cell will end up with its entire complement of metabolic products. The result would be chaotic. Assuming that these uncatalysed reactions occur at the rate of 1 molecule per unit time, t, then the addition of an enzyme with even a modest acceleration capability (say 10^9 times) would result in the formation after time t of one each of the possible products except for the chosen one, of which there would be 1,000 million molecules. As far as an analytical chemist, and probably even the cell, is concerned, the formation of the product from the catalysed reaction would be the only reaction to have effectively occurred.

In our hypothetical cell, 1,000 million molecules of reaction product may be all the cell requires for its immediate needs. Since, at least in theory, the enzyme remains unaffected after it has catalysed the reaction, it seems logical for the cell to have some means of controlling its activity. Such control may be exerted in several ways. First of all, the initial synthesis of the enzyme may be controlled by the presence of the substrate molecule. This makes sound economic sense, since it could be considered a waste of energy to synthesize an enzyme if it could not be put to immediate use. Enzymes which are synthesized in response to the presence of the substrate are termed "inducible". Secondly, the enzyme molecule can be "switched" on or off by interaction with molecules other than the substrate. Many metabolic pathways consist of long sequences of reactions which bring about gradual, step-by-step changes in the substrate molecule. In such cases the formation of one of the later products in the series may interact with an enzyme catalysing an early stage of the process and inactivate it. Conversely, an early intermediate in the series may interact positively with a dormant enzyme which catalyses a later stage and activate it so that the reaction sequence can proceed to its conclusion. A schematic representation of some of the ways in which these controls are exerted is shown in Fig. 8.2. If all of the chemical reactions of a cell are considered together, it is apparent that the subtlety of control required is enormous if metabolic processes are not to degenerate into a random state. This observation leads to a consideration of the third important property of enzymes, which is specificity. Many enzymes are so highly specific that they will catalyse a single

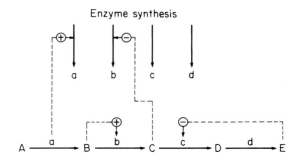

Capital letters: substrates
Lower case letters: enzymes
⊕ = Enhancement
⊖ = Inhibition

Fig. 8.2 Schematic of control mechanisms for enzyme-catalysed reactions.

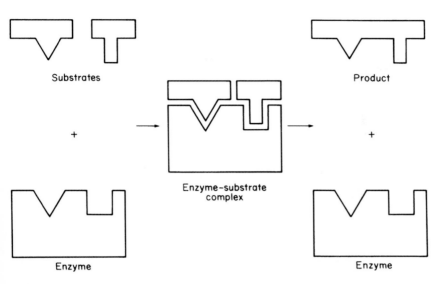

Substrates

+

Enzyme

Enzyme–substrate
complex

Product

+

Enzyme

Fig. 8.3 Schematic of enzyme action.

transformation of only one compound, and frequently only one stereoisomer of that compound. This specificity is due to the precise juxtaposition of the functional groups making up the active site. Much of the accelerating power of enzymes lies in their ability to capture and orientate the substrate in such a conformation as to provide optimum conditions for the reaction to occur. A schematic of this mode of action is shown in Fig. 8.3.

(c) Enzyme kinetics

It is convenient to obtain some quantitative measure of the rates of reactions catalysed by enzymes. The obvious way of going about this would be to add a given concentration of substrate to an enzyme in solution and to monitor the concentration of product generated with time. The type of curve shown in Fig. 8.4 would result. After a time it can be seen that the curve tails off as the rate of the reaction slows down. This could be due to the disappearance of substrate making the number of successful collisions between it and the enzyme per unit time decrease, inhibition of the enzyme by the product, inhibition by some other factor, such as pH, which might change during the reaction, or the gradual breakdown of the protein molecule itself. The initial part of the curve is linear, however, and its slope gives the initial velocity of the reaction, V_0. V_0 is a measurable quantity, but it only quantifies the rapidity of the reaction in the presence of a given substrate concentration. Repeating the experiment in the presence of different substrate

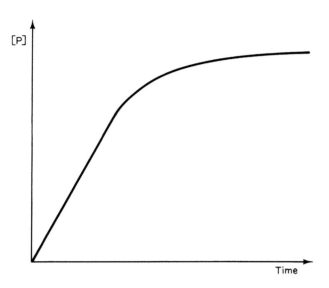

Fig. 8.4 Time course of an enzyme-catalysed reaction.

concentrations would yield a family of curves as shown in Fig. 8.5, from which a series of V_0 values could be obtained. At very high substrate concentrations, V_0 would tend towards a maximum value, which we will call

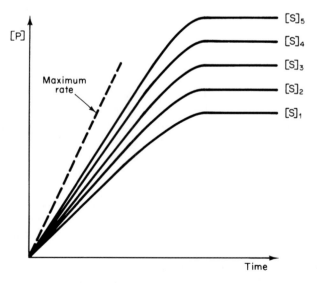

Fig. 8.5 Time course of enzyme-catalysed reactions with different substrate concentrations.

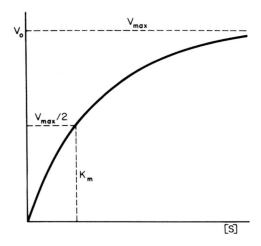

Fig. 8.6 Effect of substrate concentration on reaction velocity.

V_{max}. Further increases in the substrate concentration, [S], would not lead to higher reaction rates because all of the enzyme's active sites would be already saturated with substrate molecules. This phenomenon is more simply illustrated by plotting V_0 against [S], as shown in Fig. 8.6. The curve in Fig. 8.6 is a rectangular hyperbola and is described by the equation:

$$V_0 = \frac{V_{max}[S]}{K_m + [S]}$$

(8.1)

where V_{max} and K_m are constants for a particular set of experimental conditions (e.g. pH, temperature, etc.). This is often referred to as the Michaelis–Menten equation. At very large values of [S], $V_0 = V_{max}$. V_{max} is, quite simply, the maximum rate of reaction which can be obtained, which is a fairly useful measure. It is probable, however, that many enzymes do not function in conditions where [S] is large enough for the reaction to proceed at maximum rate. The other constant, K_m, is related to substrate concentration and gives a useful measure of the action of the enzyme at less than maximum rates. If K_m is made equal to [S] then $V_0 = V_{max}/2$. K_m is therefore defined as the substrate concentration required for the reaction to proceed at half its maximum rate. A catalysed reaction with a high K_m value would proceed with much less than maximum efficiency at low substrate concentration, and changes in the concentration of substrate would markedly affect the rate of the reaction. An enzyme with a low K_m value, however, would operate more efficiently at low substrate concentrations, and would be less affected by changes in substrate concentration at this level.

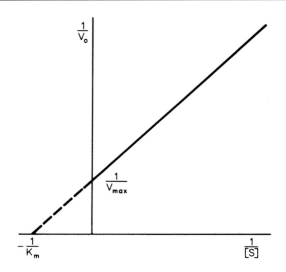

Fig. 8.7 The double reciprocal plot of equation (8.2).

An examination of Fig. 8.6 reveals that it would be fairly difficult to determine V_{max} from such a curve. Moreover, without V_{max} the value of K_m could not be determined accurately. Quite a wide range of substrate concentrations would have to be incorporated into an experiment: at the lower level in order to determine the shape of the curve in the region of K_m and at a much higher level in order to approach V_{max}. Even at a value of $[S_0]$ equivalent to $5K_m$, V_0 would be less than 85% of V_{max}. A simpler way of obtaining these constants would be to rearrange equation (8.1) by taking reciprocals of both sides.

$$\frac{1}{V_0} = \left(\frac{K_m}{V_{max}}\right)\left(\frac{1}{[S]}\right) + \frac{1}{V_{max}} \tag{8.2}$$

Equation (8.2) is of a straight line, and a plot of $1/V_0$ against $1/[S]$ is shown in Fig. 8.7 from which values of V_{max} and K_m are easy to obtain.

8.4 Polysaccharides

8.4.1 Structure

There are fewer naturally occurring monomers (monosaccharides or sugars) which are used by cells in the synthesis of polysaccharides than there are amino acids available for incorporation into proteins. There are three general types of naturally occurring monosaccharide – tetroses, pentoses and hexoses having four, five and six carbon atoms respectively.

They all have the general formula $C_nH_{2n}O_n$ but for each value of n there are a number of different monosaccharides because of a form of isomerism.

Re-examination of D-glyceraldehyde (Section 8.2), which has one asymmetric carbon shows that the insertion of another CHOH group into the carbon chain adds another centre of asymmetry. If the D-configuration is retained for the original CHOH group of glyceraldehyde, then there are two possible isomers of the four-carbon compound derived from it. These isomers are therefore D-tetroses. The insertion of another asymmetric carbon yields four possible D-pentoses, and in the same way there are eight isomers of D-hexose. The sugars which can be related structurally to glyceraldehyde in this way are all termed aldoses, and the asymmetric centres exhibit a form of stereoisomerism termed epimerism.

A similar series of sugars, termed ketoses, can be derived from dihydroxyacetone. Since dihydroxyacetone has no asymmetric carbon atom the progressive addition of one, two and three asymmetric centres will result in one D-tetrose (because the first group is fixed in the D-configuration), two D-pentoses and four D-hexoses. Fourteen L-aldoses and seven L-ketoses can be derived from L-glyceraldehyde and dihydroxyacetone with the insertion of the first CHOH group in the L-configuration. Whether a sugar belongs to the L or D series is determined by the configuration of the —H and —OH groups attached to the carbon atom next to the CH_2OH group, whereas individual monosaccharides are named according to the configurations about the other asymmetric carbons.

The five-carbon and six-carbon aldoses and the six-carbon ketoses, when in solution, adopt an equilibrium between the open chain structure and a closed ring. Five- and six-membered rings are found, but the six-membered ring in aldohexoses is the more common form. Closure of the ring at the C-1 position introduces another centre of asymmetry which is readily reversible (unlike the epimeric centres) because of the rapid interconversion between the open chain structure (C-1 not asymmetric) and the ring structure (C-1 asymmetric). Isomerism at the C-1 position is termed anomerism to distinguish it from epimerism. The two anomers are termed α, where the —OH group of C-1 protrudes from the ring in the opposite direction to the CH_2OH group on the C-5, and β, where the two groups protrude in the same direction. Chain–ring equilibrium and the formation of the two anomers is shown for D-glucose in Fig. 8.8.

Anomerism is not an important property of monosaccharides in solution, since it is freely exchangeable, but when they polymerize to form disaccharides and the larger polymers, the ring form is stabilized and the anomeric configuration is retained. Monosaccharides are joined together by glycosidic bonds which are formed by condensation between the anomeric carbon of one sugar and any of the —OH groups of another sugar. Terminology used to describe glycosidic bonds incorporates the numbers of the carbon atoms

β-D-glucose   α-D-glucose

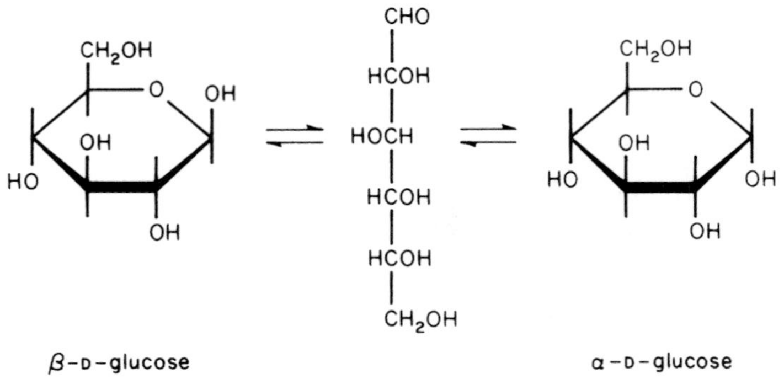

Fig. 8.8 The equilibrium between chain and ring forms of D-glucose.

joined together and the anomeric configuration, e.g. $\alpha(1-6)$ linkages, $\beta(1-4)$ linkages.

Not all possible isomers of tetroses, pentoses and hexoses occur in nature. There are only about ten naturally occurring monosaccharides. Moreover, many polysaccharides contain only one type of monosaccharide (these are termed homopolysaccharides). Some contain two or three different monosaccharides, and are termed heteropolysaccharides, while molecules with more than five or six different sugar constituents are rarely found. This suggests, therefore, that polysaccharides do not possess the same degree of complexity and specificity as the proteins in terms of their primary structure. Moreover, this relative lack of specificity extends in some polysaccharides to some variation in final chain length. Also, secondary and higher levels of structure in some polysaccharides have less complexity and more simple regularity than the corresponding levels of organization of the proteins.

8.4.2 Function

Sugars can be utilized as substrates by very many micro-organisms and since their primary use is as food it may be expected that polysaccharides have a role as food storage molecules. In this role they have the advantage of being able to form very large, relatively insoluble molecules. Insolubility is an advantage since it prevents the sugars from taking part in the general metabolism of the cell and allows their deposition in a specific site of storage within (or sometimes outside) the cell from where they cannot diffuse away.

Some bacteria, protozoa and algae have been found to possess starch, which exists in the form of insoluble granules. Starch exists in two forms, amylose which is a chain of glucose residues joined by $\alpha(1-4)$ linkages, and amylopectin, which is a branched molecule with both $\alpha(1-4)$ and $\alpha(1-6)$

linkages. The structures of these polymers are shown in Fig. 8.9. Dextrans are produced by some bacteria and yeasts and are primarily $\alpha(1-6)$ polyglucose chains with some cross-linkages. Dextrans can be hydrated to form a very sticky, viscous material, which coats the surface of some bacteria and

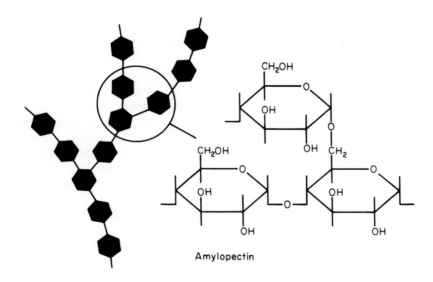

Amylopectin

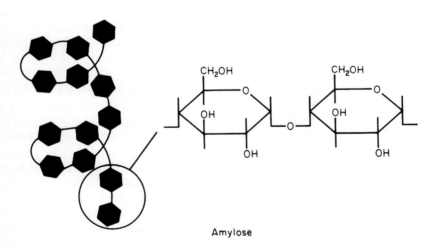

Amylose

Fig. 8.9 Structures of starch, amylose and amylopectin.

is therefore termed an extracellular polysaccharide. Extracellular polysaccharides also have a functional role.

Some extracellular polysaccharides which form a capsule are homopolysaccharides of glucose (e.g. dextran and cellulose) or other sugars, while many are quite complex, incorporating sugar derivatives (e.g. *N*-acetylhexosamines, hexuronic acids) as well as simple monosaccharides, often with three or four different constituents. Capsules are sometimes so large that they dwarf the cells embedded within them. One obvious consequence of this is that the cell may be protected against certain aggressive components of its external environment. These could include certain toxic chemicals or even predators unable to graze on larger particles. Another possible consequence is that extracellular material, by increasing the effective surface area of the cells, allows them to scavenge nutrients in a nutritionally poor environment. Certainly, whether extracellular polysaccharides have a storage or structural role, they do not appear to be an integral part of the structure of the cell itself or its metabolism, since many bacteria can lose the capsule without any effect on viability or growth.

The bacterial cell wall, which is formed outside the cell membrane, but within the extracellular polymer layer, is a far more rigid structure than both of them. All but one or two small groups of bacteria possess a cell wall. Apart from conferring considerable mechanical strength on the cell and providing a degree of protection, the cell wall is important in preventing distortion of the cell due to the often considerable osmotic pressure difference between the cytoplasm and the external environment. The osmotic pressure of bacterial cytoplasm may be as much as 600 atmospheres. The structure of the cell wall is an example of the combination of polysaccharides with other biological polymers to produce a continuous coat around the entire cell. Murein is a basic component of all procaryotic cell walls and consists of a very long chain of alternating *N*-acetylglucosamine and *N*-acetylmuramic acid units linked through glycosidic bonds. Two-dimensional rigidity is conferred on this chain by short peptide chains (containing four amino acids) attached to the muramic acid residues which themselves are cross-linked by peptide bonds or short peptide chains. The completed molecular structure is that of a mesh, and the thickness of the murein layer is controlled by the extent of further cross-linking in the third dimension. The structure of murein is shown in Fig. 8.10.

In addition to murein, Gram-positive bacteria also possess techoic acids which are complex polymers of amino sugars, sugar alcohols and amino acids. The cell walls of Gram-negative bacteria contain a lower proportion of murein and contain larger quantities of complex molecules consisting of polysaccharides and lipids. It is the different chemical components of the cell walls which account for the different staining characteristics of Gram-positive and Gram-negative bacteria.

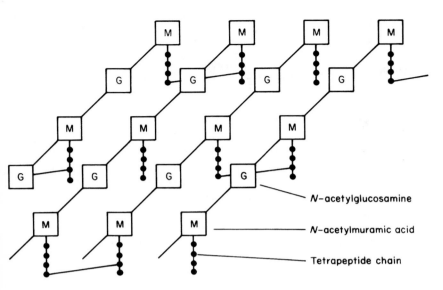

Fig. 8.10 Structure of murein.

8.5 Nucleic acids

8.5.1 Structure

The monomeric unit from which the nucleic acid polymer chain is constructed by condensation is the nucleotide. This in turn is made up of three smaller moieties. These are a purine or pyrimidine base, linked to a sugar, which in turn is linked to a phosphate group. The primary chain consists of alternating sugar–phosphate–sugar units, with the bases protruding to the side, rather like the side-chain functional groups of amino acids in proteins. The major bases found in nucleic acids are limited to two purines and three pyrimidines (plus a few minor bases occurring in small proportions) and there are only two different sugar molecules. The structures of these components and their linkage to form the polymer chain are shown in Fig. 8.11. The phosphodiester bond between the sugar and phosphate groups allows the polymer chain to assume a variety of conformations which are sometimes extremely convoluted. The bases contribute significantly to stabilizing the structures of the different classes of nucleic acids. The bases are planar (flat), due to the nature of their ring structures, and can take part in hydrophobic interactions when juxtaposed in stacks. Additionally, adjacent bases can form hydrogen bonds between each other which can also significantly stabilize the conformation of the molecule.

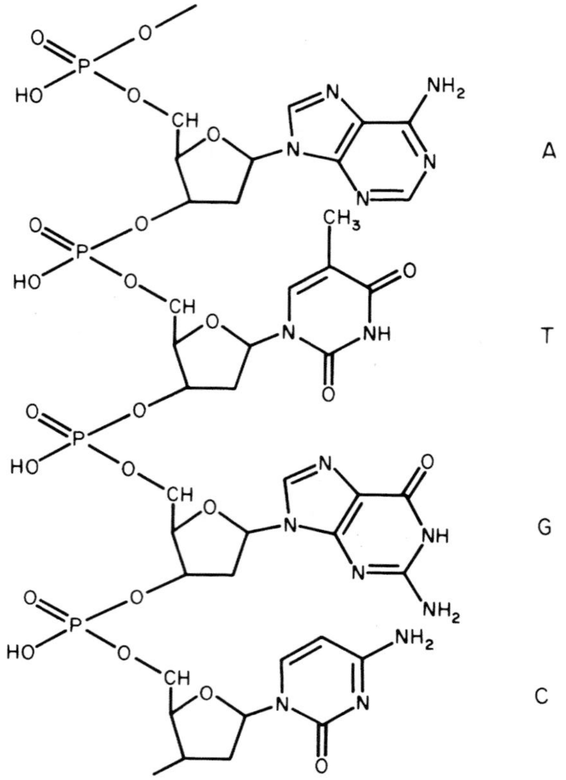

Fig. 8.11 Structure of the polymer chain of a nucleic acid.

(a) Deoxyribonucleic acid (DNA)

DNA contains only deoxyribose as the sugar component of the chain. Analysis of the purine and pyrimidine contents of DNA (i.e. adenine (A), thymine (T), cytosine (C) and guanine (G)) from a variety of eucaryotic and procaryotic sources reveals that:

A = T
G = C
A + C = G + T
and purines = pyrimidines.

These observations among others led Watson and Crick in the 1950s to propose a double-stranded structure for DNA, stabilized by hydrogen bonding between adjacent bases. The most stable conformations are the two hydrogen bonds between A and T and the three hydrogen bonds between G and C, hence the equality between the contents of purines and pyrimidines in DNA. Perhaps a useful analogy is to consider the DNA molecule to be similar to a staircase, the sugar–phosphate backbone representing the railings on each side and each hydrogen-bonded base pair representing a step in the staircase. The analogy can be taken further since the two strands are coiled around each other, thus resembling a spiral staircase. The structure is regular since the lateral dimensions of the complete A–T and G–C base pairs are identical. However, for such a structure to be possible, the two strands making up the double helix have to run in an antiparallel fashion, as shown in Fig. 8.12. A schematic of the DNA double helix is shown in Fig. 8.13.

The double helix is the normal conformation of DNA, although in some viruses single-stranded DNA is found. Although the double helix is only maintained by hydrogen bonds, a typical molecule may contain in excess of 10^7 base pairs, so that the number of these bonds is very large indeed. Another important feature of DNA is that once the sequence in one strand has been specified, the sequence in the other strand is fixed. The two strands

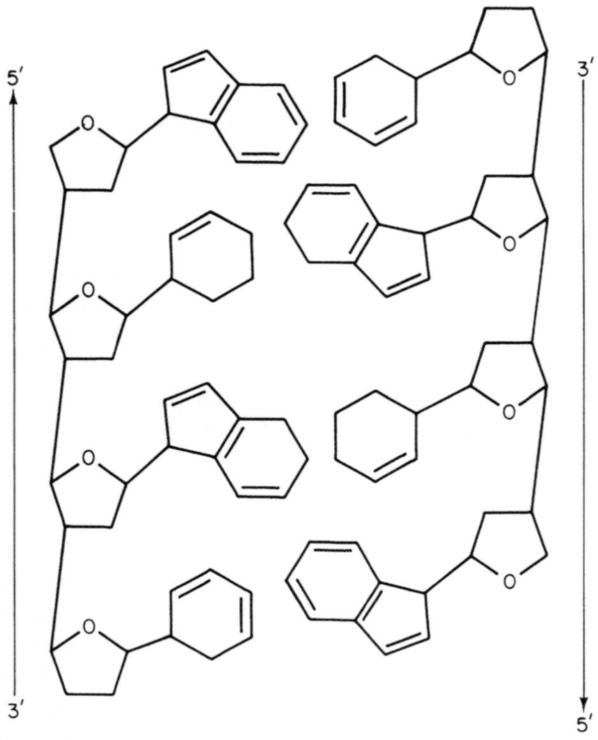

Fig. 8.12 Antiparallel arrangement of the double-stranded DNA molecule.

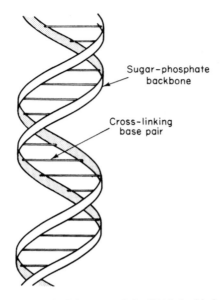

Sugar-phosphate
backbone

Cross-linking
base pair

Fig. 8.13 Schematic of the DNA double helix.

are said to be complementary. This is an important property which allows the molecule to fulfil its specific function.

(b) Ribonucleic acid (RNA)

Molecules of RNA contain exclusively ribose instead of deoxyribose. This does not permit it to form a regular double helix. Consequently RNA is normally single-stranded (although again there are a few exceptions). However, the flexibility of the polymer does permit it to fold back on itself and if complementary sequences are present considerable base pairing may occur. RNAs can assume a variety of shapes, examples of which are shown in Fig. 8.14.

8.5.2 Function

The only major function of nucleic acids is to store, maintain and transmit genetic information. It is this information which determines the inherited features of an organism. DNA is generally the material which stores this information and transmits it to the next generation during cell division. DNA is generally found in the nucleus in eucaryotes as an important component of the chromosomes. In procaryotes it forms the single circular chromosome. Several types of RNAs are involved in relaying this information

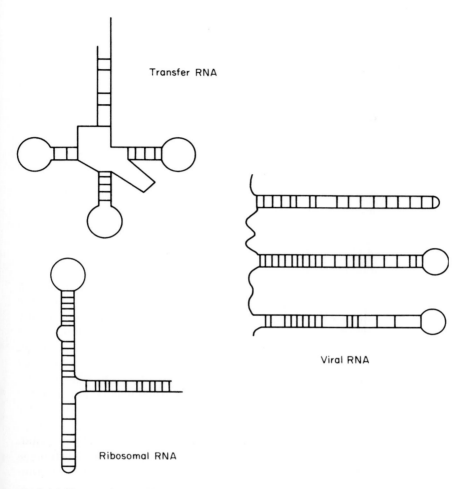

Fig. 8.14 Shapes of some RNA molecules.

from the nucleus to the sites in the cell where it is to be expressed. In order to be able to fulfil these functions the genetic material must be able to replicate itself accurately, it must be stable so that changes in hereditable characteristics (mutations) occur only at a low frequency, it must have the potential to determine the characteristics of the organism and it must be able to transmit this information.

In more precise terms, DNA contains all the information which is necessary to synthesize proteins, including the number of amino acids and their precise sequence of incorporation. It is through the activities of the proteins and enzymes that the characteristics of the cell are determined. Although the

DNA molecule has a far more specific function than merely coding for the range of proteins synthesized, all of the information embodied in every protein is incorporated into the cells' DNA complement. This information is encoded in the sequence of bases in such a way that combinations of the main four different bases can code for combinations of twenty different amino acids. The mechanisms by which the proteins are synthesized from the DNA code are described later.

8.6 Lipids

8.6.1 Structure

The lipids are a class of molecules which have considerable hydrophobic properties. These are conferred in many cases by long chain aliphatic molecules (alkanoic acids). The chain length is typically between twelve and twenty-four carbon atoms. At the carboxyl end two or three such molecules may be linked together by a bridging molecule, frequently glycerol. There may be one or two double bonds in the fatty acyl chains which can introduce a bend or "kink" into an otherwise linear structure. A number of other long chain molecules may also be found, but these superficially resemble the fatty acids, and so will not be described in detail. A generalized structure of a lipid is shown in Fig. 8.15. If all three of the hydroxyl groups on the glycerol moiety are linked to a fatty acid, then the molecule is a simple fat. If only two are linked in this way, then the third hydroxyl group may be phosphorylated to form a class of compounds known as phosphatidic acids. A number of substituents carrying an OH group may be joined to the phosphate group, including amino acids, amines, sugar alcohols, to confer varying degrees of hydrophilia on one end of the molecule.

There are many possible permutations involved in lipid structure, including some unsaturated multiple ring structures with aliphatic substituents (e.g. cholesterol). Many are linear in shape, being either partially or com-

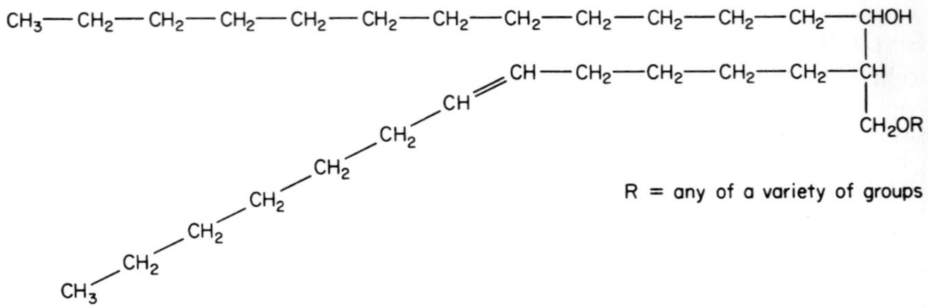

Fig. 8.15 Generalized structure of a simple lipid.

pletely hydrophobic, some appear to be branched, due to the presence of an occasional double bond, and some tend almost towards being globular.

8.6.2 Function

Fats are typically food storage molecules. Their poor solubility in water permits their deposition as droplets in the cell, from which they can easily be mobilized.

The other major function of lipids is as components of biological membranes, which normally contain approximately 50% lipid and 50% protein. Derivatives of phosphatidic acid are important components of these structures. The electron microscope has revealed that cell membranes have a characteristic double-layered structure, with two electron dense bands each approximately 20 nm in width separated by a lighter 35 nm band in the middle. The most probable structure which could account for this is a lipid bilayer with the hydrophobic fatty acid residues pointing outwards and coated on the inner and outer surfaces by electron-dense protein.

A consideration of the properties of membranes, including their selective permeability to certain substances and the existence of active transport mechanisms, would also suggest that the membrane may have pores, or hydrophilic channels, bound enzymes and cofactors and other systems connecting the inner and outer surfaces. Although much remains to be discovered about membrane structure, it would appear that the model in Fig. 2.2 is a reasonable one. A biological membrane can be thought of as proteins floating in a sea of lipid, their relative "buoyancy" being dependent on the degree of hydrophilia. In certain conditions some of the proteins can migrate laterally in the lipid bilayer, while others will be fixed relative to each other; the membrane therefore has a degree of fluidity. This model also permits the inclusion of hydrophobic or hydrophilic "channels" or pores through which certain molecules can pass. One can also imagine that some enzymes could effect the transfer of materials across the membrane.

References and further reading

Alberts, B., Bray, D., Johnson, A., Lewis, J., Raff, M., Roberts, K. and Walter, P. (1998) *Essential Cell Biology: An Introduction into the Molecular Biology of the Cell*, Garland Publishing Inc., New York.

Rawn, J. D. (1989) *Biochemistry*, Neil Patterson Publishers, North Carolina.

Yudkin, M. and Offord, R. (1980) *A Guidebook to Biochemistry*, 4th edn, Cambridge University Press, Cambridge.

Chapter 9

Microbial metabolism

9.1 Introduction

A living cell, in order to survive, grow and reproduce, has two absolute requirements. These are an external source of carbon (termed the substrate) which is assimilated and used to make new cellular material, and a source of energy which is converted and utilized in the synthesis of this material. A third requirement is for reducing power. This requirement is variable, depending on the particular mode of metabolism possessed by the organism. In the case of autotrophs, whose source of carbon is carbon dioxide, the requirements for energy and reducing power involved in the synthesis of, say, a large polysaccharide are considerable indeed. These major requirements and their relationships to biosynthetic processes are shown schematically in Fig. 9.1. This chapter is concerned predominantly with those processes which appear on the left of Fig. 9.1 and which are involved in the degradation of the substrate and the abstraction of energy. These processes are collectively termed catabolism. A brief discussion of the anabolic processes involved in the synthesis of biological macromolecules is included in Section 9.4.

Figure 9.1 also indicates that there are intermediate stages between the primary sources and their ultimate utilization in anabolic processes. These intermediate stages are of importance in allowing a degree of flexibility and control in the cell's utilization of energy. The existence of reservoirs for energy and reducing power permits the dissociation of energy-yielding metabolism and energy-consuming metabolism on both a temporal and a spatial basis.

The first part of this chapter examines the mechanisms whereby energy transfer occurs in metabolic reactions. Before considering the biochemical aspects of this, however, it is necessary to examine the nature of the driving force which makes chemical reactions occur, the associated energy changes involved (thermodynamics), and how these contribute to establishing the extent to which the reaction occurs (chemical equilibrium).

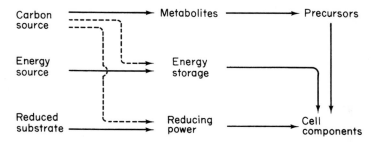

Fig. 9.1 Overview of metabolic processes.

9.2 Thermodynamics and chemical equilibrium

9.2.1 Reversible reactions

It is well known that some chemical reactions occur spontaneously. If a piece of sodium metal is added to water it rapidly and spontaneously reacts to form sodium hydroxide and hydrogen. It is not recommended that this be confirmed experimentally, however, because the reaction is violent and can be very dangerous. Hydrogen peroxide also decomposes spontaneously to oxygen and water, but this occurs very slowly (unless, of course a catalyst is added). Thus spontaneity is a phenomenon independent of rate. Another reaction, which is also spontaneous, is that which occurs when a pure crystalline sample of α-D-glucose is dissolved in water. After a time a mixture of α- and β-D-glucose is formed. The dissolution of pure β-D-glucose also leads to the same mixture. Upon "completion" of the reaction, therefore, measurable quantities of both reactants and products are present. Also, both the forward and backward reactions are spontaneous. It is probable that all chemical reactions are reversible, although in the case of the reaction of sodium with water, the composition of the reaction mixture on completion lies overwhelmingly in favour of the products, and the back reaction is undetectable. This reversibility is evident in the notation used to describe a chemical reaction:

$$A + B \rightleftharpoons C + D$$

In the general reaction of A with B, upon "completion" the reaction is said to have attained equilibrium, and the reaction mixture contains a fixed ratio of reactants to products. This is expressed in terms of the equilibrium constant, K_{eq}, which is given by:

$$K_{eq} = \frac{[C]_e [D]_e}{[A]_e [B]_e} \tag{9.1}$$

where the subscript e is used to denote concentrations at equilibrium. In the case of the glucose solution, K_{eq} is close to unity, whereas for the reaction of sodium with water it is very large.

The equilibrium attained is a dynamic one. That is, both the forward and backward reactions are still occurring simultaneously, but at rates which are equal so that there is no net change in the composition of the equilibrium mixture. We have already seen that the rate of a reaction is proportional to the concentrations of reactants, hence

$$\text{rate forward} = k[A][B] \tag{9.2}$$

and

$$\text{rate backward} = k'[C][D]$$

where k and k' are the rate constants of the forward and backward reactions respectively. It is apparent from equation (9.2) that:

$$K_{eq} = k/k' \tag{9.3}$$

Large values of K_{eq} indicate that the reaction can proceed to an appreciable extent, whereas small values indicate that the back reaction will predominate (since K_{eq} for the forward reaction equals $1/K_{eq}$ for the back reaction).

Returning to the question of spontaneity, non-spontaneous reactions do not proceed unless there is a form of energy input. For example, the reaction which follows the addition of zinc metal to copper sulphate solution is spontaneous:

$$CuSO_4 + Zn \longrightarrow ZnSO_4 + Cu$$

The back reaction which is not spontaneous can be made to occur by passing an electric current between the copper metal and the zinc sulphate solution. It is reasonable to suppose that this input of electrical energy is equivalent to the amount of energy generated by the reaction (or "lost" by the reactants) proceeding spontaneously in the forward direction. This energy output is termed free energy (ΔG). Spontaneous reactions are associated with a negative value of ΔG, since the reactants lose energy in the process. Non-spontaneous reactions are said to have positive ΔG.

As a chemical reaction proceeds towards equilibrium, the effective ΔG of the forward reaction, which is a function of the concentration of reactants, decreases in negative value, whereas the effective ΔG of the back reaction increases as the concentration of products of the forward reaction increases. At equilibrium, these effective ΔG values cancel each other out and no further net change occurs. The ΔG of a given reaction is obviously related in some way to K_{eq}. ΔG depends on the difference between the concentrations of reactants and products before the reaction commences and their concentrations at equilibrium, according to the relationship:

$$\Delta G = -RT \ln K_{eq} + RT \ln \frac{[C][D]}{[A][B]} \tag{9.4}$$

where R is the gas constant (8.31 J K^{-1} mol^{-1}) and T the absolute temperature. Since ΔG is concentration dependent it is usual to express it as the standard free energy change at 298 K with reactants and products at unit concentration ($\Delta G^{\ominus\prime}$). If [A], [B], [C] and [D] are set to unit concentration, the final term in equation (9.4) becomes zero. Therefore:

$$\Delta G^{0\prime} = -RT \ln K_{eq} \tag{9.5}$$

9.2.2 Coupling of reactions

Having seen how free energy changes and equilibria affect the spontaneity and extent to which a reaction will (or will not) occur, it is now possible to examine how free energy changes can be put to good use from the cell's point of view. Let us assume that the reaction of A and B to form C and D has a free energy change of ΔG_1^0. For another reaction:

$$C \rightleftharpoons E + F$$

the free energy change is ΔG_2^0. K_{eq1} and K_{eq2} are, respectively, [C][D]/[A][B] and [E][F]/[C]. If both reactions are allowed to occur in the same system, the net reaction will be:

$$A + B \rightleftharpoons D + E + F$$

and it is simple to show that the equilibrium constant for this reaction, K_{eq3}, is equal to the product of K_{eq1} and K_{eq2}. Therefore

$$\ln K_{eq3} = \ln K_{eq1} + \ln K_{eq2} = \frac{\Delta G_1^0 + \Delta G_2^0}{-RT}$$

and therefore:

$$-RT \ln K_{eq3} = \Delta G_1^0 + \Delta G_2^0 = \Delta G_3^0 \tag{9.6}$$

Let us now suppose that ΔG_1 is positive, and the reaction can therefore not occur under normal conditions. Provided that ΔG_2 has a negative value which is numerically larger than ΔG_1, if reactions one and two are coupled then the first reaction can occur because it is driven by the second, since for the overall reaction ΔG_3 is negative.

9.2.3 Biochemical energy coupling

It is obvious that the order and complexity of living organisms are not directly the result of spontaneous chemical reactions, and the cell therefore uses a coupling mechanism to derive energy for endergonic (energy-consuming) synthetic reactions from other exergonic (energy-yielding) reactions. The mechanism of coupling is shown schematically in Fig. 9.2. The X–Y coupling system has been found to be essentially identical in all organisms, and consists of the synthesis of adenosine triphosphate (ATP) from adenosine diphosphate (ADP) and inorganic phosphate (P_i). The structure of ATP is shown in Fig. 9.3. ATP has often been termed a "high-energy" molecule since a considerable amount of free energy is released from its hydrolysis ($\Delta G^0 = -30$ kJ mol^{-1}). This is higher than the positive value of ΔG^0 for many endergonic reactions which occur in the cell and these can therefore proceed in coupled reactions involving ATP hydrolysis.

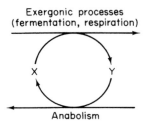

Fig. 9.2 Mechanism of coupling of endergonic and exergonic reactions.

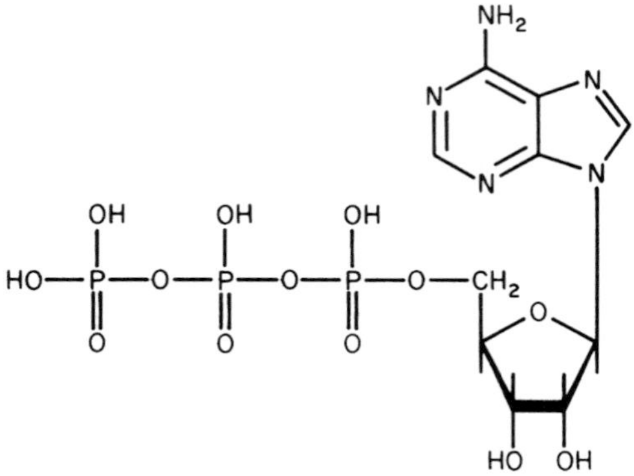

Fig. 9.3 Structure of ATP.

However, the exergonic reactions which can be used to drive the synthesis of ATP (phosphorylation of ADP) must have values of ΔG^0 greater than ΔG^0 for ATP hydrolysis. ATP is not therefore a uniquely "high-energy" molecule. Micro-organisms generally use one of three processes which can generate high values of negative ΔG^0 as sources of energy for phosphorylation. Since all chemical reactions involve the movement or transfer of electrons, then the requirements of energy-yielding metabolism are a source of electrons with a high potential energy and an electron acceptor of lower potential energy towards which they are moved with the concomitant decrease in ΔG.

9.2.4 Coupling of oxidation and reduction

It is possible, theoretically at least, that wherever a metabolic reaction involves a sufficiently large negative free energy change it could be coupled to the synthesis of ATP. Many catabolic reactions involve an oxidation of the substrate. These can be represented in simplified form as:

$$AH_2 + \tfrac{1}{2}O_2 \rightleftharpoons A + H_2O$$

The ΔG^0 for this type of reaction is typically between -120 and 200 kJ mol^{-1}, and could in theory be used in the synthesis of four to six molecules of ATP if the coupling mechanisms existed. However, the reducing power associated with the substrate will need to be preserved so that it can be used in biosynthetic reactions (as shown in Fig. 9.1). Thus, for many substrate oxidations of this type, instead of synthesizing ATP, the oxidation is coupled to the reduction of a carrier molecule, thus preserving both reducing power and much of the available free energy. The carrier molecule is either nicotinamide adenine dinucleotide (NAD) or nicotinamide adenine dinucleotide phosphate (NADP), the structures of which are shown in Fig. 9.4. Thus, the oxidation of the substrate can be represented as:

$$AH_2 + NAD \rightleftharpoons A + NADH_2$$

The oxidation of $NADH_2$ by molecular oxygen has a ΔG^0 of about -170 kJ mol^{-1}, so that in a sense it can also be considered a "high-energy" compound. The energy and reducing power preserved in $NADH_2$ or $NADPH_2$ can be diverted towards synthetic reactions (usually $NADPH_2$) or oxidative phosphorylation (usually $NADH_2$).

9.3 Energy-yielding metabolism

There are three main modes of microbial energy-yielding metabolism: respiration, fermentation and photosynthesis. In heterotrophic organisms

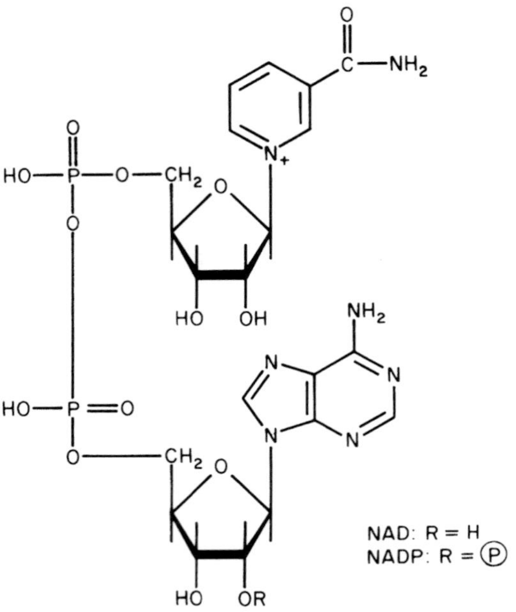

Fig. 9.4 Structure of NAD and NADP.

which rely on respiration or fermentation, the carbon substrate typically supplies energy, reducing power and the intermediate metabolites used in biosynthetic reactions. In autotrophs the sources of energy and carbon are different. However, oxidation of the substrate in heterotrophs is an important source of energy; since typical oxidations such as that of AH_2 by oxygen can release sufficient energy for the synthesis of several molecules of ATP, it is not surprising that catabolic pathways generally incorporate a large number of discrete steps. This also has the advantage of generating a large number of intermediate metabolites which act as starting points for biosynthesis.

9.3.1 Substrate level phosphorylation: glycolysis

It is not possible to describe here in detail how micro-organisms degrade the extensive variety of substrates which can be utilized. However, some of the major groups and the substrates they are capable of utilizing are shown in Table 9.1. Instead, the utilization of glucose (glycolysis) will be described as an example of the principles involved. Since carbohydrates are fairly widely used both as substrates and food storage molecules by both aerobic and anaerobic organisms, glycolysis is an important degradative pathway in the

Table 9.1 Examples of heterotrophic groups and the substrates they can metabolize

Group	Mode of metabolism	Substrates	Examples
Enteric group	Fermentative	Sugars; some can use starch or pectin	Escherichia, Shigella, Salmonella
	Respiratory	Sugars, amino acids, peptides, organic acids; some can use aromatics	
Aerobic pseudomonads and related genera	Respiratory	Extremely versatile. Pseudomonas spp. can use over fifty different compounds. Sugars, organic acids, alcohols	Pseudomonas, Aztobacter, Zoogloea
Gram-negative strict anaerobes	Fermentative	Carbohydrates, organic acids	Bacteroides, Veillonella
Clostridium	Fermentative	Genus is versatile (e.g. carbohydrates, organic acids, amino acids, ethanol, purines) but individual species are not	Clostridium
Gliding bacteria	Respiratory	Cellulose, sugars. Fairly complex requirements, e.g. for amino acids	
Actinomycetes	Respiratory	Nutritionally versatile. Produce extracellular proteases and can lyse other bacteria	Streptomyces

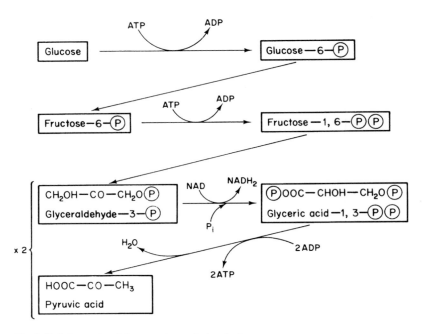

Fig. 9.5 Schematic of the reactions of glycolysis.

microbial world. A schematic of all the reactions involved in glycolysis is shown in Fig. 9.5. In order for glucose to be metabolized, it must first be activated. This involves the expenditure of energy by the cell through the hydrolysis of two molecules of ATP with the stepwise transfer of the phosphate groups to the hexose molecule. (Throughout the text ⓟ is used to denote a phosphate group and P_i to denote inorganic phosphate.) The phosphorylated hexose is cleaved to form two trioses (three-carbon sugars) which are further phosphorylated by inorganic phosphate in a reaction coupled with the reduction of NAD. The resulting compound, glycerate-1,3-diⓟ is termed a "high-energy" compound because it has a greater free energy of hydrolysis than ATP. As might be expected, then, its synthesis from glyceraldehyde-3-ⓟ has a positive value of ΔG^0. A net synthesis does occur, however, due to the way in which the cell manipulates the concentrations of reactant and product. In an equilibrium mixture which lies overwhelmingly in favour of the reactants, if their concentrations are increased, causing an increase in the magnitude of the denominator in equation (9.1), then a net synthesis of the products occurs consistent with increasing the numerator in equation (9.1) so that the value of K_{eq} remains constant when the new equilibrium is reached. Furthermore, if the concentration of products is lowered by removing them from the reaction mixture then again net synthesis occurs.

Glycerate 1,3-di℗ can directly phosphorylate ADP, and the product of this reaction, glycerate-3-℗, undergoes isomerism before a further phosphorylation occurs.

The net reaction can be represented as:

$$glucose + 2ADP + 2P_i + 2NAD \rightleftharpoons 2pyruvate + 2ATP + 2NADH_2$$

The important features of glycolysis are the net generation of ATP and the formation of an oxidized, three-carbon end-product at the expense of the cell's reserves of NAD. Another important consequence of the way in which glucose is metabolized is that the formation of a large number of intermediates provides a considerable variety of organic compounds which serve as starting points in synthetic reactions. For example, triose derivatives are used for the synthesis of lipids and some amino acids, and pyruvate and phospho-*enol*-pyruvate are required for the synthesis of other amino acids.

Glycolysis occurs by essentially the same pathway in all organisms, aerobic and anaerobic. The basic differences between aerobic respiration and anaerobic fermentation, and in particular the crucial role of molecular oxygen in aerobic metabolism, lie in the reactions which occur after the formation of pyruvate.

9.3.2 Aerobic respiration

The degradation of glucose to pyruvate yields a free energy of about 170 kJ mol^{-1}. However, this represents a much less than complete oxidation of the substrate. The major end-product of respiration is carbon dioxide, the formation of which from glucose yields about 3000 kJ mol^{-1}.

$$glucose + 6O_2 \rightleftharpoons 6CO_2 + 6H_2O$$

The energy yield from the complete combustion of glucose is almost twenty times that from its metabolism to pyruvate. Again, the cell does not perform the complete oxidation in a single highly exergonic step, but in a series of reactions. Neither does it utilize molecular oxygen directly in the oxidation of the organic carbon derived from pyruvate.

(a) Tricarboxylic acid cycle

In order to obtain carbon dioxide from pyruvate ($CH_3COCOOH$) an obvious requirement is for dehydrogenation. First, however, an acetyl group ($CH_3 \cdot CO-$) is split from pyruvate and transferred to a carrier molecule termed coenzyme A (CoA) in a reaction coupled to the reduction of NAD:

$$CH_3COCOOH + NAD + CoASH \rightleftharpoons CH_3COSCoA + CO_2 + NADH_2$$

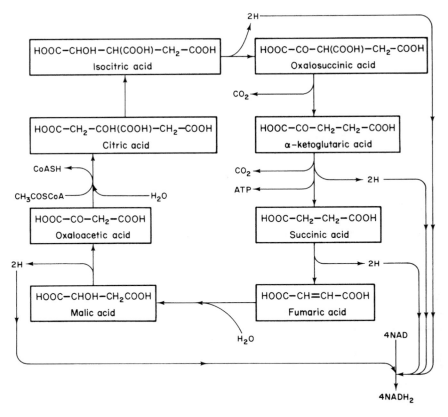

Fig. 9.6 The tricarboxylic acid cycle.

The acetyl group is transferred to oxaloacetate, a four-carbon compound, to form citrate:

$$CH_3COS-CoA + HOOCCOCH_2COOH + H_2O$$
$$\Updownarrow$$
$$HOOCCH_2(COH(COOH))CH_2COOH + CoASH$$

In a series of reactions (Fig. 9.6) involving eight intermediates there are four dehydrogenations (i.e. the reduction of four molecules of NAD), two hydrations (addition of H_2O) and two decarboxylations (removal of CO_2) leading to the regeneration of oxaloacetate. Hence, the reaction sequence is cyclical and the intermediate compounds serve as carriers of the two-carbon fragment derived from acetyl CoA as it is gradually modified to CO_2. The overall cycle can be represented in a simplified form as:

$$\text{oxaloacetate} + 2H_2O + \text{acetyl CoA} + 4NAD + ADP + P_i$$

$$\xrightarrow{} \xrightarrow[\text{intermediates}]{} \xrightarrow{} \xrightarrow{} \text{oxaloacetate} + \text{CoASH} +$$

$$NADH_2 + ATP + 2CO_2$$

This method of representing the cycle suggests that the role of oxaloacetate is akin to that of a catalyst because it is not consumed during substrate oxidation; coenzyme A is also regenerated from the reaction sequence which generates CO_2 from pyruvate, and the formation of ATP is a positive advantage to the cell. The only product of the complete oxidation of the substrate which requires regeneration is $NADH_2$, and this is where the requirement for molecular oxygen is significant in aerobic organisms.

(b) The respiratory electron transport chain

Although the substrate has been completely oxidized to CO_2, the involvement of molecular oxygen has not yet occurred at this stage. There remains a considerable quantity of free energy available to the cell from the respiratory electron transport chain which mediates in a number of stages the reoxidation of $NADH_2$ and the ultimate reduction of oxygen. The significance of the electron transport chain is therefore twofold. It brings about the reoxidation of $NADH_2$ and in so doing generates a considerable amount of free energy for use in the synthesis of ATP.

The first stage in the process involves the transfer of hydrogen to a carrier protein, called flavoprotein (FP).

$$NADH_2 + FP \rightleftharpoons NAD + FPH_2$$

Subsequently electrons are generated from intermediate hydrogen transfer processes:

$$2H \rightleftharpoons 2H^+ + 2e^-$$

These are transferred to another series of carriers called cytochromes, which are proteins containing a large non-protein ring structure within which a reducible ferric ion is held:

$$\text{Cyt—}Fe^{3+} + e^- \rightleftharpoons \text{Cyt—}Fe^{2+}$$

The cytochrome system is consistent and fairly well characterized in eucaryotes, but in bacteria less is known about it. There are probably at least three different cytochromes, however. The end of the cytochrome chain is directly coupled to the reduction of oxygen to form water. The entire process, illustrating the path of the electrons from one carrier to another is shown in Fig. 9.7.

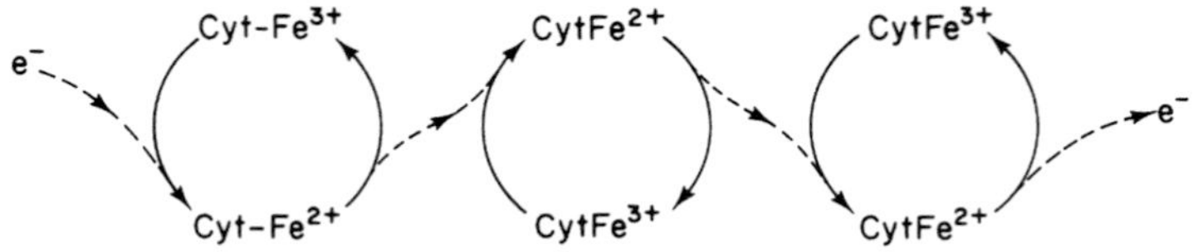

Fig. 9.7 The electron transport chain.

Unlike the reactions involved in glycolysis, reactants, products and intermediates do not exist in solution; in fact, no real chemical change in the normal sense is discernible experimentally during electron transport. The respiratory chain is, therefore, an important example of the close structural integration of components of a metabolic pathway from which intermediates are not released into the cytoplasm. This structural integration occurs at sites on unit membranes within the cell. In eucaryotes, the mitochondrial membranes are the site of electron transport, whereas in procaryotes the process occurs in invaginations of the cell membrane (the mesosomes). Unless the structural integrity of the membrane is maintained electron transport cannot occur.

Much of what is known of the sites of synthesis of ATP comes from studies of eucaryotes. There are three steps within the chain involving a sufficiently large free energy change for the synthesis of ATP. These are during the oxidation of $NADH_2$ by flavoprotein, the transfer of an electron to the initial stage of the cytochrome system and the oxidation of the terminal cytochrome by oxygen. The free energy changes involved in this process are illustrated schematically in Fig. 9.8.

9.3.3 Anaerobic fermentation

It is important to consider the fate of pyruvate in organisms which do not possess a respiratory electron transport chain, or which cannot use the chain in the absence of molecular oxygen. These organisms carry out fermentation where complete oxidation of the substrate does not occur because the cells do not possess a mechanism for reoxidation of the large quantities of $NADH_2$ which would be formed. Instead of further oxidizing pyruvate, it is reduced in reactions coupled to the reoxidation of $NADH_2$ generated during the initial substrate breakdown (glycolysis).

The lactic acid bacteria achieve this reoxidation in a single step with the formation of lactate:

$$2CH_3COCOOH \underset{\xrightarrow{\hspace{1cm}}}{\overset{NADH_2 \qquad NAD}{\rightleftharpoons}} 2CH_3CHOHCOOH$$

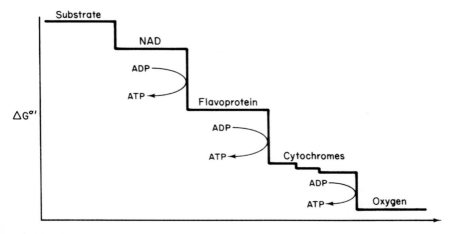

Fig. 9.8 Free energy changes during electron transport.

The molecular formula of lactate ($C_3H_6O_3$) is exactly half that of glucose. Hence there is no change in the oxidation–reduction level stoichiometry between the initial substrate and its fermentative end product. Accordingly, there is no net change in the cellular complement of NAD during fermentation. All that has happened is that the substrate has undergone a rearrangement.

Other organisms achieve the same end in several stages with the formation of different end products as, for example, in alcoholic fermentations:

$$2CH_3COCOOH \xrightleftharpoons{\quad CO_2 \quad} 2CH_3CHO \xrightleftharpoons{\quad NADH_2 \quad NAD \quad} 2CH_3CH_2OH$$

In this case, ethanol is considerably more reduced than lactate, but the CO_2 generated is considerably more oxidized. This is a clear illustration of the fundamental characteristic of fermentative metabolism; the transfer of electrons necessary to generate energy in the form of ATP does not require the existence of an independent external acceptor. The electrons are transferred from one part of an organic molecule to another and the average oxidation level of the products is identical with that of the substrate. The energy is derived from the rearrangement of the electrons. Some bacteria possess a more complex mechanism: they utilize two different organic compounds as substrates (usually a pair of amino acids), oxidizing one at the expense of the other, which is reduced. The principle of fermentation, however, remains the same.

9.3.4 Implications of respiratory and fermentative metabolism

The first major difference between the two modes of metabolism is the ATP yield. Fermentation (essentially glycolysis in this case) yields two moles of ATP for each mole of glucose fermented. In respiration the passage of a pair of electrons through the electron transport chain yields 3 moles of ATP. Four pairs are generated from each turn of the tricarboxylic acid cycle, which also generates one mole of ATP itself, and the cycle turns twice for each mole of glucose degraded to pyruvate. Hence 28 moles of ATP can be formed from the oxidation of one mole of glucose.

The second difference is in the range of organic compounds which can be used as electron donors. Because fermentative organisms have to maintain a strict oxidation-reduction balance in both anabolic and catabolic processes due to constraints imposed on the capacity for reoxidation of $NADH_2$ they are limited to compounds which are neither highly oxidized nor highly reduced. Moreover, these substrates must be capable of taking part in pathways which are used for substrate level phosphorylations. Aerobes, in contrast, can generally use a greater variety of organic compounds for energy because if they can enter the cell's metabolism at any point prior to the completion of the tricarboxylic acid cycle considerable energy can be obtained from the donation of their electrons to the electron transport chain.

9.3.5 Photosynthetic energy-yielding metabolism

A popular conception of the importance of photosynthesis on a global scale is that green plants consume carbon dioxide and generate oxygen, which in turn is used by animals for the oxidation of food, yielding carbon dioxide once again. Moreover, it is evident that green plants rely on sunlight to be able to generate oxygen and that carbon dioxide is the sole source of carbon in these organisms. They are, therefore, photoautotrophic. If we employ a very simplistic empirical formula for cellular carbon compounds, denoted (CH_2O) (the empirical formula of carbohydrates), then the metabolism of the green plants can be denoted as:

$$CO_2 + H_2O \rightleftharpoons (CH_2O) + O_2$$

In order to be able to perform this conversion, there are two obvious requirements. Since the carbohydrates synthesized are very much larger molecules than carbon dioxide there will be a considerable net requirement for ATP. Also, since carbon dioxide is a fairly highly oxidized compound there will be a requirement for substantial quantities of $NADPH_2$ for use in reductive biosyntheses.

It is evident that the only source of energy available to the green plants is sunlight. In 1937, Hill found that $NADPH_2$ could be formed in the presence of light entirely independently of carbon dioxide fixation:

$$2H_2O + 2NADP \xrightarrow{\text{light}} 2NADPH_2 + O_2$$

Later, it was also discovered that ATP could be formed in the presence of light, again without the involvement of carbon dioxide:

$$ADP + P_i \xrightleftharpoons{\text{light}} ATP$$

Thus, these light reactions can be considered entirely separately from the biosynthetic reactions which utilize carbon dioxide.

A brief description of the way in which light energy is used to form ATP and $NADPH_2$ is best afforded by reference to Fig. 9.9. Eucaryotes possess two pigment systems, PS 1 and PS 2. Each system consists of several hundred molecules of pigments including chlorophyll and carotenoids (Fig. 9.10). These systems are orientated in the stacks of membranes found within chloroplasts, and appear to act as small antennae for the collection of light. The light energy is channelled towards a specialized molecule of chlorophyll at the centre of each system. In PS 1 the light energy is used to promote an electron to a high-energy state so that PS 1 can reduce an electron carrier which under normal conditions it would not be able to do. The electron can then take one of two courses. It can either return to PS 1 in a transfer process which liberates sufficient energy for the formation of a molecule of ATP, or it can be transferred to NADP. If it takes this latter course it leaves PS 1 deficient in an electron which needs to be replaced. This electron is either obtained from water, in the case of all eucaryotes, or another electron donor such as hydrogen sulphide or an organic compound in the case of the photosynthetic bacteria. The bacterial photosynthesis machinery is simpler, being limited to only type 1 pigment systems coupled to the oxidation of the electron donor. In the eucaryotes, however, a second pigment

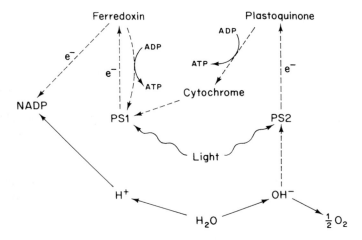

Fig. 9.9 The mechanism of photosynthesis.

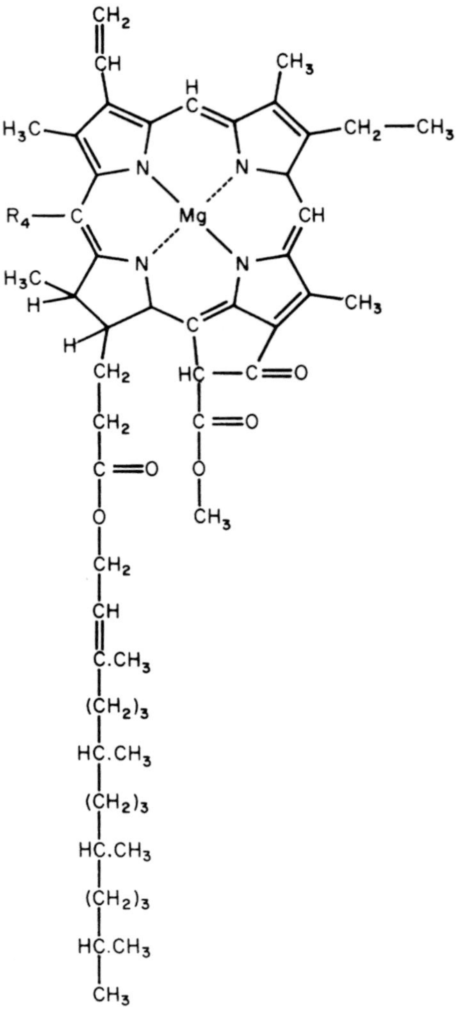

Chlorophyll a

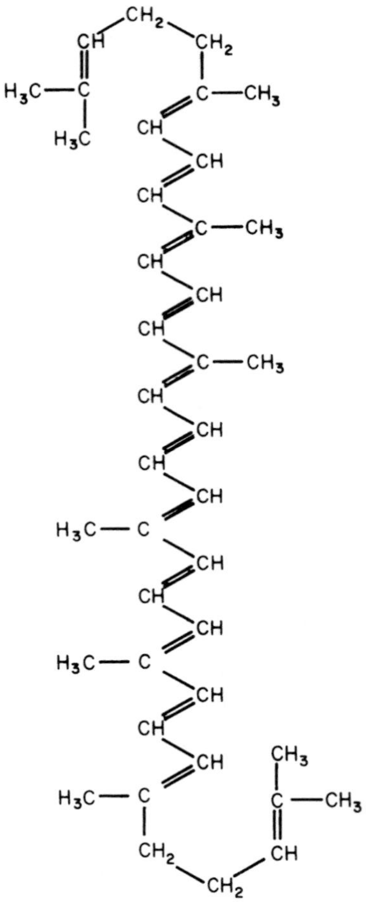

A carotenoid

Fig. 9.10 Structures of photosynthetic pigments.

system is involved in the oxidation of water. Water is an energetically unfavourable electron donor and a second light reaction is required to promote an electron via PS 2 to an acceptor molecule which then transforms it through an electron transport chain to PS 1 with the concomitant synthesis of another molecule of ATP.

9.3.6 Anaerobic respiration

Unfortunately, the division of heterotrophs into groups which are fermentative, and therefore anaerobic, and respiratory, and therefore aerobic, is an oversimplification. Certain groups are capable of using oxidized inorganic ions instead of molecular oxygen as terminal electron acceptors. This ability is confined to bacteria which utilize nitrate, sulphate and carbonate. An important group is the denitrifying organisms which reduce nitrate to nitrogen gas.

$$2NO_3^- + 10e^- + 12H^+ \rightleftharpoons N_2 + 6H_2O$$

The enzyme system which mediates this reduction is distinct from the normal respiratory chain, and most denitrifiers possess both systems. Hence, they are facultatively anaerobic and only generate nitrogen gas in the absence of air. In general, the organisms are heterotrophic but some can use inorganic compounds as electron donors, for example:

$$5S + 6NO_3^- + 2H_2O \rightleftharpoons 5SO_4^{2-} + 3N_2 + 4H^+$$

9.3.7 Other metabolic modes

So far we have examined heterotrophic metabolism in both respiration and fermentation (the use of organic compounds and energy sources), and phototrophic metabolism (the use of light as an energy source, frequently photoautotrophic, sometimes photoheterotrophic). There are in addition to these groups a number of bacteria which can grow chemoautotrophically using reduced inorganic compounds as energy sources. An important group is the nitrifying organisms which oxidize ammonia and nitrite to nitrate. Many of these chemoautotrophs can also grow heterotrophically. Notable among these are the hydrogen bacteria which use molecular hydrogen as an energy source and CO_2 as the carbon source; these organisms are more properly categorized as belonging to the Pseudomonadaceae. One group which may be considered intermediate between the autotrophs and heterotrophs are the methanogens, which use hydrogen as an electron donor and carbon dioxide as an electron acceptor, resulting in the formation of methane. These organisms are strict anaerobes, and can utilize carbon dioxide, formate and possibly acetate as carbon sources. Examples of micro-organisms which grow autotrophically using a variety of energy sources and electron donors are summarized in Table 9.2.

9.4 Biosynthesis

It has already been stated that the stepwise degradation of substrates by micro-organisms is advantageous in that the release of energy is thereby

Table 9.2 Examples of obligate and facultative autotrophic metabolism

Energy source	Major carbon source	Electron donor	Relation to oxygen	Examples
Light	CO_2	H_2O	Aerobic	Algae (some can use organic compounds as secondary sources of carbon).
Light	CO_2	H_2S, H_2	Anaerobic	Green and purple sulphur bacteria (some can use organic compounds as secondary sources of carbon in the presence of CO_2).
Light/organic	CO_2/organic	Organic	Facultative	Purple non-sulphur bacteria. Facultatively heterotrophic.
NH_3, NO_2^- $S, H_2S, S_2O_3^{2-}$	CO_2/organic	NH_3/NO_2^- $S, H_2S, S_2O_3^{2-}$	Aerobic Aerobic	Nitrifying bacteria. Sulphur-oxidizing bacteria.
H_2	CO_2/organic	H_2	Aerobic	Hydrogen bacteria. Nutritionally versatile; probably related to non-autotrophs such as some *Pseudomonas* spp.
H_2	CO_2/formate/acetate	H_2	Anaerobic	Methanogenic bacteria. Use CO_2 as terminal electron acceptor.

achieved in increments which permit its storage in the form of ATP (or $NADH_2$). Another advantage of the numerous steps of the major degradative (catabolic) pathways is that the intermediates formed act as the starting points for the synthesis of new cellular material. For example, pyruvate acts as the starting point for the synthesis of some amino acids, and the intermediates of the tricarboxylic cycle are essential starting points for amino acids, cytochrome precursors and pyrimidine synthesis amongst others. All of these low molecular weight compounds are known as the central intermediary metabolites and from them the cell will synthesize the several hundred compounds (at least) it requires for normal purposes. Although there are a large number of different low molecular weight compounds present in the cell, on a weight basis most of the cell consists of the four types of macromolecules, the structures and functions of which were described in Chapter 8: proteins, polysaccharides, lipids and nucleic acids.

In Chapter 8, brief mention was made of the involvement of the condensation reaction in biological polymerization. The proteins, polysaccharides and nucleic acids are all synthesized through condensation reactions occurring at one end of the growing polymer chain. In this respect, then, the chemical changes involved in biosynthesis are fairly straightforward. So too is the mechanism whereby the chemical energy stored in the form of ATP is used in promoting the thermodynamically unfavourable polymerization process. In all cases the bond energy is used to activate the monomers prior to polymerization. For example, glucose-6-phosphate may be activated by ATP to form ADP-glucose with the release of inorganic pyrophosphate. Amino acids are also initially activated by reaction with ATP. Adenosine monophosphate (AMP) is one of the monomers of nucleic acid, so that the triphosphate acts as its own activator, with GTP, UTP and CTP performing the same role in the case of the other ribonucleotides.

The following sections focus on the wider biological implications of polymer biosynthesis rather than the detailed chemistry.

9.4.1 Polysaccharide synthesis

Monosaccharides are the raw materials from which polysaccharides are synthesized. The biochemistry involved is reasonably straightforward, with the monomers being added to one end of the growing linear polymer chain. Branching is a secondary event.

In photosynthetic organisms, polysaccharide synthesis can be considered the latter stage in a reaction sequence which extends back to the level of carbon dioxide fixation. Carbon dioxide is fixed by reaction with a pentose sugar, ribulose 1,5-diphosphate, to give two triose phosphates. The fixation of three molecules of CO_2 results in a net gain of one triose, the other five trioses being rearranged to form three pentoses to initiate the cycle again. This cyclical reaction sequence is called the Calvin cycle.

9.4.2 Nucleic acid synthesis

The mechanisms of synthesis of the nucleic acids reflect the crucial importance of their two roles in storing and transmitting genetic information.
During reproduction, replication of the DNA complement of a cell prior to
distribution to the progeny must proceed accurately to ensure that a faithful
copy of the genetic information is passed on to the next generation.
Replication of double-stranded DNA occurs by the dissociation of the two
complementary strands, each of which can then act as a template for the
missing strand which is synthesized *in situ* to form a new double-stranded
molecule. This process is shown schematically in Fig. 9.11. The mechanism
is termed semi-conservative, because in each replication half of the parental
DNA is passed intact to each daughter molecule.

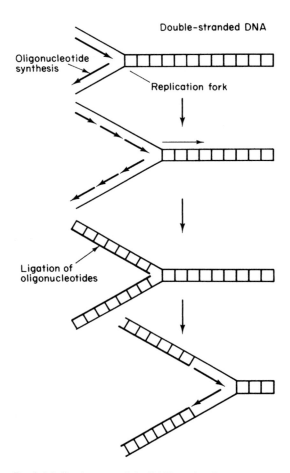

Fig. 9.11 Replication of the DNA molecule.

The expression of the genetic information carried by the cell's comple-ment of DNA involves as a first step the synthesis of RNA. The synthesis of RNA proceeds in a manner analogous to that of DNA, with one strand of the DNA molecule acting as a template for the synthesis of a complementary strand of RNA. This process is termed transcription. The RNA so formed is called messenger RNA (mRNA) and acts as the intermediate carrier of the

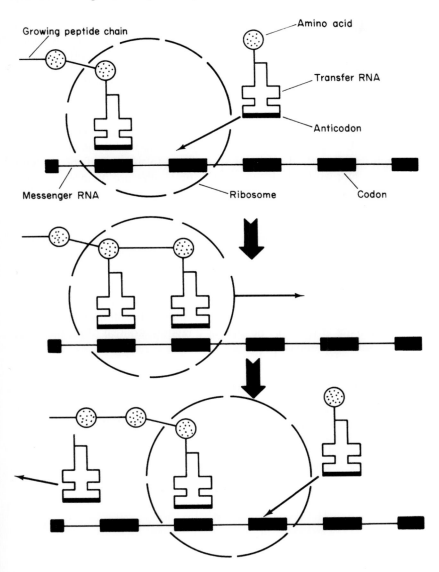

Fig. 9.12 Schematic of protein synthesis.

genetic information to the site in the cell where it will be expressed. This expression is through the synthesis of protein.

9.4.3 Protein synthesis

The functional characteristics of proteins depend entirely upon the precise sequence of amino acids in the peptide chain. The code which specifies this sequence exists in the sequence of bases on the DNA molecule which in turn specifies the sequences of three bases coding for particular amino acids. There are sixty-four possible combinations of the four nucleotides which can make up a codon (a trinucleotide sequence), which is more than enough to code for the twenty amino acids normally found in proteins. Consequently, many amino acids are coded for by more than one base sequence.

The sites of protein synthesis in the cell are the ribosomes. The ribosome binds the mRNA in such a way that each codon is exposed to the carrier molecule which transfers the individual amino acid to the end of the growing peptide chain. Not surprisingly, the carrier molecules are themselves composed of RNA. Each of these transfer RNAs (tRNAs) carry on one end of the molecule a trinucleotide base sequence complementary to the codon on the mRNA molecule, called the anticodon, and on the other end is bound the amino acid specified by the corresponding codon. Thus there are different tRNAs, each acting as carrier for one amino acid. The reaction sequence involved in protein synthesis is shown in Fig. 9.12.

In summary, we can say that the genetic characteristics of a cell are carried in the form of a triplet code in the DNA of the chromosomes. This code is expressed exclusively through the synthesis of proteins which in turn have an immediate structural role and, as enzymes, control of all the other metabolic reactions of the cell.

References and further reading

Atkins, P. W. (1998) *Physical Chemistry*, 6th edn, Oxford University Press, Oxford.

Stanier, R. Y., Ingraham, J. L., Wheelis, M. L. and Painter, P. R. (1987) *General Microbiology*, 5th edn, Macmillan Education, London.

Yudkin, M. and Offord, R. (1980) *A Guidebook to Biochemistry*, 4th edn, Cambridge University Press, Cambridge.

Chapter 10

Microbial growth

10.1 Introduction

Growth is a characteristic of all living organisms. In higher organisms the growth of an individual is manifested as an increase in size, which is accompanied by the complex processes of physiological maturation including cell differentiation. Growth can also be considered at the level of the population, and in animals this is governed by, amongst several factors, the complex mechanisms of sexual reproduction. The growth of an individual organism is, of course, a consequence of cell division. In the bacteria, however, cell division by binary fission directly affects the population size since each new cell constitutes a new individual. Growth of individual cells does occur, of course, but as with many studies on bacteria, phenomena at the population level are generally of most interest. This discussion will be largely confined to the growth of microbial populations, reproducing by binary fission. Most of the experimental work on the growth of such populations has been conducted using bacteria, and the following sections will generally be confined to this group.

10.2 Mathematical description of growth

Under favourable conditions, in the presence of carbon and energy sources and essential nutrients, growth which proceeds by the division of one cell into two, followed by the division of each of these into two more and so on results in an increase in the number of cells in the population according to the series 1, 2, 4, 8, 16, 32, ... Provided that the culture conditions remain favourable for growth over a reasonable period of time, cell division will occur regularly and the population will double in size at regular intervals. If the cells were all to behave in the same manner discrete stepped increases in population size would be observed, but in practice, cultures containing a large number of cells rapidly become desynchronized so that a gradual increase in cell numbers (N) or biomass concentration (X) with time is observed, as shown in Fig. 10.1.

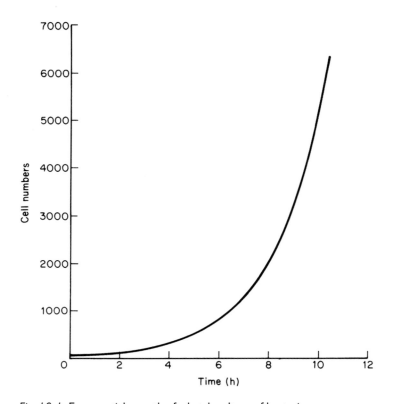

Fig. 10.1 Exponential growth of a batch culture of bacteria.

The type of growth shown in Fig. 10.1 is termed exponential or log-arithmic. The series 1, 2, 4, 8, 16, 32, etc. can be expressed as $2^0, 2^1, 2^2, 2^3, 2^4, 2^5 \ldots$ Therefore, if $\log_2 N$ is plotted against time, the straight line relationship shown in Fig. 10.2 is obtained. This is a more con-venient graphical representation of the data, since the gradient of the line gives the exponential growth rate, K:

$$K = \frac{\log_2 N_t - \log_2 N_0}{t} \tag{10.1}$$

If tables of logarithms to the base 2 are unavailable the data can be plotted in terms of $\log_{10} N$: since $\log_{10} N = 0.301 \log_2 N$, K is given by

$$K = \frac{\log_{10} N_t - \log_{10} N_0}{0.301t} \tag{10.2}$$

Growth rate can also be expressed in terms of the mean doubling time, $1/K$. In the example given in Fig. 10.2 the exponential growth rate is $0.67\ \text{h}^{-1}$, and the mean doubling time is 1.5 h.

Another term is often used to describe the rate of change in cell numbers. This is the specific growth rate constant (μ), sometimes called the instantaneous growth rate constant. This term is particularly important in describing growth in continuous culture systems, as will be seen in Section 10.5. The constant μ is defined as follows. The rate of increase in cell numbers is proportional to the initial cell concentration:

$$\frac{dN}{dt} = \mu N \qquad (10.3)$$

The cell concentration after a time t can be related to the initial cell concentration by integration of equation (10.3):

$$N_t = N_0 e^{\mu t} \qquad (10.4)$$

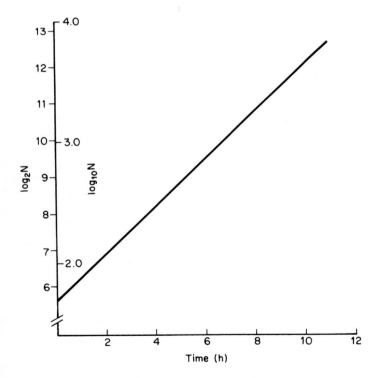

Fig. 10.2 Logarithmic plot of exponential growth.

During exponential growth μ is a constant since, from equation (10.4)

$$\ln N_t = \ln N_0 + \mu t \qquad (10.5)$$

This can be rearranged to yield

$$\mu = \frac{\ln(N_t/N_0)}{t} \qquad (10.6)$$

If a value of t equal to $1/K$, the doubling time of the culture, is substituted into equation (10.6) then:

$$\mu = K \ln 2$$
$$= 0.693K \qquad (10.7)$$

10.3 Growth curve

In a batch culture, the conditions necessary for exponential growth to occur will obviously not obtain indefinitely. Thus, the curves in Figs 10.1 and 10.2 only represent in part the events which would follow the inoculation of a suitable culture medium.

Following inoculation, there may be a period of apparent inactivity during which no growth occurs. This is called the lag phase and its occurrence and duration are controlled by a number of factors. A lag phase generally occurs if the inoculum is taken from a medium of different composition, and the delay is caused by the time taken for the cells to synthesize the enzymes necessary to metabolize the new substrate. Enzymes which are not always present in the cell, being synthesized only in the presence of the substrate, are termed inducible. Enzymes which are generally always present are termed constitutive. If the new substrate is metabolized by constitutive enzymes, the lag phase may be very much reduced in duration or virtually absent altogether. The lag phase would also be expected to be very short or absent in cultures inoculated from an exponentially growing population in the same culture medium, whereas an inoculum from a population which had ceased exponential growth may require a significant lag before the onset of growth in the new culture.

As the cells begin to grow, the population moves into the exponential phase and divides at a constant rate. The length of the exponential phase is limited by the availability of the substrate. Growth can only occur if all of the essential nutrients required by the organism are present, but will cease if only one of them is exhausted. The particular component which becomes exhausted before the others is termed the growth-limiting nutrient. Thus cultures can be grown, for example, under conditions of carbon, nitrogen or phosphorus limitation, or even limited by a trace nutrient, such as iron or magnesium.

As the limiting substrate approaches exhaustion, the growth rate declines and the population enters the maximum stationary phase, characterized by a growth rate of zero. The population may remain in this phase for a significant time without losing viability; if the exhausted nutrient were to be replenished, or an inoculum transferred to a fresh medium, the cells would start to divide once again (sometimes after another short lag phase). Although no net growth occurs during the maximum stationary phase, some residual cellular activity continues, utilizing internal reserves, for the maintenance of viability. This is discussed at greater length in Section 10.5.2. Viability can only be maintained for a limited period, and the cells will begin to die as some cellular functions are irretrievably lost. Ultimately, the cells will also lose their structural integrity and start to break up, a process known as lysis. This decline in viability and cell mass following the stationary phase is termed the death phase. A complete curve showing all of these phases of growth for a typical bacterial culture is shown in Fig. 10.3.

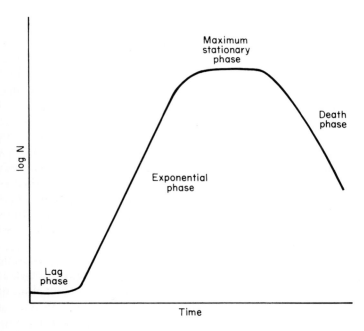

Fig. 10.3 The growth curve.

10.4 Factors affecting growth

10.4.1 Substrate concentration

(a) Effect on growth rate

Consider that in batch culture, when the concentration of limiting substrate [S] is high, growth proceeds exponentially, and only begins to decline as [S] becomes very low. In effect, μ is markedly affected by [S] when the latter is low, but at increasingly higher values of [S] this effect diminishes and disappears altogether as μ attains a maximum value (μ_{max}) while [S] remains finite. Consider also that virtually all of the metabolic processes of the cell are controlled by enzymes. Thus, the rate of substrate utilization, the rate of synthesis of cellular components, and hence the rate of growth may be expected to be related to the kinetics of enzyme-catalysed reactions in some way.

In fact, the effect of substrate concentration on the growth rate in batch culture is consistent with both considerations and can be described by the relationship

$$\mu = \frac{\mu_{max}[S]}{K_s + [S]} \tag{10.8}$$

where K_s is the substrate concentration at which μ is half of μ_{max}.

The similarity between equation (10.8) and the Michaelis–Menten equation (see Chapter 8) which relates the rate of an enzyme-catalysed reaction to the substrate concentration, is immediately apparent. The curve obtained from a plot of μ against [S] is similar in shape to the plot of V_0 against [S] for enzyme kinetics. Exponential growth in a bacterial culture and the rate of an enzyme-catalysed reaction are not directly analogous, however; the precise effects of [S] on the rate parameters in each case are different. The Michaelis–Menten equation was expressed in terms of the *initial* substrate concentration. This is considered valid since the values of V_0 plotted are those obtained during the initial period of the enzyme reaction, where, because the enzyme concentration is normally very low compared to the substrate concentration, the actual substrate concentration is almost equal to the initial substrate concentration. In growing bacterial cultures, however, during the exponential phase in which μ is supposedly constant, very large changes in [S] occur. In order for equation (10.8) to be valid if [S] is the initial concentration, therefore, it would be necessary for the growth rate to be somehow fixed according to this initial substrate concentration and maintained even after a large proportion of the substrate has been utilized. This is indeed the case in many bacterial cultures and equation (10.8) can therefore be expressed in terms of the initial substrate concentration. Under certain conditions, however, μ is affected by even small changes in $[S]$ where the latter is initially low and true exponential growth characterized by a constant value of μ over a measurable time does not occur.

(b) Effect on cell yield

In a batch culture, the maximum stationary phase is reached just as the substrate utilization rate reaches zero. When this point is reached, the maximum conversion of substrate into cells has been attained. This can be expressed in terms of the cell yield coefficient (Y), given by:

$$Y = \frac{X_t - X_0}{[S_0] - [S_t]}$$

(10.9)

X, denoting the biomass concentration, is used here instead of N because cell yield is normally expressed in terms of cell mass, while $[S]$ may be expressed in grammes, gramme moles, BOD, COD or other units, depending on the nature of the substrate and the application. Typical yields based on carbon equivalents given by Heijnen and Roels (1981) for a wide range of bacteria and substrates are in the range 0.4–0.6, but may be as high as 0.8 or as low as 0.1. Obviously, Y depends on the type of organism, the substrate and the cultural conditions. However, for a given culture, it is a constant over the entire substrate utilization period as shown in Fig. 10.4.

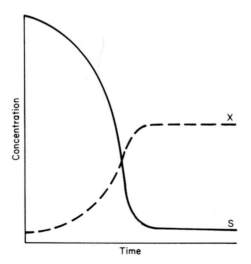

Fig. 10.4 Change in substrate concentration during bacterial growth.

10.4.2 Temperature

The temperature range in which microbial growth can occur lies between the freezing point of water and approximately 80°C, although survival is possible at temperatures considerably outside the range. However, no single species can grow at all temperatures within this range. It is possible to define three major groups of bacteria based on the temperatures at which they can grow. These are given in Table 10.1.

In the range between the minimum and maximum tolerable temperatures, the typical effect of temperature on growth is shown in Fig 10.5. At the lower temperatures, a gradual increase in growth rate with increasing temperature is due to the general enhancement of metabolic reactions in much

Table 10.1 Classification of temperature ranges which permit the growth of procaryotes (after Stanier et al., 1987)

Group	Optimum temperature (°C)	Temperature range (°C)
Thermophiles	55–75	40–80
Mesophiles	30–45	10–45
Psychrophiles		
(obligate)	15–18	Freezing point–20
(facultative)	25–30	Freezing point–35

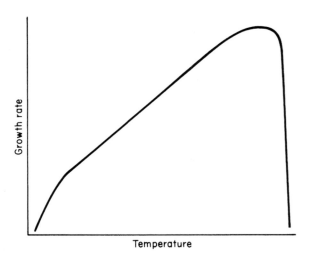

Fig. 10.5 Effect of temperature on growth.

the same way as very many chemical reactions are accelerated by heating. The sharp decline in growth rate at the upper limit of the temperature range is generally due to thermal denaturation of proteins. At this level, temperature can induce a conformational change in the structure of an enzyme leading to a significant loss of activity. The sharp decline may represent the loss of activity of only a few essential enzymes out of the many present in the cell.

10.4.3 pH

The effect of pH on enzyme activity generally follows the pattern given in Chapter 8. Active sites and functional groups which contribute to the tertiary structure of proteins frequently contain ionic groups whose dissociation is particularly sensitive to changes in pH. However, unlike its temperature, the pH of the cytoplasm can be controlled by many species almost independently of the external pH through the selective permeability of the membrane to certain inorganic ions. Thus some bacteria, notably the *Thiobacilli*, can grow in acid media having a pH of 2 or less, while others are tolerant of significantly alkaline conditions. For many species however, the optimum pH for growth is 7 ± 1 or 2 units.

10.5 Continuous cultivation

The curve in Fig. 10.3 described microbial growth in batch culture and is in many ways of limited value in obtaining an understanding of the behaviour of natural populations. In many systems growth of the population does not constitute a batch culture or even a succession of batch cultures. In such open systems growth may proceed more or less continuously, with a balance between the input of substrate and substrate utilization and between the growth of cells and their depletion due to death, predation or migration from the system. Another important difference between laboratory batch cultures and open, natural systems is that the substrate concentration in the former is often very much higher than in the latter.

A culture which approximates more closely to a natural system may be established by use of a chemostat. As its name suggests, this is a device which permits the control of culture parameters at a steady state, allowing the continuous growth of a microbial population (although it can equally well be used to study the effects of unsteady state parameters). In its simplest form, a chemostat consists of a vessel equipped with a stirrer into which culture medium can be pumped at a controlled rate to a growing culture whose volume and biomass concentration are maintained constant by the incorporation of an overflow device. For each volume increment of fresh culture medium added to the vessel an equivalent volume of the culture itself leaves the system. The steady state biomass concentration, X, can only be maintained if the growth rate of the population is sufficient to offset the continuous loss of cells in the overflow. A schematic diagram of a simple chemostat is shown in Fig. 10.6.

10.5.1 Control of growth rate by dilution rate

The rate of addition of fresh culture medium to the chemostat may be expressed in terms of the dilution rate, D, given by:

$$D = \frac{Q_1}{V_2} \tag{10.10}$$

where Q_1 is the influent flow rate and V_2 the volume of the vessel (reactor). The instantaneous growth rate of the population is given by:

$$\frac{dX_2}{dt} = \mu X_2 \tag{10.11}$$

while the loss of biomass in the overflow from the reactor is

$$\frac{dX_2}{dt} = -DX_2 \tag{10.12}$$

Hence the net change in biomass concentration is given by

$$\frac{dX}{dt} = \mu X_2 - DX_2 \tag{10.13}$$

$$= (\mu - D)X_2$$

If the biomass is growing, then it does so at the expense of the supplied

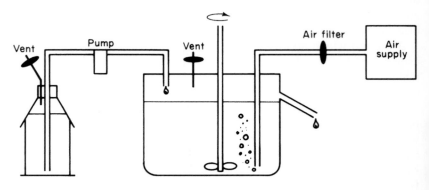

Fig. 10.6 A simple chemostat.

substrate, which has a concentration in the reservoir $[S_1]$. Continuous cultures are normally operated under conditions of complete mixing. This means that every drop of growth medium supplied is immediately and instantaneously dispersed throughout the culture volume, so that the substrate concentration in the reactor is the same as that in the effluent from the reactor, denoted by $[S_2]$. According to equation (10.8), the growth of the culture is dependent on the substrate concentration $[S_2]$ (in this case). If, initially $\mu > D$, X_2 will increase and if $[S_1]$ is held constant the amount of substrate utilized will increase. This will result in a lower value of $[S_2]$ in the reactor. But, a lower value of $[S_2]$ will cause a decrease in μ, such that the higher X becomes, the lower μ becomes, tending gradually towards D. Similarly, if $\mu < D$, X decreases, S increases, with μ tending towards D once more. At the point where $\mu = D$, $dX/dt = 0$, so that X_2 and therefore $[S_2]$ attain constant steady state values.

The relationships between X, $[S_1]$, $[S_2]$, μ and D at steady state are as follows. Since $\mu = D$, equation (10.8) can be written:

$$D = \frac{\mu_{max}[S_2]}{K_s + [S_2]} \tag{10.14}$$

whence

$$[S_2] = \frac{K_s D}{\mu_{max} - D} \tag{10.15}$$

The relationship between $[S_2]$, $[S_1]$ and X_2 can be derived from a substrate mass balance. The rate of change in substrate concentration in the reactor is the rate of substrate input minus the rate of output, minus the substrate used for growth, i.e.,

$$\frac{V_2 d[S_2]}{dt} = Q_1[S_1] - Q_1[S_2] - \frac{\mu X V}{Y} \tag{10.16}$$

Dividing throughout by V_2 and setting $d[S_2]/dt = 0$ and $\mu = D$ for steady state conditions yields:

$$X = Y([S_1] - [S_2]) \tag{10.17}$$

There are two important points to be noted about equations (10.15) and (10.17). The first is that by controlling the dilution rate any steady state value of $[S_2]$ which is less than $[S_1]$ can be obtained, irrespective of the value of $[S_1]$. The second is that the biomass concentrations can be controlled by manipulation of the influent substrate concentration.

A continuous culture can be maintained in a chemostat at any desired growth rate simply by setting the appropriate dilution rate. However, if D is set to a value greater than μ_{max} the rate of loss of cells exceeds their growth rate and a condition known as washout occurs. As D increases towards μ_{max} washout does not occur at the point at which it just becomes greater than μ_{max}, since the maximum growth rate in any given culture depends on the steady state substrate concentration, and may therefore be less than μ_{max}. The critical dilution rate at which washout occurs (D_c) can be obtained by substituting $[S_1]$ for $[S_2]$ in equation (10.14). It can be seen that D_c is lower than μ_{max} by a factor of $[S_1]/(K_s + [S_1])$.

10.5.2 Cell maintenance

In Section 10.4.1 it was stated that the growth yield (Y) is a constant for all phases of growth in batch culture. However, in continuous culture at very low dilution rates, the biomass concentration may be less than that expected for a given value of $[S_2]$.

The assumption that Y is constant suggests that a fixed proportion of the substrate is always utilized for growth processes (or that all of the substrate is used with a constant efficiency). But it seems reasonable to suppose that a small quantity of the substrate utilized must be used for processes which do not directly contribute to the growth and division of the cell, such as repair of damaged or denatured components. The quantity of substrate channelled into these maintenance processes cannot also be used for growth. In batch culture, the requirement for cell maintenance is seen in the phase immediately following the maximum stationary phase. The initial stages of the decline in biomass concentration are often accompanied by a continuing utilization of oxygen and the maintenance of constant viability. This is known as endogenous respiration and is due to the utilization by the cells of some of their own mass for maintenance purposes, in the absence of the external supply of substrate. It can therefore be considered to be a form of auto-oxidation.

Even in the presence of an external supply of substrate, there will still be a maintenance requirement to offset the decay of cellular functions. At very low growth rates in continuous culture, with low substrate utilization rates, it seems likely that a greater proportion of the substrate turnover will be required for maintenance than at high growth rates, and that the apparent cell yield will decrease.

All phenomena relating to maintenance, endogenous respiration, etc., which tend to reduce the biomass concentration, can be incorporated into the single term called the specific decay rate b. This term can be incorporated into cell and substrate mass balances to yield the following modifications to equations (10.15) and (10.17). At steady state,

$$\mu = D + b \tag{10.18}$$

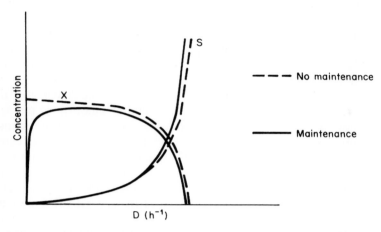

Fig. 10.7 Effect of the decay coefficient on substrate and biomass concentrations in continuous cultures.

Hence

$$S_2 = \frac{K_s(D + b)}{\mu_{max} - (D + b)}$$ (10.19)

and

$$X = \frac{YD([S_1] - [S_2])}{D + b}$$ (10.20)

According to Atkinson and Mavituna (1990), who gave maintenance requirements in terms of g substrate per g biomass per h, values of b vary considerably for different organisms utilizing different carbon and energy sources both aerobically and anaerobically. For a typical carbon-limited culture, however, b might be expected to be between 10 and 100 times lower than μ_{max}.

The magnitude of the effect of b on both X and $[S_2]$ is shown in Fig. 10.7 in terms of D, for a b equivalent to $0.02\mu_{max}$. It is clear that b influences the behaviour of continuous cultures most at low growth rates, which are typical of biological wastewater treatment processes. (The important consequences of b within this context are discussed more fully later on.)

10.6 Growth on two substrates

During his extensive studies on bacterial growth, Monod noticed that in batch cultures supplied with two limiting substrates acting as both carbon and energy sources, in many cases one was preferentially utilized and growth on the second only commenced after exhaustion of the first, normally after a short lag phase. This phenomenon is known as diauxie, and the type of growth curve obtained is shown in Fig. 10.8. Monod tested this effect on *Bacillus subtilis* and *Escherichia coli* supplied with various sugars and sugar alcohols and found that glucose was frequently the preferred substrate.

Diauxie is caused by a combination of two factors. The first is that the enzyme or enzymes which mediate the uptake and metabolism of the preferred substrate are constitutive, whereas those which oxidize the second substrate are inducible. This in itself would not be sufficient to cause diauxie, unless the lag phase prior to utilization of the secondary substrate were to be longer than the time taken for growth on the preferred substrate to attain stationary phase. The second factor, which effectively prevents induction of the enzymes for the metabolism of the second substrate until the first is exhausted, is that a separate mechanism in addition to induction needs to be activated prior to enzyme synthesis. This mechanism involves a promoter gene, which will only activate transcription of the region of DNA coding for the necessary enzymes if a certain metabolite is present in sufficient concentration: it has been found that this metabolite is maintained only at low concentrations during metabolism of the primary substrate.

It might be expected intuitively that in continuous culture, since the steady state substrate concentration is constant, and the cells are continuously metabolizing it, utilization of the second substrate would never occur. This is indeed the case at high dilution rates, where the relationship between

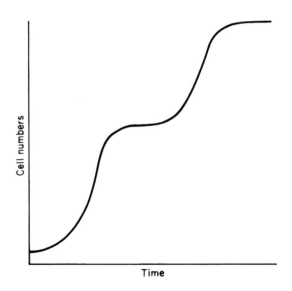

Fig. 10.8 Diauxic growth.

$[S_1]$ and $[S_2]$ for the preferred substrate obeys equation (10.17) and $[S_1] = [S_2]$ for the second substrate. At very low dilution rates, however, the substrates may be utilized concurrently. At very low dilution rates, $[S_2]$ also becomes very low, and it would appear that inhibition of enzyme synthesis for metabolism of the second substrate cannot occur below a certain low, but finite value of $[S_2]$.

10.7 Mixed cultures

Except in specific industrial applications, pure microbial cultures rarely exist outside the laboratory. Most natural environments are constantly subject to invasion by a wide variety of micro-organisms and consequently support the growth of mixed populations. Not all species which enter a mixed population survive to become a stable component of that system. Their survival and growth depend on the nature and extent of interactions with the other species present, the availability of essential nutrients and the ways in which substrate concentrations are influenced by other members of the population. The types of interactions which can occur between different species in mixed cultures are summarized in Table 10.2.

Table 10.2 Types of interaction between different microbial species in mixed populations

Interaction	Definition	Effects
Neutralism	No interaction	Growth of two species in mixed culture exactly the same as in two separate pure cultures
Competition	Joint dependence on a common growth-limiting factor (substrate)	Faster growing species will outgrow the slower growing one unless growth rates are identical
Commensalism	Growth of one species promoted by the presence of the other	More prolific growth of one species in the mixed culture than in pure culture
Mutualism	The growth of each species is promoted by the presence of the other	More prolific growth of both species in mixed culture than in separate pure cultures
Symbiosis	Mutual obligate dependence of two species on each other	Growth of both species in mixed culture. No growth in separate pure cultures
Amensalism	Growth of one species repressed by an inhibitor produced by the other	Growth of one species is less prolific in mixed culture than in pure culture
Predation	One organism uses the other	Complex. Density of each species controlled by density of the other

Observations which demonstrate neutralism in mixed culture are fairly rare. Although different species can often utilize different major carbon sources, under certain conditions they may have a common dependence on other limiting factors, such as phosphorus, nitrogen, trace elements or even the availability of suitable attachment sites in the case of organisms which tend to grow as films on solid surfaces.

Competition is a much more common interaction. This occurs when two (or more) species utilize the same growth-limiting substrate. In batch culture, the faster growing species will dominate the slower growing ones with the result shown in Fig. 10.9. In continuous culture, the faster growing species utilizes the limiting substrate at a rate which causes it to drop below the level required to maintain the growth rate of the slower growing species at a value equal to D. Thus the slower growing species is gradually washed out of the culture. An important exception to this general rule is the case of an organism which has a low μ_{max} for a given substrate, together with a high affinity for the substrate (low K_s), in competition with an organism which can attain relatively high growth rates but has a low affinity for the substrate. Such a situation is shown in Fig. 10.10. In this case, organism A will be washed out of the culture at low dilution rates, and organism B will be washed out at high dilution rates. Thus, in mixed cultures, control of the dilution rate cannot only be used to control the steady state substrate concentration and hence the biomass concentration, but it will also control the dominance of different species to some extent.

For the same reasons that natural populations consisting of only a single bacterial species are rare, in mixed populations species of higher organisms may also be present. From the public health engineering standpoint the pro-

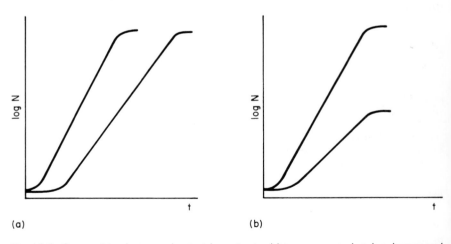

(a)   (b)

Fig. 10.9 Competition between bacterial species in: (a) two separate batch cultures; and (b) a two-membered culture.

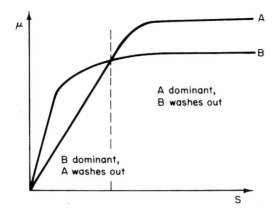

Fig. 10.10 Effect of substrate concentration on the outcome of competition in continuous culture.

tozoa are probably the most important group, in particular the classes Mastigophora, Ciliata and Sarcodina since they consume bacteria as a source of nutrients. In many mixed continuous cultures of predatory protozoa and bacteria, both species can often survive. In many cases oscillations in the population densities of both species occur. This is probably due to the consumption of bacteria by the predator to such an extent that the bacterial numbers fall to a level at which the predator population can no longer be supported. This causes a decrease in the numbers of predators and in response to the reduced rate of predation the bacterial population increases once more.

References and further reading

Andrews, J. F. (1983) Kinetics and mathematical modelling, in *Ecological Aspects of Used Water Treatment* (eds C. R. Curds and H. A. Hawkes), Vol. 3, Academic Press, London, pp. 113–172.

Atkinson, B. and Mavituna, F. (1990) *Biochemical Engineering and Biotechnology Handbook*, 2nd edn, Macmillan Publishers Ltd, Basingstoke.

Gaudy, A. F. and Gaudy, E. T. (1980) *Microbiology for Environmental Scientists and Engineers*, McGraw-Hill Book Company, New York.

Heijnen, J. J. and Roels, J. A. (1981) A macroscopic model describing yield and maintenance relationships in aerobic fermentation. *Biotech. Bioeng.* **23**, 739–741.

Mudrack, K. and Kunst, S. (1986) *Biology of Sewage Treatment and Water Pollution Control*, Ellis Horwood Ltd, Chichester.

Stanier, R. Y., Ingraham, J. L., Wheelis, M. L. and Painter, P. R. (1987) *General Microbiology*, 5th edn, Macmillan Education, London.

Chapter 11

Water quality chemistry

11.1 Introduction

To a certain extent, the quality of water will depend on its end use. For example, water used for human consumption will conform to superior water quality standards than water used for agricultural irrigation. These water quality standards are regulated legislation and are used to manage water quality. This chapter discusses water quality chemistry and describes new water quality standards due for implementation in the new EC directive relating to water intended for human consumption.

11.2 The water molecule

Water possesses a number of special characteristics that make this molecule unique, mainly its ability to form hydrogen bonds. Figure 11.1 illustrates the water molecule.

As oxygen is more electronegative than hydrogen, the hydrogen atoms are

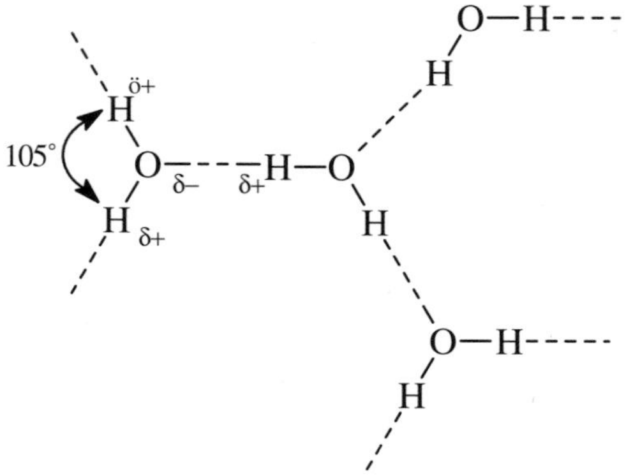

Fig. 11.1 Hydrogen bonding in water.

left with a partial positive charge ($\delta+$) and the oxygen atom attains a partial negative charge ($\delta-$). This phenomenon results in the production of a *dipole moment*. The water molecules can aggregate and form bonds that are called hydrogen bonds (see Fig. 11.1). It is these hydrogen bonds that make water unique when compared with other dihydrides in the same group (e.g. H_2S), as they are gases at room temperature.

Owing to the molecules polarity, water has excellent solvent properties. For example, with ionic solids, water molecules surround both the positive and negative ions which neutralizes the forces between the ions. Water is said to hydrate each ion and the ions become solvated. Water also has an extremely high dielectric constant in comparison with other liquids which indicates a high solubility for ionic compounds.

Water possesses a very high heat capacity, i.e. a large amount of heat is required to change the temperature of a body of water significantly. This ability protects the flora and fauna in a water body from sudden temperature changes.

The maximum density of water as a liquid is achieved at 4°C. This is very significant as it means that ice will float on a body of water. If ice were denser than water in its liquid form, water bodies would freeze from the bottom upwards which would have disastrous consequences for aquatic organisms.

11.3 Oxygen demand

A normal healthy river or stream will be aerobic. This means that it will contain a certain level of dissolved oxygen (DO). This is essential to aquatic life, including plants, animals and fish, and is also used by aerobic bacteria which break down any organic matter that enters the stream. This organic matter is said to exert an oxygen demand when it is degraded. Hence, the concentration of dissolved oxygen is a critical parameter relating to the overall quality of any water course.

There is a limit to the amount of oxygen that can be dissolved in water. This varies with temperature, but at normal temperatures it is about 10 mg l^{-1}. This upper limit is called the *saturation concentration*. As the temperature of the water increases, the DO concentration will decrease. This is seen in Fig. 11.2.

The minimum desirable DO concentration for maintaining the aquatic life population is about 5 mg l^{-1}. Thus, even a small oxygen demand could have a significant effect on a water body.

11.3.1 Biochemical oxygen demand (BOD)

In a natural ecosystem, such as a river or a biological wastewater treatment works, substances which can be oxidized do not exert their total oxygen

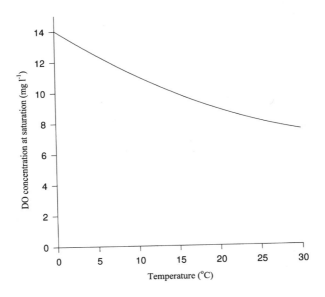

Fig. 11.2 The effect of temperature on DO saturation.

demand during breakdown by aerobic bacteria. This is because they are not completely degraded to carbon dioxide, water, nitrate, etc. Some of the organic compounds will be used as food and therefore are not completely oxidized. The amount of oxygen actually utilized in biological degradation is called the biochemical oxygen demand (BOD). The BOD is defined as the amount of oxygen required by bacteria while stabilizing decomposable organic matter under aerobic conditions.

BOD values generally have a numerical subscript, such as BOD_5 or BOD_{20}. This indicates the time period in days over which the test was performed. BOD_5 is most commonly used, being indicative of the short to medium term biodegradability of the sample.

Table 11.1 shows some typical BOD values of wastewaters, effluents and rivers which contain a complex mixture of organic substances.

Table 11.1 Range of typical BOD values

Sample	BOD_5 $(mg\ l^{-1})$
Raw sewage	300
Good quality sewage effluent	<20
Good quality river water	<3
High strength vegetable processing wastewater	10,000

From Table 11.1, it can be seen that a good quality sewage effluent has a BOD greater than the DO saturation value. Thus, on its own, such an effluent sample would use up all of the DO available and become anoxic (no molecular oxygen present). This would have severe implications if this effluent were discharged into a river system and appropriate dilutions by the receiving river water must be taken into account in order to prevent the river from becoming anoxic.

The BOD of the river generally must not exceed a limit of 4 mg l^{-1} which is the value incorporated into recommendations published by the Royal Commission on Sewage Disposal. This would reduce the DO from saturation to 5–6 mg l^{-1}, which is still capable of supporting aquatic life.

11.3.2 Reaeration

Oxygen will continue to dissolve in a river as it is depleted in DO. This is called reaeration. Reaeration can occur by diffusion of oxygen into the water from the air. Other major routes of dissolved oxygen in waters are given in Table 11.2. The example is given for the Thames Estuary.

Two sets of figures for photosynthesis are given, as the extent of photosynthesis will vary greatly. This process can account for 0% to 50% of the DO in the Thames Estuary. From Table 11.2, it can be seen that diffusion from the air is a major source of DO.

When an effluent containing oxidizable matter is discharged to a stream, its BOD will not be exerted instantaneously. Oxidation will take place over a period of time. If we look at the combined effects of reaeration and exertion of oxygen demand, a curve is produced. This is illustrated in Fig. 11.3, along with the oxygen demand and reaeration curves.

The solid curve is called the oxygen sag curve. The two processes of oxidation and reaeration together cause a drop in the DO to a minimum value which then recovers again.

Table 11.2 Sources of dissolved oxygen in the Thames Estuary

Source	Percentage of DO	
	Minimal photosynthesis	Maximal photosynthesis
Upper reaches and tributaries	10.6	5.0
Rainfall	0.3	0.1
Effluent discharges	0.8	0.4
Sea	10.4	4.9
Air	77.8	37.1
Photosynthesis	0.1	52.5

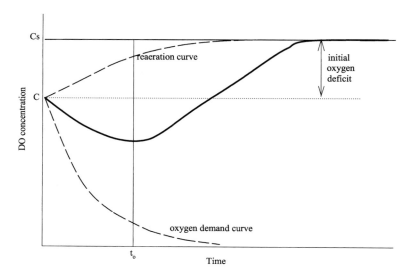

Fig. 11.3 The oxygen sag curve (Cs = saturation concentration).

11.3.3 Chemical oxygen demand (COD)

Unlike the BOD test, the chemical oxygen demand measures the total quantity of oxygen required to oxidize substances to carbon dioxide and water. As a result of this, COD values tend to be greater than BOD values. COD values can be vastly greater if large amounts of biologically resistant organic matter are present, e.g. a high lignin content. COD concentrations range from $\leqslant 20$ mg l^{-1} O_2 in unpolluted surface waters to >200 mg l^{-1} in effluent rich waters. Industrial wastewaters may have COD levels in the range 100 to 60,000 mg l^{-1} O_2.

One advantage of the COD procedure is the shorter time required to produce results (~3 hours) compared with the BOD (5 days). For scientists and engineers with considerable experience in these measurements, the COD values can be related to the BOD values, thus saving analysis time. In both oxygen demand procedures, blank determinations must also be undertaken.

One problem with the COD determination is that certain inorganic ions (e.g. Cl^-) can be oxidized, which results in very high values. Interference is controlled by the use of mercuric sulphate which forms a complex with the chloride.

The COD test is frequently used in the analysis of industrial wastes and particularly in surveys for determining and controlling losses in sewer systems. If BOD data are also taken into account, this can indicate possible toxic conditions and any biologically resistant organic substances that may be present.

11.4 Acidity and alkalinity

11.4.1 Acidity

This is classed as the ability of water to neutralize a strong base. It is controlled by weak acids (e.g. carbonic, humic and fulvic), strong mineral acids (e.g. HCl, HNO_3, H_2SO_4), and hydrolizing metal salts of iron and aluminium.

Acidity can be further classed into two types: mineral acidity and carbon dioxide acidity. Determination of these is carried out by titration with standard basic/alkaline reagents. Mineral acidity is measured up to pH4 and CO_2 acidity is measured up to pH8.3 using different indicators to observe the endpoints (methyl orange and phenolphthalein respectively).

The corrosive constituent in most waters is carbon dioxide, with mineral acidity being prevalent in industrial wastes.

Carbon dioxide acidity determinations are important in the area of public water supplies. Groundwater supplies can contain considerable amounts of CO_2, resulting from bacterial oxidation of organic matter. Concentrations of CO_2 can be as high as 30 to 50 mg l^{-1}. The amount of CO_2 present determines which choice of treatment will be used, e.g. aeration or neutralization with lime.

Industrial effluents containing mineral acidity are required to be neutralized before discharge to the environment. The amount of chemicals and the process involved will be chosen based upon the evaluation of the acidity data produced.

11.4.2 Alkalinity

This is a measure of the acid neutralizing capacity (ANC) of water and is expressed in mmol l^{-1}, as are acidity measurements. The major contributors of alkalinity in waters are bicarbonate, carbonate and hydroxides. However, borates, silicates and phosphates may also affect alkalinity determinations due to their basicity.

Alkalinity is due to the formation of salts of weak acids and strong bases. These act as buffers resisting a pH change on addition of acid (e.g. acid rain). Thus, alkalinity can be used as a measure of the buffer capacity of a water.

Alkalinity is measured by titration with standard acidic reagents. As with acidity measurements there are two titration stages. The first stage involves titration to a pH of 8.3. This is classed as the free alkalinity. Further titration to pH4 gives the total alkalinity in the water sample. At this pH all the carbonate and bicarbonate ions have been converted to carbonic acid, H_2CO_3. It is sometimes necessary to determine the kind of alkalinity present, i.e. hydroxide, bicarbonate and carbonate. This value is required in water softening processes where lime is added to soften water by precipitation.

Alkalinity data are important in the treatment of water and wastewater, and their determination is necessary before methods such as chemical coagulation, water softening, corrosion control and treatment of industrial wastes can be correctly executed.

11.4.3 Carbon dioxide and the buffering of natural waters

The carbonate system is the most important acid–base system in natural waters, and is mainly responsible for the natural buffering of waters. The species that constitute the carbonate system are: $CO_{2(g)}$ and $CO_{2(aq)}$, H_2CO_3 (carbonic acid), HCO_3^- (bicarbonate) and CO_3^{2-} (carbonate). These can be expressed in the following equilibrium equations for natural waters:

$$CO_{2(g)} \leftrightarrow CO_{2(aq)} \tag{11.1}$$

$$CO_{2(aq)} + H_2O \leftrightarrow H_2CO_3 \tag{11.2}$$

$$H_2CO_3^* \leftrightarrow HCO_3^- + H^+ \tag{11.3}$$

$$HCO_3^- \leftrightarrow CO_3^{2-} + H^+ \tag{11.4}$$

where $H_2CO_3^*$ represents $[CO_{2(aq)}]$ and $[H_2CO_3]$.

Carbonates and bicarbonates react with metal ions such as Ca^{2+} and Mg^{2+} which can result in the formation of precipitates such as $CaCO_3$. The formation of such a precipitate is one of the processes utilized in water softening. Due to these reactions, the type of chemicals used to treat water will vary.

11.5 Hardness and metal ions in water

The hardness of water is largely dependent on the geological strata with which it has been in contact. Hard water originates from chalk and limestone areas and soft water originates from granite and other areas that contain impermeable rock strata.

Water hardness generally represents the total concentration of calcium (Ca^{2+}) and magnesium (Mg^{2+}) ions in solution, other multivalent metal ions such as strontium (Sr^{2+}), ferrous iron (Fe^{2+}) and manganese (Mn^{2+}) may contribute to the overall water hardness if present in significantly high concentrations. Aluminium (Al^{3+}) and ferric (Fe^{3+}) ions are unlikely to have any influence on water hardness due to their negligible ionic concentration at the pH of natural waters. Table 11.3 shows the major ions associated with hardness, the predominant ones being calcium and magnesium.

Table 11.3 Principal cations causing water hardness
and the anions associated with them

Cations	Anions
Ca^{2+}	HCO_3^-
Mg^{2+}	SO_4^{2-}
Sr^{2+}	Cl^-
Fe^{2+}	NO_3^-
Mn^{2+}	SiO_3^{2-}

Therefore water hardness data (Ca and Mg) are usually reported in mg l^{-1} as $CaCO_3$. Table 11.4 gives some classification on the degree of hardness based on these values.

The hardness of water is important in industrial processes where efficiency may be reduced due to scaling of pipes and boilers. In some cases it is left to the consumer to treat their own supply rather than the water company treating all the water supply, which is costly. If the hardness is greater than 150 mg l^{-1} for public water supplies, water softening treatment may be required.

Water hardness can be sub-divided depending on the type of water hardness present. Calcium and magnesium constitute the greatest amount of hardness found in natural waters. This is often referred to as the total hardness in waters, and can be determined using a simple titrimetric method. Individual values for calcium and magnesium can also be evaluated. Generally the amount of calcium hardness is determined by titration and this is subtracted from the total hardness to reveal the amount of magnesium hardness present.

Carbonate hardness is represented in the amount of hardness produced by the carbonate and bicarbonate salts of calcium and magnesium. It is often referred to as temporary hardness, i.e. hardness that is removed by boiling. This is shown in the following equation:

$$Ca(HCO_3)_2 \xrightarrow{\text{heat}} CaCO_{3(s)} + CO_2 + H_2O \tag{11.5}$$

Table 11.4 Classification for water hardness

Concentration $CaCO_3$ (mg l^{-1})	Degree of hardness
0–75	Soft
75–150	Moderately hard
150–300	Hard
>300	Very hard

Non-carbonate hardness is associated with any salts of calcium and magnesium except carbonates and bicarbonates, i.e. associated with chloride, sulphate, nitrate anions. This type of hardness can also be termed permanent hardness, which refers to hardness that cannot be removed by boiling.

Above concentrations of $500 \, mg \, l^{-1}$, hardness produces an unpleasant taste in waters. The EC drinking water directive has no standards set concerning water hardness. However, if water is subject to softening processes, it must have a hardness concentration of no more than $150 \, mg \, l^{-1}$ and a minimum concentration of at least $60 \, mg \, l^{-1}$ Ca.

11.5.1 Metal ions

Many metals are present in natural waters. As well as metals being present due to natural phenomenon, i.e. weathering of rocks and soils, metal concentrations can also increase due to discharges from various anthropogenic sources. This is discussed in detail in Chapter 13. Within the aquatic environment, metals can be transported from one compartment to another, i.e. from waters to sediments and biota. Of particular concern is metal accumulation in biota, as this can represent a potential health risk to humans. The adsorption and accumulation of metals on to sediments can result in much higher metal concentrations compared with that in the water column. This can lead to other pollution and water quality problems.

Metal ions that are of major concern are the heavy metals which might have an effect on public health. These are primarily Al, As, Cd, Cr, Cu, Pb, Hg, Fe, Ag and Zn. Other toxic heavy metals such as thallium (Tl) may also be monitored if it is thought that they are likely to be present in the water body.

Iron and manganese are also important metal ions in terms of water quality. However, this is from an aesthetic point of view as the metal concentrations found in waters are unlikely to pose a threat to human health. Iron concentrations are generally higher in groundwaters than surface waters. This is due to the prevalent environmental conditions. In groundwaters, reducing (anaerobic conditions) result in iron being converted from insoluble ferric (Fe^{3+}) to the soluble ferrous (Fe^{2+}) state. Upon aeration, the Fe^{2+} is oxidized to Fe^{3+} and an orange precipitate is produced, $Fe(OH)_3$. Aeration and coagulation are the main processes used to remove iron at the water treatment plant. This is necessary as iron in low concentrations ($0.3 \, mg \, l^{-1}$) will give the water an unpleasant taste, making it unsuitable for drinking water supplies.

Manganese also exhibits similar chemistry to that of iron, i.e. it is soluble in reducing conditions, and in oxidizing conditions Mn(II) is converted to Mn(IV) to produce an insoluble precipitate. Mn also has similar aesthetic problems to iron, i.e. discoloration, taste and staining. Manganese can have physiological and neurological effects at high concentrations, but this is unlikely at the concentrations observed in UK waters.

11.6 Nutrients

All living organisms require a wide range of nutrients for growth and maintenance. The major nutrients are carbon, hydrogen, oxygen, nitrogen, phosphorous and sulphur. With regards to water quality, it is the nitrogen and phosphorus compounds that are of interest, as high levels can lead to potential health and eutrophication problems.

11.6.1 Nitrogen

Nitrogen occurs in the environment through a variety of natural and anthropogenic sources, e.g. decomposition of organic and inorganic matter, excretion by biota, reduction of nitrogen gas directly from the atmosphere, industrial processes and use of agricultural fertilizers. The last is a main source for increasing nitrogen (or more specifically nitrate) concentrations in water resources.

Nitrogen can exist in a number of organic and inorganic forms. Figure 11.4 illustrates the decomposition of organic nitrogen compounds to molecular nitrogen (N_2).

Organic nitrogen is present in dead plants and animals. This is converted to ammonia by bacterial action in aerobic or anaerobic conditions. In solution, ammonia (NH_3) is in equilibrium with the ammonium ion (NH_4^+). At high concentrations (2 to 3 mg l^{-1} as NH_4^+), ammonia/ammonium values may indicate organic pollution in a water system from sources such as domestic sewage, industrial waste or fertilizer runoff.

Under aerobic conditions, ammonia is oxidized by nitrifying bacteria according to the following equation to give nitrites:

$$2NH_3 + 3O_2 \xrightarrow{\text{bacteria}} 2NO_2^- + 2H^+ + 2H_2O \qquad (11.6)$$

Nitrite concentrations in freshwaters are usually very low. However, in drinking water distribution systems, nitrites can also be produced due to the microbial breakdown of chloramine, which may be used as a disinfectant.

$$H_2NCONH_2 \xrightarrow{\text{aerobic/anaerobic}} NH_3 \xrightarrow{\text{aerobic}} NO_2 \xrightarrow{\text{aerobic}} NO_3 \xrightarrow{\text{anoxic}} N_2$$

Urea | Ammonia | Nitrite | Nitrate | Nitrogen gas

Fig. 11.4 Breakdown of organic nitrogen to nitrogen gas.

High nitrite concentrations in waters tend to indicate the presence of industrial effluents and unsatisfactory microbiological water quality.

These nitrites can be further oxidized under aerobic conditions to produce nitrates:

$$2NO_2^- + O_2 \xrightarrow{\text{bacteria}} 2NO_3^- \tag{11.7}$$

It is nitrates that are the most common form of nitrogen found in natural waters. Natural background levels of nitrate (no greater than 0.5 mg l^{-1} NO_3^-) originate from rocks, land drainage and plant and animal matter. The most significant source is the use of nitrate fertilizers in the agricultural industry. As soil cannot retain nitrates well, the excess nitrate which is not utilized by the plant will be washed out of the soil by water percolating through it. This can result in groundwaters containing high concentrations of nitrates.

The monitoring of nitrites and nitrates in water quality programmes provides a general indication of the nutrient status and level of organic pollution. Nitrate concentrations in the UK are expressed as mg l^{-1} of NO_3^-. In some reports nitrate is represented as mg l^{-1} N present as NO_3^-. It is important to be able to distinguish between the two. For example, 50 mg l^{-1} NO_3^- is equivalent to 11.3 mg l^{-1} N as NO_3^-.

The EC directive on drinking water states a maximum value of 50 mg l^{-1} NO_3^- for waters. Nitrites had a maximum value of 0.1 mg l^{-1} NO_2^- which has been increased to 0.5 mg l^{-1} NO_2^- under the new EC drinking water directive.

Nitrates and nitrite concentrations in drinking water are monitored and controlled due to the development of methaemoglobinaemia in infants. Nitrate, which is easily reduced to nitrite, binds with haemoglobin to form methaemoglobin. This is unable to carry oxygen. Thus, excess levels of methaemoglobin will lead to oxygen deprivation. When the concentration of methaemoglobin in the blood is greater than 10%, an infant's skin will exhibit a blue tinge. If untreated, symptoms that follow this blue skin condition are stupor, coma and ultimately death. Research has indicated that at double the maximum value for nitrate (100 mg l^{-1}), increased levels of methaemoglobin are observed, but these are still within normal physiological parameters. Infantile methaemoglobinaemia is virtually non-existent in Europe because the majority of the population has mains piped water which is treated to remove bacteria. However, acute cases of this condition have been reported in well waters high in nitrate (in excess of 100 mg l^{-1}), and in waters containing bacteria. It is thought that the bacteria enhance the conversion of nitrate to nitrite.

11.6.2 Phosphorus

Phosphorus is an essential nutrient and can exist in water in both dissolved and particulate forms. Natural sources of phosphorus arise from the weathering of rocks and the decomposition of organic matter. High concentrations of phosphorus in waters are due to domestic wastewaters, industrial effluents and agricultural runoff. Phosphorus can also be released from sediments by remobilization due to bacteria. Owing to the developments in detergents which contain appreciable amounts of polyphosphates (up to 50%), domestic wastewaters can contain high concentrations of phosphorus compounds.

All polyphosphates are eventually hydrolysed to produce the ortho form, for example:

$$Na_4P_2O_7 + H_2O \longrightarrow 2Na_2HPO_4 \tag{11.8}$$

The rate of hydrolysis is increased by temperature, decrease in pH and bacterial enzyme action. For determining phosphate with any accuracy, all the polyphosphates are first converted to orthophosphates which are then measured colorimetrically.

High phosphorus concentrations are largely responsible for the eutrophication of a water body, as phosphorus is the limiting nutrient for algal growth. Eutrophication is discussed in detail in Chapter 18.

Owing to the above factors, phosphorus measurements are incorporated in water quality surveys and monitoring programmes as these values will be required in order to control the algal growth in lakes or reservoirs. In drinking water, the maximum parametric value in the UK is 5 mg l^{-1} as P_2O_5.

11.7 Water quality standards

The European Community (EC) directive relating to the quality of water intended for human consumption was approved by the Council of Ministers on 15th July 1980. It is more commonly known as the drinking water directive and relates to all waters intended for consumption, and use in food production. The directive does not include natural mineral waters, but does include bottled and containered waters.

The impact of this directive has been significant and it is generally recognized that it has been the driving force behind the overall improvement in drinking water quality which has taken place in the EC over the last decade. In 1995, the Commission published a proposal to update the directive; this is currently being considered by Council and by Parliament.

The original directive consists of 62 parameters used to monitor drinking water quality and to take the necessary steps to ensure compliance with established values. These parameters were divided into six categories: organoleptic, physicochemical, parameters concerning substances undesirable in excessive amounts, toxic substances, microbiological, and the minimum required concentration for softened water intended for human consumption. The directive has two tiers of standards for most parameters;

a guide level which is a non-binding value, and a Maximum Admissible Concentration (MAC) – this value must not be exceeded.

The revised directive has dispensed with these parameters and adopted parametric values (PVs) which are essentially the same as MACs. The new directive also splits the parameters into two classes: *mandatory* and *indicator*. The PVs for the mandatory parameters are binding values that are health related and consist of 26 chemical parameters and two microbiological parameters. Indicator parameters are non-binding and are analysed in water quality monitoring programmes. They consist of three microbiological, eleven chemical and four aesthetic.

11.7.1 Microbiological standards

Recommended guidelines for bacteriological water quality in the UK are generally in keeping with those of the World Health Organisation and the European Community. In all waters intended for drinking, no coliform organisms should be detectable in any 100 ml sample taken.

Since it should be the aim of any treatment works to produce water of such quality at all times, conformity with the guidelines will depend upon the taking of samples on a fairly frequent basis.

Water undertakings are only practically responsible for the quality of water leaving the treatment works and entering the distribution system. Once in the system, the quality may deteriorate to some degree and this is allowed for in the UK guidelines. The tolerances which may be allowed in samples taken from the distribution system, including the consumer's tap, are as follows:

1 *E. coli* should not be detectable in any sample of 100 ml.
2 Coliform organisms should not be detectable in any two consecutive samples of 100 ml from the same or closely related sampling points.
3 For any given distribution system, coliform organisms should not occur in more than 5% of routine samples, provided that at least fifty samples have been examined at regular intervals.

The minimum action necessary when any coliform organisms are found is to check the disinfection process at the works and to take another sample immediately from the same point. Additional samples should also be taken in an attempt to locate the source of contamination. Where coliform organisms persist in successive samples, or where *E. coli* is found, this constitutes a definite indication that undesirable material is gaining access to the system. Its location and removal should be sought immediately and at the same time it may be considered necessary to increase the disinfectant dose at the treatment works in order to maintain a more effective residual (see Section 20.3).

Provided that sufficient samples have been taken on a routine basis, the water quality in a distribution system can be reviewed periodically and classified on the basis of the criteria in Table 11.5.

The requirement of fifty samples per annum is a minimum requirement. In the UK recommendations for a monthly sampling frequency for treated water in the distribution system are based on the size of the population served. These guidelines are summarized in Table 11.6. Based upon these, a population of 100,000 would necessitate twenty samples per month, 200,000 would require forty samples per month, 400,000 would require seventy samples per month and 600,000 would need eighty-five samples per month.

EC directives also give guidelines for raw water quality (i.e. surface water to be abstracted for treatment) which specify levels of faecal indicator organisms and salmonellae. The new EC drinking water directive has only two mandatory parameters for drinking water, no *E. coli* and Enterococci should be detectable in any 100 ml sample. However, for bottled or containered water, several other parameters are required to be determined and these are shown in Table 11.7.

The indicator parameters provide information on the microbiological quality of the supplied water as well as checking the effectiveness of the water treatment processes. These are given in Table 11.8.

Table 11.5 Categories of water quality in a distribution system

Category	Quality of supply	Counts from routine samples (cells/100 ml)	
		Total coliforms	E. coli
1	Excellent	0	0
2	Satisfactory	1–3*	0
3	Intermediate	4–9*	0
4	Unsatisfactory	10 and/or	>1

* No coliforms should be detectable in consecutive samples or in >5% of routine samples; if coliforms are found at such levels the supply is classed as unsatisfactory.

Table 11.6 Guidance on sampling frequencies for water in a distribution system (Government of Great Britain, 1983)

Population size	Monthly sampling frequency
<300,000	One sample per 5,000 head of population
300,000–500,000	As above plus one sample per additional 10,000 head of population
>500,000	As above plus one sample per additional 20,000 head of population

Table 11.7 Mandatory microbiological standards for bottled and containered waters

Microbiological parameter	Parametric value
Escherichia coli (E. coli)	0/250 ml
Enterococci	0/250 ml
Pseudomonas aeruginosa	0/250 ml
Colony count 22°C	100/ml
Colony count 37°C	20/ml

Table 11.8 Indicator microbiological standards for drinking waters

Microbiological parameter	Parametric value
Clostridium perfringens (including spores)	0/100 ml*
Colony counts 22°C	No abnormal change
Coliform bacteria	0/100 ml†

*Only measured if the water originates from or is influenced by surface waters.
†For bottled or containered water the value is 0/250 ml.

Many agencies do not specify tolerable limits on the number and types of viruses in treated waters, although where such limits are specified they tend to be at least as stringent as the limits on faecal coliforms (e.g. the World Health Organisation limit of one plaque-forming unit per litre). Enumerating viruses in water supplies is fairly difficult. From what is known about their fate in water treatment processes, however, the absence of indicator bacteria is generally taken to indicate that the viral quality of the water is also acceptable.

11.7.2 Chemical standards

The 26 mandatory chemical parameters which operate under the new EC drinking water directive are given in Table 11.9.

On comparison of the new guidelines with those of the WHO, it can be observed that many of the parametric values are comparable or lower for the new directive. Generally, WHO guidelines are the basis for EC and USEPA legislation. Since their implementation, the WHO guidelines have been subject to two revisions, the latest in 1993.

There are ten new chemical parameters in the new directive. These have been added on the basis of their potential occurrence in drinking water and their toxicity, all are considered to be potential carcinogens. Acrylamide, epichlorhydrin and vinyl chloride are all present in materials used in the

Table 11.9 Mandatory chemical parameters

Chemical parameter	Parametric value ($\mu g\,l^{-1}$)	WHO guideline value (1993) ($\mu g\,l^{-1}$)
Acrylamide	0.1	0.5
Antimony	5	5
Arsenic	10	10
Benzene	10	10
Benzo(a)pyrene	0.01	0.7
Boron	1,000	300
Bromate	10	25
Cadmium	5	3
Chromium	50	50
Copper	2,000	2,000
Cyanide	50	70
1,2-dichloroethane	3	30
Epichlorhydrin	0.1	0.4
Fluoride	1,500	1,500
Lead	10	10
Mercury	1	1
Nickel	20	20
Nitrate	50,000	50,000
Nitrite	500	3,000
Pesticides	0.1	Individual values
Pesticides – total	0.5	Individual values
Polycyclic aromatic hydrocarbons (PAH)	0.1	–
Selenium	10	10
Tetrachloroethene and trichloroethene	10 (in total)	40 and 70 respectively
Trihalomethanes – total	100	Individual values
Vinyl chloride	0.5	5

water industry. Tri- and tetrachloroethene are used as dry-cleaning solvents and their presence can be sometimes be detected in contaminated groundwaters. Currently, under the UK Water Supply Regulations (1989), they have been assigned maximum values of 10 and $30\,\mu g\,l^{-1}$ respectively. The new directive reduces this to a parametric value of $10\,\mu g\,l^{-1}$ for both compounds.

Trihalomethanes (THMs) consist of bromoform, chloroform, chlorodibromomethane and dichlorobromomethane. They are generated as a result of interaction between naturally occurring organic compounds (e.g. humic and fulvic acids) and chlorine-based disinfectants. The present THM guideline is $100\ \mu g\,l^{-1}$, which is determined over a three-month average. For the first five years during the new directive, THMs will be assigned a PV of $150\ \mu g\,l^{-1}$. After this period, full compliance will be required to the $100\ \mu g\,l^{-1}$ value. Similarly to the procedure for THMs, bromate will have an intermediate value of $25\ \mu g\,l^{-1}$ which will apply from five until ten years after entry into the directive. The value will then be set at the proposed PV of $10\ \mu g\,l^{-1}$.

Other changes in the directive relate to the tightening of standards to bring them in line with the latest WHO (1993) guideline values. Arsenic, antimony, copper and nickel have had their PVs decreased to correspond with the WHO guidelines. Lead has also been reduced to equal the value set by WHO ($10\ \mu g\,l^{-1}$). However, full compliance for lead is required within a 15-year transition period. Meanwhile, an interim standard of $25\ \mu g\,l^{-1}$ will come into effect five years after entry. Monitoring of lead must be adequate at the tap to be representative of a weekly average value ingested by the consumer. It also takes into account any peak levels that may have an adverse effect on human health. Breaches of standards which are due to domestic plumbing (e.g. lead piping) are the homeowners' responsibility. Estimates of the cost of replacing all the lead piping in the UK are at £2 billion for the water industry and £8 billion for domestic replacements.

The PAH standard of $0.1\ \mu g\,l^{-1}$ relates to the sum of the concentration of four PAHs, as opposed to the six PAHs in the current directive. These are benzo-3,4-fluoranthene, benzo-11,12-fluoranthene, benzo-1,12-perylene and indeno(1,2,3-cd)pyrene. Fluoranthene and benzo(a)pyrene have been removed. Benzo(a)pyrene which is the most potent carcinogen of the PAHs is now regulated separately.

The standards for individual and total pesticides have not changed. The term pesticides in this context incorporates pesticides, herbicides, insecticides, fungicides and PCBs. In the new directive, this has been extended to include metabolites, degradation and reaction products. For four individual pesticides, aldrin, dieldrin, heptachlor and heptachlor epoxide, a standard of $0.03\ \mu g\,l^{-1}$ has been imposed.

Nitrite is regulated because of its ability to produce blue baby syndrome. This has been discussed in Section 11.6.1. The current nitrite standard of $100\ \mu g\,l^{-1}$ can cause difficulties when chloramine is used as a disinfectant. The new standard of $500\ \mu g\,l^{-1}$ will allow the continued use of chloramination processes. Where chloramination is not used, the current standard of $100\ \mu g\,l^{-1}$ will still apply. This is to ensure that the drinking water supply will not contain higher nitrite concentrations than the current levels.

Table 11.10 Indicator chemical parameters

Parameter	Parametric value
Chemical	
Aluminium	200 μg l^{-1}
Ammonium	0.5 mg l^{-1}
Chloride	250 mg l^{-1}
Conductivity	2500 μS cm^{-1} at 20°C
Iron	200 μg l^{-1}
Manganese	50 μg l^{-1}
Oxidizability	5 mg l^{-1} O$_2$
pH	≥6.5 and ≤9.5
Sulphate	250 mg l^{-1}
Sodium	200 mg l^{-1}
Total organic carbon (TOC)	No abnormal change
Radioactivity	
Total dose	0.1 mSv per year
Tritium	100 Bq l^{-1}

The WHO approach has been adopted where the levels of nitrite and nitrate must not exceed 1 mg l^{-1}. This can be calculated using the following equation:

$$\frac{[\text{Nitrate}]}{50} + \frac{[\text{Nitrite}]}{3} \leqslant 1 \tag{11.9}$$

Indicator parameters given by the new directive are shown in Table 11.10.

The proposed total dose for radioactivity (0.1 mSv) from drinking water coincides with the WHO guideline. This compares with an individual's total dose from all sources which is 2.5 mSv. Water companies will need to develop laboratory facilities for radioactivity monitoring, rather than compliance *per se*.

Since the last drinking water directive was enforced, it is clear that many of the current standards do not reflect the most up to date knowledge on toxicology. In order to correct this, the new directive specifies that the Commission will review the parameters and parametric values, and the monitoring and analytical specifications every five years. This will allow for proposals for amendments to be made along with the scientific and technical progress within that time.

References and further reading

Chapman, D. (1992) *Water Quality Assessments: A Guide to the Use of Biota, Sediments and Water in Environmental Monitoring*, Chapman & Hall, London, 564 pp.

De Zuane, J. (1990) *Handbook of Drinking Water Quality-Standards and Controls*, Van Nostrand Reinhold, New York, pp. 30–35.

EC (1980) Council Directive relating to the quality of water intended for human consumption (80/778/EEC), *Off. J. Eur. Commun.*, **L229** (30 August 1980), pp. 11–29.

Gray, N. F. (1994) *Drinking Water Quality: Problems and Solutions*, John Wiley & Sons, UK, 309 pp.

Government of Great Britain (1983) *The Bacteriological Examination of Drinking Water Supplies 1982*, Reports on Public Health and Medical Subjects No. 71; Methods for the Examination of Waters and Associated Materials, HMSO, London.

Sawyer, C. N., McCarty, P. L. and Parkin, G. F. (1994) *Chemistry for Environmental Engineering*, 4th edn, McGraw-Hill, Singapore, pp. 485–601.

Chapter 12

Environmental water chemistry

12.1 Introduction

Water sources (surface and groundwaters) have become increasingly contaminated due to increased industrial and agricultural activities. The protection of these sources involves an understanding of the ways in which water behaves (from a chemical perspective) and its movement through the hydrologic cycle.

This chapter examines the hydrologic cycle in detail, describing the various pathways and movement of water in the environment. Basic aspects of surface water, groundwater and precipitation (rain and snow) chemistry are also discussed.

12.2 The hydrologic cycle

Water beneath the land, on the surface and in the atmosphere is in a constantly moving cycle, known as the hydrologic (water) cycle. This water supply has been reused for millions of years and is a complex process which is driven by solar energy. The mechanisms of movement are complicated by a variety of air–water–land interface characteristics across which the cycle operates. For example, water can return from the land to the atmosphere by evaporation, but this will be affected by the type of soil from which the water is being evaporated. Soil type will also affect the water retention capacity of the soil.

Spatial and temporal climate conditions also affect the movement of water in this cycle. For example, the flow rate of water in a river is not dependent just on rainfall. Some of the water will be lost by evaporation, the rate of which will be affected by climatic conditions. Figure 12.1 shows the movement mechanisms involved in the hydrologic cycle. It is important to remember that the cycle is a closed system, i.e. the total amount of water in the cycle will remain constant, only it will be present in different forms. The aspects of the hydrologic cycle are discussed below.

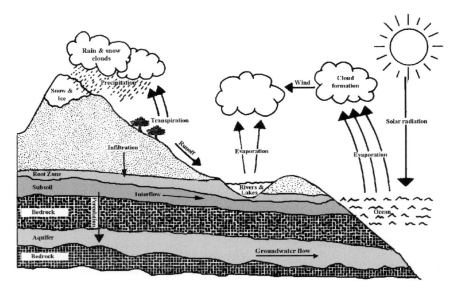

Fig. 12.1 The hydrologic cycle.

Evaporation – This occurs at the air–liquid interface when water under-goes a change of state from a liquid into vapour. To achieve this, energy is required and is provided directly or indirectly by solar radiation. Other factors that are important in controlling the evaporation rate are as follows.

Humidity – An increase in the percentage of water vapour in the air (a high relative humidity) will reduce the evaporation rate.

Wind speed – This increases the evaporation rate by removal of saturated air from just above the air–liquid interface.

Soil and surface characteristics – The rate of evaporation is dependent on soil saturation, soil depth and soil type. For example, sandy soils saturate quickly and will thus have a relatively high rate of evaporation compared to loamy or clay soils.

Type and density of plant cover – During precipitation periods (rain or snow-fall), some water will land on leaf and plant surfaces, a process called interception. In certain circumstances, such as in forested areas, up to 4% of the incoming precipitation may be intercepted. This encourages rapid evaporation.

Transpiration – This is a process which is unique to vegetation. Water moves through the soil at a rate dependent on various soil properties (e.g. porosity). Water still near the surface of the soil (sometimes referred to as the plant root zone) is utilized by the plant's root system and subsequently moves through the plant via osmosis. Eventually, water reaches the leaves' surfaces where diffusion to the atmosphere takes place. The transpiring water is said to pass along the *soil–plant–air continuum*. The water vapour produced is a natural by-product of photosynthesis. Plant root systems may reach otherwise inaccessible water sources and thus transpiration will increase the amount of water returned to the atmosphere.

Evaporation and transpiration together account for all the water returning to the atmosphere from land surfaces. As these processes are indistinguishable, they are both described as *evapotranspiration*.

Water vapour produced by evapotranspiration may condense and be transported by circulating atmospheric currents. As warm air rises, because it is less dense than the surrounding cooler air, it will experience a decrease in air pressure which causes expansion. This, together with a drop in temperature of approximately 10°C per 100 metre rise, produces a body of air which is saturated with water vapour that has condensed around anthropogenic residues. This process produces condensation nuclei and cloud droplets. Further condensation results in the formation of raindrops and ultimately precipitation. About 10^6 initial droplets must condense together to form one raindrop.

Small cloud droplets (1–100 μm) can coalesce to form larger rainwater droplets (1 mm or greater). Once the droplet growth has exceeded a critical mass to surface area ratio, large droplets begin to fall rapidly. This increases droplet collision and results in precipitation.

UK precipitation is a mixture of rain, hail and snow and is unevenly distributed both spatially and temporally, e.g. London has a mean annual rainfall of 500 mm compared with the Lake District which can have up to 2,500 mm. Some precipitation may evaporate before hitting the ground and that which reaches the ground can evaporate from anywhere (soil, plant surfaces and water body surfaces). Over a long period of time, the amount of precipitation must balance with the amount of evapotranspiration due to the constant cycling of water and the fact that it remains a closed system.

Infiltration is a term used to describe the entry of water into soils. The rate of infiltration can vary widely and depends on soil properties such as water content, texture, density, organic matter content, permeability (hydraulic conductivity) and porosity. Soil surface conditions will also affect infiltration. For example, a compacted soil surface, vegetation density, topography and human activities will play an important role in infiltration processes.

Percolation is defined as water moving down the soil profile by gravity after it has entered the soil. Water will eventually move past the plant root zone towards the bedrock. This process is called deep percolation and is responsible for "recharging" the supply of groundwater.

The portion of water that flows over and through the soils, eventually reaching surface waters, is called *runoff*. This comprises surface runoff, interflow and groundwater flow.

Surface runoff occurs if the precipitation rate is greater than the rate of infiltration. The soil will become saturated and any further precipitation will be contained in surface depressions (e.g. lakes). This is called surface storage. After this stage, water will begin to move down slopes over the land as surface runoff.

If water cannot percolate through a soil or rock formation, then it will move laterally along this impermeable layer until it is eventually discharged into a body of surface water (i.e. stream, lake or ocean). This lateral movement of water is called interflow.

Approximately 4% of the water contained in the hydrologic cycle is groundwater. This groundwater is continously moving at a slow flow rate towards the sea. In England and Wales, approximately 35% of the potable water supply comes from aquifers (water-bearing rocks). It is therefore vital that these groundwater supplies are protected from contaminants.

12.3 Surface water chemistry

Rivers, lakes and estuaries/oceans can all be classed as surface waters. The chemistry of these water bodies differ because of the physical and geological characteristics of the water body and the surrounding area. Before discussing the chemistry involved, the physical properties of several different surface waters are discussed.

Rivers can be characterized by water flow downstream under the influence of gravity. Any chemical substance disperses in a river as it moves downstream. This is due to turbulent diffusion and the different water velocities which occur across a river. The velocity of river water generally reaches a maximum near the centre of the river just below the surface. Water near to the banks and the river bed will move at a slower rate due to friction processes.

Lake water movement is restricted to the influence of the wind. The resulting water currents are responsible for the turbulent diffusion (transport) of chemical compounds. Water movement in a lake is also influenced by the shape of the lake basin, density variations in water and inflowing streams and rivers. Stratification of lakes (i.e. a creation of water layers) may occur because of the differences in the densities of waters. However, in the summer months, the surface water temperature rises due to increased solar radiation and can produce a three layer system. This is known as *thermal stratification*. The upper warmer water layer is called the *epilimnion*, the lower colder layer is the *hypolimnion*. The zone dividing these

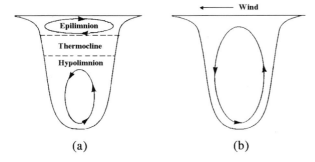

Fig. 12.2 Diagram showing: (a) thermal stratification, and (b) no stratification, in a lake.

layers is known as the *thermocline*. This is defined as a sharp temperature gradient which indicates a boundary between the water masses of different temperatures. These layers are illustrated in Fig. 12.2.

Oceans and seawater represent a complex heterogeneous system with many sources and sinks. The fundamental properties of seawater are temperature, pressure and salinity. These parameters control the density of the water which determines pressure gradients responsible for oceanic currents. Estuaries are responsible for the supply of material to the oceans. An estuary is a zone where river water and seawater are mixed. It is characterized by sharp ionic strength gradients and chemical composition. Water flow in estuaries is influenced by tides, inflow of rivers and density differences between salt water and freshwater.

12.3.1 Chemical characteristics of freshwaters

The chemical composition of several surface waters are given in Table 12.1. The data in Table 12.1 show that the composition of surface waters can exhibit a wide degree of variability. Rivers and lakes (hereafter referred to as freshwaters) attain their chemical characteristics by dissolution and chemical reactions with solids, liquids and gases they have had contact with during water movement in the hydrologic cycle. The following sections outline the processes involved.

12.3.2 Weathering processes

Weathering may be defined as the mechanical (physical) fracturing or chemical decomposition of rocks *in situ* by natural agents at the surface of the Earth. Physical weathering involves the fragmentation of parent rock with no change in chemical composition, and is brought about by two main

Table 12.1 Chemical composition of surface waters (adapted from Snoeyink and Jenkins, 1980; Harrison and deMora, 1996)

Constituent	Average river water (mg l^{-1})	Niagara River, New York (mg l^{-1})	Reservoir water, California (mg l^{-1})
Na^+	6.3	6.5	2.6
K^+	2.3	1.2	0.6
Mg^{2+}	4.1	8.1	1.1
Ca^{2+}	15	3.6	4.0
Fe^{3+}	0.67	0.02	0.07
Al^{3+}	0.01	–	–
Cl^-	7.8	13	2.0
SO_4^{2-}	11.2	22	1.6
HCO_3^- (alkalinity)	58.4	119	18.3
SiO_2	13.1	1.2	9.5
NO_3^-	1.0	0.1	0.041
Total dissolved solids (TDS)	126	165	34

processes: temperature and crystallization (freeze–thaw action). The continuous heating and cooling of rocks due to diurnal temperature variations may cause fractures. For example, a rise in temperature of 83°C in a sheet of granite 30 m in diameter would lead to an expansion of 25 mm in the granite. Thus, a temperature drop of the same amount would result in a contraction of similar size. If water should be present in these fractures (due to chemical weathering), the rate of fracturing will be increased due to the continual freeze–thaw processes in the fracture. This occurs due to the specific volume changes produced by water (i.e. crystallization). In the absence of any fractures, this freeze–thaw process will not operate as it cannot produce fractures, only alter them.

Another possible physical weathering mechanism is plant root growth which may increase the size of the fractures. Chemically this mechanism may be enhanced by the release of organic acids into the soil solution during plant growth. Erosion processes also provide a contribution to physical weathering. Rock particles in suspension can attack the exposed rock, resulting in abrasion and fragmentation.

Chemical weathering occurs by the reaction of the parent rock with freshwaters. The reaction may result in complete dissolution of the rock material with the formation of new products. There are several mechanisms of chemical weathering.

Dissolution occurs when soluble material present in rocks and soils is slowly dissolved in water. There are two types of dissolution processes: congruent and incongruent. With *congruent dissolution*, the cations and anions are dissolved in relation to their stoichiometry in the solid phase, for example, NaCl:

$$NaCl_{(s)} \leftrightarrow Na^+_{(aq)} + Cl^-_{(aq)} \tag{12.1}$$

Incongruent dissolution produces a new mineral phase as differential leaching of ions from the rock material takes place, for example, dissolution of serpentine:

$$4Mg_3Si_2O_5(OH)_4 + 12H^+ \rightarrow Mg_6Si_8O_{20}(OH)_4 + 6Mg^{2+} + 12H_2O$$
$$\text{serpentine} \tag{12.2}$$

Dissolution processes can be pH dependent and therefore microbial activity will affect the solubility of chemical compounds due to the formation of CO_2 and hence carbonic acid (H_2CO_3).

Hydration and hydrolysis are closely related chemical weathering processes. Hydration involves water molecules being incorporated into the structure of a chemical compound. These water molecules do not react with the chemical and are sometimes termed *water of crystallization*. An example, relating to chemical weathering involves the surface adsorption of H_2O molecules on haematite (Fe_2O_3):

$$Fe_2O_{3(s)} + 3H_2O \rightarrow Fe_2O_3 3H_2O_{(s)} \tag{12.3}$$

Hydrolysis, in this context, is the chemical reaction of water with the rock material. Hydrolysis processes can produce acidic or basic solutions. For example, the hydrolysis of $Fe_2(SO_4)_3$ produces an acidic solution:

$$Fe_2(SO_4)_{3(s)} + 6H_2O \rightarrow 2Fe(OH)_{3(s)} + 3H_2SO_{4(aq)} \tag{12.4}$$

The hydrolysis of CaO results in a basic solution:

$$CaO_{(s)} + H_2O \rightarrow Ca(OH)_{2(aq)} \rightarrow Ca^{2+}_{(aq)} + 2OH^-_{(aq)} \tag{12.5}$$

Hydrolysis can also affect the physico-chemical speciation of chemicals (mainly metals) in solution. This refers to the physical and chemical nature of the element. Elements in waters are generally differentiated into particulate and non-particulate fractions. This is achieved by filtration of the water using a 0.45 μm filter. The addition of $Fe(ClO_4)_3$ to water can produce the following chemical species due to hydrolysis:

$$Fe^{3+}, Fe(OH)^{2+}, Fe(OH)_2^+, Fe(OH)_4^-, Fe_2(OH)_2^{4+}$$

Although the distinction between hydration and hydrolysis is not a clear one, hydrolysis refers to a more extensive and less reversible reaction of the solute with the water molecule.

Carbonation can be defined as the chemical weathering of rock by reaction with carbon dioxide (CO_2). Carbonic acid is produced by reaction of CO_2 with H_2O and it is readily dissociated as shown:

$$CO_{2(aq)} + H_2O_{(l)} \leftrightarrow H_2CO_{3(aq)} \qquad (12.6)$$
$$\text{carbonic acid}$$

$$H_2CO_{3(aq)} \leftrightarrow H^+_{(aq)} + HCO_3^-{}_{(aq)} \qquad (12.7)$$

The dissociation of carbonic acid will produce some acidity in freshwater systems. Examples of carbonate processes are given below:

$$CaCO_{3(s)} + CO_{2(g)} + H_2O \leftrightarrow Ca^{2+}{}_{(aq)} + 2HCO_3^-{}_{(aq)} \qquad (12.8)$$
$$\text{calcite}$$

$$CaMg(CO_3)_{2(s)} + 2CO_{2(g)} + 2H_2O \leftrightarrow Ca^{2+}{}_{(aq)} + Mg^{2+}{}_{(aq)} + 4HCO_3^-{}_{(aq)}$$
$$\text{dolomite}$$
$$(12.9)$$

In many lake waters, the precipitation and dissolution of $CaCO_3$ play a significant part in pH buffering and water composition.

On the surface layer, the oxygen content is relatively high due to photosynthesis and equilibrium with the atmosphere. At depth, consumption of oxygen due to biological activity may produce an anoxic (oxygen free) environment, whereby reducing conditions will prevail. In anoxic waters, the elements found in minerals and soils will be in their reduced form. Metals will be in low oxidation states, e.g. Fe^{2+}, Mn^{2+} and NO_3^-, NO_2^-, SO_4^{2-} will be used as biochemical oxidants. On interaction with oxygenated waters, minerals are solubilized followed by oxidation.

This oxidation process is important for elements which form insoluble sulphides. For example, FeS_2 (pyrite) is oxidized to produce soluble iron and sulphate:

$$2FeS_{2(s)} + 7O_{2(g)} + 2H_2O \rightarrow 2Fe^{2+}{}_{(aq)} + 4SO_4^{2-}{}_{(aq)} + 4H^+{}_{(aq)} \qquad (12.10)$$

$$4Fe^{2+}{}_{(aq)} + O_{2(g)} + 4H^+{}_{(aq)} \rightarrow 4Fe^{3+}{}_{(aq)} + 2H_2O \qquad (12.11)$$

$$Fe^{3+}{}_{(aq)} + 3H_2O \rightarrow Fe(OH)_{3(s)} + 3H^+{}_{(aq)} \qquad (12.12)$$

$$FeS_{2(s)} + 14Fe^{3+}{}_{(aq)} + 8H_2O \rightarrow 15Fe^{2+}{}_{(aq)} + 2SO_4^{2-}{}_{(aq)} + 16H^+{}_{(aq)} \qquad (12.13)$$

Esssentially, these minerals (sulphides) are oxidized in air to form H_2SO_4

and concentrations of dissolved metals are produced that may be toxic to the aquatic biota.

All the reactions discussed so far represent a type of physical or chemical weathering. They are not independent processes and are likely to occur together to a greater or lesser extent.

12.3.3 Precipitation

The composition of water can and will be altered by precipitation of minerals. Ultimately, these precipitates will be accumulated in sediments in freshwaters. Precipitation occurs when the solubility product of a particular compound is exceeded.

Precipitates are formed by two different so-called nucleation processes. Fine particles on which precipitation may occur are present in freshwaters. *Homogeneous nucleation* occurs if the precipitate is formed from its component ions in solution (i.e. no solid phase is initially present). If other fine particles (foreign nuclei) are present, then the process is referred to as *heterogeneous nucleation*. In undergraduate experimental organic chemistry, if a precipitate is "unwilling" to form, the inside of the glass vessel which contains the solution is rubbed with a glass rod. This causes very small glass particles to be present in solution which act as the initial nuclei on which precipitation first takes place. This is an example of heterogeneous nucleation and is considered to be the major formation pathway of precipitates in freshwaters.

As precipitation processes are creating new structures, this process requires energy. A solution will generally have to be supersaturated (or in greater concentrations than equilibrium conditions) before homogeneous nucleation will take place. The amount of supersaturation tends to be larger for homogeneous nucleation than for heterogeneous nucleation as the free energy required to form a precipitate on a fine particle is less.

The initial solid phase produced by precipitation processes may be metastable compared with a thermodynamically stable solid phase. Over time, the precipitate structure is changed to be more thermodynamically stable. This process is known as aging. As the more stable precipitate has a lower solubility than the initial metastable solid phase, additional precipitation may occur which may also increase particle size. Other factors such as temperature, substance concentration and pressure (to a lesser extent) will also affect precipitation and rate of formation.

If groundwater entering freshwater bodies is calcite ($CaCO_3$) saturated, there is a strong possibility that precipitation will take place. This precipitation will be incorporated into the sediments of surface waters. Algal species form shells of silica (SiO_2). This can remove amorphous silica from the water and eventually it will become part of the sediment. Metals such as Fe and Al are present in surface water mainly as humic acid complexes.

They exist in suspended (colloidal) matter which eventually increases in size due to coagulation and aggregation processes before undergoing sedimentation.

12.3.4 Eutrophication

The eutrophication process can occur in all waters and can be defined as the nutrient enrichment of a water body. Eutrophication can occur naturally as part of the aging of a lake/waterbody, but generally occurs through anthropogenic processes which greatly increase the eutrophication rate. For most freshwaters, phosphorus is the limiting nutrient. In marine and estuarine waters, nitrogen may be the limiting nutrient. These nutrients can be released from fertilizers and detergents and eventually make their way into water courses. Eutrophication can result in increased biological activity with increased plant growth and blue-green algal blooms. Eventually the water will be "choked" by the eutrophication process. Eutrophication is discussed in more detail in Chapter 18.

12.3.5 Chemical characteristics of seawater

Seawater is a heterogeneous solution of gases and solids containing both inorganic and organic material in suspension. Table 12.2 gives the concentrations of the major constituents in seawater. Major constituents are those taken to be at concentrations greater than 1 mg kg^{-1}.

Seawater also contains other constituents such as Sr, SiO_2, N, Li, I and heavy metals, but in much smaller amounts (of the order 1×10^{-5} to 10 mg kg^{-1}). The pH of seawater is around 8 and has a high ionic strength which is controlled by NaCl. In contrast, rivers tend to be slightly acidic in nature and have a low ionic strength, with $Ca(HCO_3)_2$ being the major salt.

Table 12.2 Major constituents of seawater

Constituent	Concentration in mg kg^{-1} (ppm)
Na^+	10,635
K^+	1,320
Mg^{2+}	406
Ca^{2+}	390
Cl^-	19,177
SO_4^{2-}	2,700
HCO_3^-	142
Br^-	65
F^-	1.3
Total dissolved solids (TDS)	34,500

There are essentially two major concepts regarding control of the composition of seawater that complement each other. These are:

1   control by chemical equilibria between seawater and sediments;
2   kinetic regulation by the rate of supply of individual components and the interaction between biological and mixing cycles.

Equilibrium processes also exist at the air–sea interface. To some extent, biological activity will also affect the amount of dissolved gases in the seawater. Gas concentrations play an important role in seawaters, e.g. O_2 determines the redox potential in seawater and will thus affect speciation; dissolved CO_2 is important in controlling the pH of seawaters.

All equilibrium processes play a part in controlling the composition of seawater. These processes affect dissolved gas concentrations, physicochemical speciation and the solubilities of elements and compounds.

12.4 Precipitation chemistry

Precipitation (i.e. rain, snow and hail) is a major pathway in which natural and anthropogenic compounds are removed from the atmosphere. The fact that precipitation effectively "washes" these chemicals out of the atmosphere is particularly evident in highly air polluted areas (i.e. major cities).

Large scale cooling is required in order for significant precipitation to take place. This cooling occurs by the upward movement of air currents and there are three main ways by which this occurs:

1   frontal (*cyclonic*), in which a warm air mass meets and rises above a cooler air mass;
2   *orographic*, where warm air is deflected upwards (e.g. over mountains);
3   *convective*, caused by differences in local solar heating.

Precipitation consists of a heterogeneous solution of dissolved chemical compounds. It will contain atmospheric gases (N_2, CO_2, O_2), impurities in the air originating from the Earth's surface (sea spray, volcanic emissions, dust, volatile organic compounds), and traces of other gases (e.g. ozone) derived from chemical reactions within the atmosphere. Apart from the effects of atmospheric pollution, the composition of rainfall is determined mainly by seawater and dust.

Precipitation naturally contains 10 to 20 mg l^{-1} of dissolved solids and has a pH of around 5.6 due to its content of atmospheric gases, mainly CO_2. This lower pH is due to the fact that it is poorly buffered and cannot maintain a neutral pH. Chemical analysis of precipitation shows the presence of nine major ionic compounds:

1 Na^+, Mg^{2+}, Cl^- – from sea spray;
2 K^+, Ca^{2+} – soil derived;
3 H^+ – from strong acids;
4 NH_4^+ – from neutralization of strong acids;
5 SO_4^{2-} and NO_3^- – from SO_2 and NO_2 oxidation processes.

Carbonate CO_3^{2-} may also be present, but this is removed as $CO_{2(g)}$ at lower pH values. Table 12.3 shows the general composition of precipitation.

Precipitation near the sea contains more SO_4^{2-}, Cl^-, Na^+ and Mg^{2+} than that which falls inland. It can be seen from Table 12.3 that precipitation is an extremely dilute solution of dissolved compounds. Owing to its acidic nature, precipitation forms part of a worldwide acid–base titration in which the rainwater acids are involved in titration of the bases of rocks.

12.4.1 Sampling

There are generally two accepted methods of sampling precipitation. One method involves collecting precipitation in an open rain gauge. One problem with this is that samples will also contain a portion of the dry deposition. The other method requires the use of automatic collectors which are only open during rainfall. The latter method obviously produces a more accurate sample of precipitation. However, due to their simple nature, open rain gauges are generally utilized.

12.4.2 Dry deposition

As the name implies, this process involves the removal of chemicals from the atmosphere without the intervention of rainfall. Dry deposition includes

Table 12.3 Chemical composition of precipitation (adapted from Snoeyink and Jenkins, 1980 and Harrison and de Mora, 1996)

Constituent (mg/l)	"Average" rain	Menlo Park, California, rain	Station 526U, Belgium, rain	Nevada, snow
Na^+	1.98	9.4	0.97	0.6
K^+	0.30	0	0.23	0.6
Mg^{2+}	0.27	1.2	0.36	0.2
Ca^{2+}	0.09	0.8	3.3	0
Cl^-	3.79	17	2.0	0.2
SO_4^{2-}	0.58	7.6	6.1	1.6
HCO_3^-	0.12	4	0	3
pH	5.7	5.5	4.4	5.6

the removal by sedimentation, impaction on obstacles such as leaves and vegetation, and absorption at the Earth's surface. The concentration of a chemical ($\mu g \ m^{-3}$) and the rate of deposition ($\mu g \ m^{-2} \ s^{-1}$) are measurable quantities. The ratio of these two quantities is called the *deposition velocity*:

$$\text{deposition velocity } (V) = \frac{\text{deposition rate}}{\text{air concentration}} \qquad (12.14)$$

Deposition velocity has units of $m \ s^{-1}$. Different surfaces (earth, water) will have different deposition velocities. It should be noted that removal on to a wet or dry surface can be termed dry deposition. In fact, transfer to a wet surface is generally faster than that to a dry surface. Values of deposition velocities of some chemical compounds in the atmosphere which are subject to dry deposition on various surfaces are given in Table 12.4.

Gases can be absorbed by plant and soil surfaces. Owing to their large surface area and ability to absorb water soluble gases, plants provide a major sink for dry deposition.

It can be seen from Table 12.4 that there is a great degree of variability in these data. This relates to the chemical in question and the deposition surface. The data show that nitric acid (HNO_3) vapour represents a major dry deposition sink mechanism.

Dry deposition represents a downward flux of material (i.e. from air to land surfaces). However, a few substances are capable of being in a state of dynamic equilibrium and will also have upward fluxes. For example, ammonium (NH_4^+) in the soil is in dynamic equilibrium with ammonia gas (NH_3):

$$NH_4^+{}_{(aq)} + H_2O \leftrightarrow NH_{3(g)} + H_3O^+{}_{(aq)} \qquad (12.15)$$

Table 12.4 Deposition velocities of some atmospheric chemicals (Harrison, 1992)

Chemical	Surface	Deposition velocity ($m \ s^{-1}$)
SO_2	Ocean	0.005
	Soil	0.007
	Grass	0.01
	Forest	0.02
O_3	Wet grass	0.002
	Dry grass	0.005
	Snow	0.001
HNO_3	Grass	0.02
CO	Soil	0.0005
Aerosols ($< 2.5 \ \mu m$)	Grass	0.0001

Hence, when atmospheric ammonia concentrations exceed equilibrium conditions, there will be a net flux of ammonia downwards.

12.4.3 Particulate matter

Aerosols can be classed as very fine particulate material present in the atmosphere and are generally in the size range 0.1 to 20 μm. The deposition of this particulate matter depends on the particle size. Thus, the larger the particles, the more rapidly they will fall. Table 12.5 shows deposition velocities for particles of density 2 g cm^{-3}.

Aerosols ($\leqslant$5 μm) have low sedimentation velocities and the turbulence of air determines their movement. Therefore, these aerosols are removed very slowly by dry deposition (e.g. NO_3^- and SO_4^{2-}). Intermediate particles (5 to 20 μm) can be removed due to impaction on to leaves and other surfaces. Particles > 150 μm are falling at around 1 m s^{-1} and their residence time in the atmosphere is so short that any chemicals in this category do not concern us as far as air pollution is concerned. With the finer particles, removal by wet deposition is more likely.

12.4.4 Wet deposition

This term describes the removal of chemical compounds by precipitation. There are two processes involved in wet deposition and these are dependent on the point at which the chemical is absorbed by the water droplets. Absorption of compounds within the clouds followed by precipitation is termed *rainout*; absorption of compounds by falling precipitation is called *washout*. It follows that the rate of removal of compounds by washout will be dependent on the rate of rainfall.

Aerosols between 0.1 and 1 μm contain most of the nitrates and sulphates, and are removed very slowly by dry deposition due to low deposition velocities (~ 10^{-3} m s^{-1}). Their removal from the atmosphere is most likely to be achieved by rainout. Washout does not provide an efficient means of removal, although it can be quite effective for larger particles.

Table 12.5 Sedimentation velocities for particles (Harrison, 1992)

Particle diameter (μm)	Deposition velocity (m s^{-1})
5	0.0015
10	0.0061
20	0.024
50	0.14
100	0.46
150	0.8

Rainout of gases can occur in three different ways:

1   solution formation in cloud water according to Henry's law, e.g. N_2O and CH_4;
2   solution formation with subsequent reversible hydration and dissociation, e.g. CO_2 and NH_3;
3   solution formation and irreversible reaction with other materials in cloud water, e.g. SO_2 and NO_2. This process is responsible for the production of "acid rain".

Fog or mist (cloud droplets) can be removed from the atmosphere by vegetation which effectively "scavenges" chemicals. This process is termed *occult deposition* and becomes significant on hilltops or mountains where there is plenty of moisture in the air.

12.4.5 Acid rain

This term is recognized by the majority of the population as an environmental problem. Considering the possible damage caused by acid precipitation to other environmental media, the acidity or pH of precipitation is of great interest. Rainwater is naturally slightly acidic (pH ≈ 5.6) due to the reaction of water with atmospheric CO_2 (see equations (12.7) and (12.8)). Formation of strong acids (nitric and sulphuric) due to pollutants undergoing oxidation (SO_2 and NO_2) causes the pH to drop to a value below 5, but in extreme cases, pH values as low as 2 may be observed.

In low humidity conditions, sulphur dioxide reacts with the hydroxyl radical:

$$SO_2 + OH^{\cdot} \rightarrow HOSO_2 \tag{12.16}$$

$$HOSO_2 + O_2 \rightarrow SO_3 + HO_2 \tag{12.17}$$

$$SO_3 + H_2O \rightarrow H_2SO_4 \tag{12.18}$$

Reactions in water droplets to produce an acid also occur in wetter climates:

$$SO_{2(aq)} + 2H_2O \leftrightarrow H_3O^+_{(aq)} + HSO_3^-_{(aq)} \tag{12.19}$$

$$HSO_3^-_{(aq)} + H_2O \leftrightarrow H_3O^+_{(aq)} + SO_3^{2-}_{(aq)} \tag{12.20}$$

Sulphurous acid (H_2SO_3) which is a relatively weak acid will be produced. This can be oxidized by atmospheric oxygen to produce H_2SO_4, a strong acid. This reaction is slow and can be catalysed by the presence of transition metal ions (Cu^{2+}, Fe^{3+}). Other oxidation processes include oxidation of bisulphite by ozone and sulphite reactions with hydrogen peroxide.

$$HSO_3^- + O_3 \rightarrow HSO_4^- + O_2 \tag{12.21}$$

$$SO_3^{2-} + H_2O_2 \rightarrow SO_4^{2-} + H_2O \tag{12.22}$$

There are two mechanisms for HNO_3 production from NO_2 and NO. The first involves the reaction with the hydroxyl radical:

$$2NO + O_2 \rightarrow 2NO_2 \tag{12.23}$$

$$NO_2 + OH^{\cdot} \rightarrow HNO_3 \tag{12.24}$$

The second process involves the formation of $NO_3^{\cdot}$ (peroxy-nitrite radical), a reactive intermediate. These substances are highly reactive and have very short lifetimes.

$$NO_2 + O_3 \rightarrow NO_3^{\cdot} + O_2 \tag{12.25}$$

$$NO_3^{\cdot} + NO_2 \leftrightarrow N_2O_5 \tag{12.26}$$

$$N_2O_5 + H_2O \rightarrow 2NO_3^{\cdot} \tag{12.27}$$

These highly reactive intermediates can also produce nitric acid by reaction with an alkane:

$$R\text{-}H + NO_3^{\cdot} \rightarrow HNO_3 + R \tag{12.28}$$

Both types of acid produced can be neutralized by reaction with ammonia, NH_3:

$$H_2SO_4 + 2NH_3 \rightarrow (NH_4)_2SO_4 \tag{12.29}$$
$$\text{ammonium sulphate}$$

$$HNO_3 + NH_3 \rightarrow NH_4NO_3 \tag{12.30}$$
$$\text{ammonium nitrate}$$

The most obvious effect that acid rain produces is the potential acidification of lakes and other water bodies. This may affect the aquatic biota in the water system, and can increase amounts of chemicals and metals such as aluminium, which are leached from the soil into surface runoff.

Acid rain is also known to attack stonework and buildings constructed of limestone. SO_2 deposition in both wet and dry forms is involved. In wet conditions, H_2SO_4 is produced which produces $CaSO_4$ from calcium carbonate

(CaCO$_3$). As CaSO$_4$ is more soluble than the carbonate, dissolution takes place. The rate of dissolution is dependent on the deposition rate of SO$_2$.

Knowledge of the chemical composition of precipitation can provide an indicator of the degree of air pollution. It is for this reason, amongst others that the World Meteorological Organization has decided to include the measurement of the chemical composition of precipitation in its background air pollution monitoring programme.

12.5 Groundwater chemistry

Chemical processes in groundwater are more or less the same processes that occur for surface waters. However, the chemistry of groundwaters is affected by the nature of the subsoils and rocks through which it passes. Chemicals will enter groundwater due to transfer processes between subsurface strata containing groundwater, and via infiltration and percolation of water from the surface. In contrast to surface waters, due to the soil and water ratio being large for groundwater systems, transfer processes will favour the soil phase. Thus, chemicals strongly bound to soil or rocks will be present in groundwaters in very low concentrations unless subject to dissolution processes.

A general relationship between the composition of groundwater and the nature of the soil and rocks will be expected. In predominately limestone rock and subsoils, groundwater often contains high concentrations of calcium and magnesium ions, whereas in areas of underlying volcanic rock or sandstone a softer water is produced. Owing to this filtration process through the subsoil and rock, groundwater is generally considered to be safe for drinking purposes.

With industrial and domestic waste being disposed of in landfill sites, it is important that the chemical (and microbiological) processes of groundwaters are evaluated. This is of paramount importance in the case of sites which contain hazardous material (e.g. a low level radioactive waste landfill site). The composition of groundwater at such a site could mobilize these radionuclides, resulting in unwanted transport of these hazardous pollutants.

This section looks at the basic concepts of groundwater chemistry. Further reading on this subject and water chemistry in general can be found in Snoeyink and Jenkins (1980) and Stumm and Morgan (1981).

12.5.1 Groundwater storage

Water moves underground through infiltration and percolation processes. An aquifer can be defined as a water-bearing rock formation situated below ground level. The water contained in these aquifers is known as groundwater. A slow flow of this groundwater is continually moving towards the sea. Figure 12.3 shows the various soil zones in which this water travels. Note that the water table is the level of water available in an ordinary well.

In aquifers, it is assumed that the voids are occupied with water. Thus, the porosity of the rock material is an important parameter. It is defined as the actual volume available in the soil/rock to hold water. It is given by the equation:

$$\text{porosity } (n) = \frac{V_v}{V_t} \qquad (12.31)$$

where V_v = void volume and V_t = total volume.

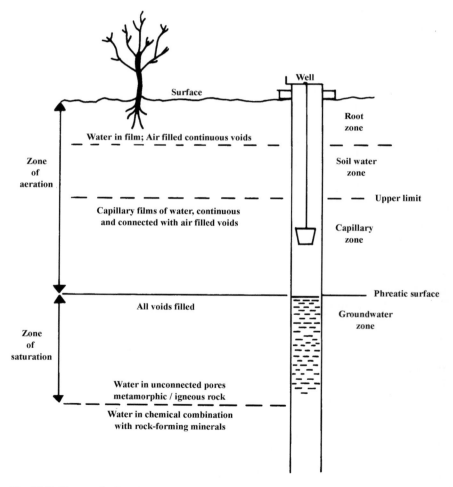

Fig. 12.3 Zones of subsurface water.

Table 12.6 Porosities and permeability data for geological materials

Material	Porosity (% void space)	Permeability range ($\times 10^{-3} \, m^3 \, day^{-1} \, m^{-2}$)
Clay	50–60	0.0005–0.05
Boulder clay	20–40	0.05–500
Alluvial sands	30–40	500–500,000
Alluvial gravels	25–35	500,000–50,000,000
Shale	5–15	0.00005–0.0005
Sandstone	5–25	0.05–5,000
Limestone	0.1–10	0.005–500
Basalt	0.001–50	0.005–500
Granite	0.001–14	0.00005–0.005
Slate	0.001–1	0.000005–0.005

Another important parameter is permeability which is the rate at which a substance allows fluids to pass through it. Table 12.6 gives some porosity and permeability data for certain geological materials.

These data show that clay, sands and gravels have the greatest capacity as water bearing materials, and hard rocks such as granite and slate have a low water holding capacity. Although gravels hold less water than clays, the high permeability value indicates that they will transmit water at a much higher rate.

Normally, aquifers consist of sands, gravels and certain rocks which have an enlarged water capacity due to chemical reaction (e.g. limestone). Rock strata which exclude water due to their impermeable nature are called *aquicludes*. There are two main types of aquifer as illustrated in Fig. 12.4.

1 *Unconfined aquifers* are bounded below by aquicludes, but are not bound above, i.e. their upper limit is the natural water table line.
2 *Confined aquifers* are sandwiched between two impermeable layers (aquicludes). The water contained in these aquifers is often under considerable pressure. Confined aquifers rely on recharge by infiltration at their exposed edges in order to maintain pressure. This pressure is used in the Artesian well principle where water is forced upwards into the well above the normal level created by the water table. This new "water table line" is known as the *piezometric surface*. A flowing well is created on piercing a confined aquifer when the point of bore entry is below the piezometric surface.

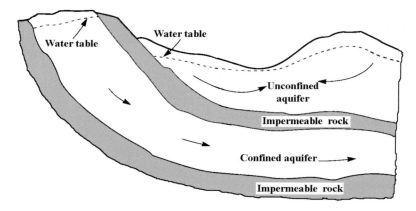

Fig. 12.4 Types of aquifer.

In Britain, aquifers are found in four major rock types:

1   carboniferous limestones mainly in northern England, Midlands and around Bristol;
2   triassic rocks which are found in the Midlands;
3   Cretaceous chalk which occurs in Yorkshire and some of the West Country;
4   recently-formed rocks including sands and gravels, also form aquifers.

12.5.2 Chemical characteristics of groundwater

Table 12.7 shows the composition of groundwaters. Compared with the data

Table 12.7 Chemical composition of groundwater (adapted from Snoeyink and Jenkins, 1980)

Constituent (mg l⁻¹)	Well water, Dayton, Ohio	Groundwater from chalk aquifer	Groundwater, Davis, California	Thorpe Borewater, Norwich, UK
Na^+	8.2	6.2	67	38.5
K^+	1.4	1.2	2.0	3.3
Mg^{2+}	34	1.2	66	9.6
Ca^{2+}	92	119	34	121
Fe^{3+}	0.09	Trace	–	0.005
Cl^-	9.6	8.5	39	64
SO_4^{2-}	84	2.5	57	77.4
HCO_3^-	339	–	224	225
SiO_2	10	11	–	15.5
NO_3^-	13	6.2	13.9	13.1
TDS	434	–	523	589

from Table 12.1, groundwaters have higher dissolved mineral concentrations than surface waters due to greater contact time between the water, soils and rock formations for dissolution processes to take place.

The major ions in groundwater are Na^+, Mg^{2+}, Cu^{2+}, Cl^-, HCO_3^- and SO_4^{2-}. The total concentration of these six ions constitutes more than 90% of the total dissolved solids (TDS) in water. The TDS in groundwater consists of all the inorganic constituents and minute amounts of organic matter. This parameter can vary over orders of magnitude and will be generally larger for surface waters as they may contain suspended solids (organic and inorganic matter). Freshwater generally has a TDS of 0 to 1000 mg l^{-1} compared with seawater which has a TDS of approximately 35,000 mg l^{-1}. Water containing a TDS above 2000 mg l^{-1} is considered too salty for drinking water.

Dissolved organic matter (DOC) is ubiquitous in all water environments. DOC comprises hydrophobic acids, bases and neutrals. The majority of DOC is composed of humic and fulvic acids. These are heterogeneous organic macromolecules which play a significant role in the binding of metals and pesticides, the result of which affects transport, toxicity and speciation of these compounds. Humic substance concentrations range from 20 μg l^{-1} in groundwaters to over 30 mg l^{-1} in surface waters. The chemistry of humic substances is beyond the scope of this book and the reader should refer to Aiken *et al.* (1985) for a more detailed account.

Groundwater also contains dissolved gases due to exposure of surface water to the atmosphere prior to infiltration, soil gas contact during infiltration, and gas production by reactions of groundwater with minerals and biological activity. The most abundant gases are those that constitute the Earth's atmosphere, namely O_2, N_2 and CO_2. Methane (CH_4), H_2S and N_2O are also present in groundwater due to biogeochemical processes that occur in non-aerated subsurface zones. These gases may pose a problem in groundwater use. For example, an accumulation of CH_4 could become a possible explosion hazard and H_2S, the gas that smells of rotten eggs, above concentrations of 1 mg l^{-1} renders the groundwater unfit for human consumption.

Radon (^{222}Rn) is also found in groundwaters. Radioactive decay of uranium and thorium which are common in rocks and soils produces radon which can accumulate in dwellings. The decay series for uranium is shown below.

$$^{238}U \longrightarrow {}^{234}Th \longrightarrow {}^{234}Pa \longrightarrow {}^{234}U \longrightarrow {}^{230}Th \longrightarrow {}^{226}Ra \longrightarrow {}^{222}Rn$$

This radon may contribute to indoor levels when high radon concentrations are observed in groundwater. The decay products of radon can also be hazardous to human health.

Groundwater processes vary from surface water processes in terms of their redox and biological activity. The chemistry of a groundwater system can vary from oxidizing conditions near the surface, which produce $Ca(OH_3)_2$ water chemistry, to reducing conditions at depth, which result in NaCl water chemistry. In general, surface water will be highly oxygenated and reducing conditions are unlikely, except where there may be a high oxygen demand (e.g. marshes or peat bog wetlands where reducing conditions are inches from the surface). In groundwater, an oxygen demand due to biological activity will decrease the dissolved O_2 concentration. Eventually, some bacteria utilize other electron acceptors (e.g. NO_3^-) thereby reducing these species and changing the groundwater chemistry. Table 12.8 shows these alternative electron acceptors and the order in which species tend to be reduced.

Nitrate (NO_3^-) can be readily reduced in groundwater by biological activity. However, nitrate may become stable in deeper water if there are insufficient heterotrophic bacteria (due to the lack of available organic carbon) to promote oxidation.

The metal compounds iron (Fe^{III}) and manganese (Mn^{IV}) oxides are the next to be reduced. This results in increasing amounts of metals in groundwaters. However, metal concentrations are related to the amount of available bisulphide (HS^-). Bisulphide is produced by the reduction of sulphate by bacteria. As iron and manganese oxides are present in the soil in percent by mass, the reducing conditions must be extreme before sulphate reduction takes place. If high bisulphide concentrations are present, iron (and possibly manganese) precipitates are formed, thus decreasing these metal concentrations in groundwaters.

It should be noted that most reduction processes are alkaline, therefore leading to an increase in pH. It follows that most oxidizing processes are acidic in nature (e.g. nitrification).

Figure 12.5 illustrates the chemical changes that take place in groundwater with changing depth.

The groundwater will become more alkaline with a sodium bicarbonate ($NaHCO_3$) composition which will allow more fluoride and phosphate to dissolve, because calcium is the controlling cation for these anions. Carbon dioxide produced by bacterial activity contributes to the CO_3^{2-} and HCO_3^- concentrations in the water, thus assisting it in achieving a $NaHCO_3$ composition.

Table 12.8 Redox reactions in groundwater systems (adapted from Ward and Elliot, 1995)

Reaction	Oxidized species	Reduced species
$O_{2(g)} + 4H^+ + 4e \rightarrow 2H_2O$	O_2	O^{2-}
$2NO_3^- + 12H^+ + 10e^- \rightarrow N_{2(g)} + 6H_2O$	NO_3^-	N_2
$MnO_{2(s)} + 4H^+ + 2e^- \rightarrow Mn^{2+} + 2H_2O$	MnO_2^-	Mn^{2+}
$FeOOH_{(s)} + 2H^+ + 2e^- \rightarrow Fe^{2+} + H_2O$	$FeOOH$	Fe^{2+}
$SO_4^{2-} + 10H^+ + 8e^- \rightarrow H_2S_{(g)} + 4H_2O$	SO_4^{2-}	H_2S

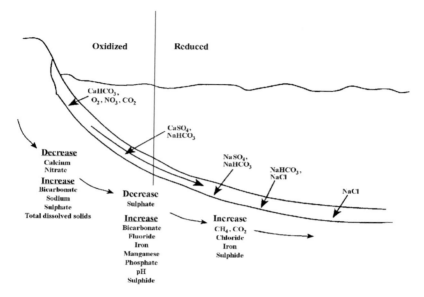

Fig. 12.5 Chemical changes in groundwater with increasing depth (adapted from Pyne and David, 1995).

12.5.3 Ion exchange

The interaction of groundwater with the soil can be seen in terms of a simple exchange of cations and anions. Ion exchange processes are particularly prevalent in soils that have a high clay content, as the soil solution will contain a large proportion of colloidal particles (10^{-3} m to 10^{-6} m) in suspension. These colloids generally have a negative charge which is large relative to the surface area. This charge is balanced at the surface by an accumulation of ions of opposite charge, called counterions. The layer of counterions can have a changeable composition so long as the overall charge on the colloid is balanced. Thus, an ion exchange reaction may be defined as involving the replacement of one ionic species in a solid compound by another ionic species taken from aqueous solution in contact with the solid, for example:

$$CaCO_{3(s)} + Mg^{2+}_{(aq)} \leftrightarrow MgCO_{3(s)} + Ca^{2+}_{(aq)} \tag{12.32}$$

It is generally the clay minerals that are responsible for ionic substitution, and will affect the composition of soil and groundwaters due to these ion exchange processes.

Of the ion exchange mechanisms, the cation exchange capacity (CEC) is the parameter most commonly measured. The CEC is the summation of exchangeable cations (including hydrogen ions). Generally expressed in milliequivalents per 100 grammes (meq 100 g^{-1}) of material, these values are also pH dependent.

References and further reading

Aiken, G. R., McKnight, D. M., Wershaw, R. L. and MacCarthy, P. (eds) (1985) *Humic Substances in Soil, Sediment and Water: Geochemistry, Isolation and Characterisation*, Wiley, New York,

Goody, R. (1985) *Principles of Atmospheric Physics and Chemistry*, Oxford University Press, New York, p. 219.

Harrison, R. M. (ed.) (1992) *Understanding our Environment: An Introduction to Environmental Chemistry and Pollution*, The Royal Society of Chemistry, Cambridge, 318 pp.

Harrison, R. M. and deMora, S. J. (1996) *Introductory Chemistry for the Environmental Sciences*, 2nd edn, Cambridge University Press, Cambridge, pp. 290–326.

Pyne, R. and David, G. (1995) *Groundwater Recharge and Wells: A Guide to Aquifer Storage Recovery*, Lewis Publishers, pp. 169–216.

Snoeyink, V. L. and Jenkins, D. (1980) *Water Chemistry*, Wiley, New York, 451 pp.

Stumm, W. and Morgan, J. J. (1981) *Aquatic Chemistry: An Introduction Emphasizing Chemical Equilibria in Natural Waters*, 2nd edn, Wiley, New York, 756 pp.

Ward, A. D. and Elliot, W. J. (eds) (1995) *Environmental Hydrology*, CRC Press, New York, pp. 1–17.

Chapter 13

Water pollution

13.1 Introduction

Pollution can be defined as "the introduction by man into the environment substances or energy liable to cause hazards to human health, harm to living resources and ecological systems, damage to structures or amenity, or interference with legitimate uses of the environment".

Environmental pollutants in general can be taken up by humans through breathing, skin contact or the consumption of foodstuffs. The important pollutants are the heavy metals (e.g. lead, cadmium or mercury), dangerous gases such as sulphur dioxide, carbon monoxide, nitrogen oxide and ozone, N-nitroso compounds and chlorinated hydrocarbons (e.g. DDT or vinyl chloride). The most common sources of environmental pollutants are industrial or production facilities, power stations, motor vehicles, agriculture and domestic heating, which release pollutants in the solid, liquid or gaseous forms.

Pollutants can cause immediate (acute) or long-term damage (e.g. through carcinogenic, mutagenic or embryotoxic substances). Particularly dangerous pollutants are characterized by a good rate of absorption into the human body (resorption), slow degradation (long biological half-life) and high stability (persistence). Thus, they can remain in the natural circulatory system for a long time and unfold their full damaging effects.

Toxic chemicals in the water courses pose the greatest threat to the safety of water supplies in the developed nations. Specific water pollutants include chlorinated hydrocarbons, heavy metals, bacteria, saline water and municipal and industrial wastes. Examples of water pollution are pesticide runoff from agricultural land and industrial discharges into surface waters. Table 13.1 illustrates some examples of water pollutants.

In order to be able to control these pollutants, it is important to have an understanding of the sources, interactions and effects of water pollutants. This chapter sets out to provide the reader with information on the types and potential sources of water pollution.

Table 13.1 Types of water pollution

Pollutant	Potential effects
Trace elements	Health, aquatic biota
Heavy metals	Health, aquatic biota
Metal-organic compounds	Metal transport
Inorganic pollutants	Toxicity, aquatic biota
Nutrients	Eutrophication
Acidity, alkalinity, salinity	Water quality, aquatic life
Organic pollutants	Toxicity
Polychlorinated biphenyls (PCBs)	Possible biological effects
Pesticides	Toxicity, aquatic biota, wildlife
Petroleum wastes	Wildlife, aesthetics
Sewage wastes	Water quality, oxygen levels
Biochemical oxygen demand	Water quality, oxygen levels
Pathogens	Health
Detergents	Eutrophication, wildlife, aesthetics
Taste, odour and colour	Aesthetics
Sediments	Water quality, aquatic biota, wildlife

13.2 Sources of pollutants

There are several points within the water cycle where the water can become polluted. The most significant is the point at which the water is removed (i.e. abstracted) by humans from rivers, lakes and wells and used for drinking, cooking, cleaning, industrial processes and a whole variety of purposes. During use, the water is converted into wastewater. What is done with the wastewater largely determines the extent of the water pollution which exists. The options available range from putting it back to source or treating it to a very high standard (i.e. restoring it to its former condition). Wastewater treatment processes are discussed in Chapters 14, 15 and 17.

Sources of pollution fall into two main categories:

1   point sources;
2   non-point sources (or diffuse sources).

Table 13.2 Point and non-point sources of pollution

Point sources	Non-point sources
Discharges from sewage treatment works to rivers	Runoff and underdrainage from agricultural land into rivers
Discharges of industrial wastewaters to rivers	General contamination of recharge rainfall to outcropping aquifers
Discharges of farm effluents to rivers	Septic tank soakaways into permeable strata
Discharges from small domestic sewage treatment plants to rivers	Wash-off of litter, dust and dry fallout, from urban roads to rivers
Discharges by means of well or borehole into underground strata	General entry of sporadic and widespread losses of contaminants to rivers
Discharges of collected landfill leachate to rivers	Seepage of landfill leachate to underground strata and rivers

Table 13.2 summarizes the more common point and non-point sources.

13.2.1 Point sources

A point source is a defined point of entry of polluting material into a water-course. This could be an effluent discharge pipe, a storm water overflow or a known point where waste is habitually dumped into the water. Point sources are far easier to control than non-point sources. For this reason, non-point sources are converted as far as possible to point sources. In London, during the 19th century, this was done through the construction of inter-ceptor sewers which carried all of the capital's wastewater, previously dumped into cesspools or discharged direct to local streams, to two sites downstream where it could be discharged to the Thames on the ebb tide. Subsequently, treatment works were built on these sites so that pollution of the river from most of London's wastewater was brought under close control.

(a) Sewage effluents

Point sources will naturally vary according to the particular area being investigated. In a river like the Thames, sewage treatment works will con-stitute the main sources of pollution. All of the wastewater and runoff gen-erated within the catchment area will be directed to the treatment works, where the pollution load will be reduced by well over 90%. The fact that sewage treatment works constitute the greatest point source of pollution in the river illustrates the success of the authorities in minimizing other sources.

Storm waters are another significant point source. However, they can also be considered to be non-point sources as they come from a very large catchment area and their deposition cannot be directly controlled.

(b) Industrial effluents

In certain areas, industrial discharges have in the past caused severe pollution problems. It is characteristic of many industrial effluents that they are very much stronger than domestic wastewater so that the effect of an equivalent volume discharged can be much greater. Today there is a much greater emphasis on the treatment of industrial wastewaters prior to their discharge. In many cases these wastewaters will be discharged to a sewer and will be treated in a mixture with domestic sewage at the sewage treatment works. The Water Authority will levy a charge on the industry which chooses this option, based on the volume and strength of the wastewater. As a result of this, many factories partially treat their effluents before discharging to the sewer in order to cut down on their costs. Some industries can treat their wastewaters to produce effluents of the same standard as those discharged from the sewage works and can therefore discharge these directly to the river with the consent of the Water Authority.

One other point source is the discharge from power stations. There are, in fact, two types of power station discharges. One type is the scrubbing liquors which arise from the scrubbing of stack gases to remove sulphur dioxide. These can be treated prior to discharge on the same basis as the industrial effluents mentioned above. The second type of discharge is the cooling waters. River water is used for cooling purposes and is discharged back into the river at a higher temperature. This can lead to thermal pollution which will be discussed in more detail later in this chapter.

13.2.2 Non-point sources

Pollution from point sources for the most part enters the water cycle at a relatively late stage in the terrestrial leg of the cycle, i.e. lowland rivers and estuaries. In contrast, pollution from non-point sources tends to enter at an earlier stage during precipitation or overland flow.

The actual point where the pollutants enter depends on the type of source, its location and on the physical form of the pollutants. If the pollutants are gases or fine airborne particles, they can fall to the ground directly as rain. The rain can also wash particles which have been deposited on surfaces into nearby watercourses. If the pollutants are soluble they can be transported large distances in the water. During storms, quite large particles, including soil, can be washed down from the land into water bodies. These can sometimes have pollutants such as pesticides attached to them.

Non-point source pollution transport processes in urban areas are likely to be different from those in rural areas. There are several reasons for this:

1 Quite a large part of urban areas are covered by impermeable materials. Thus, quite a lot of the rainwater will become runoff and eventually enters drains and sewers.
2 In urban areas less soil is exposed and so less erosion and hence transport of soil particles into surface waters can be expected.
3 In urban areas, pollutant loadings are mainly affected by the accumulation of litter, fallout and road traffic. In rural areas most of the pollution is due to the erosion of soils.
4 In the long term, almost all of the pollutants deposited on impermeable surfaces in urban areas which are not removed by external processes (e.g. street cleaning) will eventually end up as surface runoff. In rural areas, deposits can be incorporated into the soil where their removal rate can be reduced.

(a) Acid rain

Acid rain is an environmental problem that has been given wide coverage in the media. The pollutants responsible for this phenomenon are in fact air pollutants. Acid rain is a clear case of the pollution of one part of the environment directly or indirectly resulting in the pollution of another part. This illustrates the pitfalls of treating the environment as if it were made up of numerous separate compartments. The chemistry of acid rain has been discussed in Chapter 12.

(b) Mine water

Mining for coal and mineral ores can have serious implications for water pollution. There are essentially two types of mining operation:

1 strip mining;
2 deep mining.

Strip mining involves exposing bare land surfaces. It can result in severe erosion which leads to high sediment loads in rivers as well as increased concentrations of the ore minerals themselves.

Water which gets into mines can dissolve sulphur-bearing minerals, converting them into sulphuric acid. The leachate that drains out of the mines is termed acid mine drainage. pH values of acid polluted water can fall below 3 which is deadly to most forms of aquatic life except some bacteria.

Pyrite, FeS_2, is oxidized to Fe^{2+} by the following reaction:

$$2FeS_2 + 4H_2O + 6O_2 \rightarrow 4H_2SO_4 + 2Fe^{2+} \qquad (13.1)$$

Oxidation to Fe^{3+} then occurs slowly due to the low pH values:

$$4Fe^{2+} + 4H^+ + O_2 \rightarrow 4Fe^{3+} + 2H_2O \tag{13.2}$$

Below pH4, the iron oxidation is catalysed by a variety of bacteria. The Fe^{3+} ion can also dissolve pyrite:

$$FeS_2 + 14Fe^{3+} + 8H_2O \rightarrow 15Fe^{2+} + 16H^+ + 2SO_4^{2-} \tag{13.3}$$

Although amorphous $Fe(OH)_3$ can be precipitated, it is the production of sulphuric acid that causes concern.

Controlling pollution from these sources involves reducing the amount of water entering the mines either by sealing them or diverting surface runoff, collecting the leachate and treating it, and using greater care in the disposal of mine spoil. Effective cleaning up after the termination of mining operations is also important.

(c) Hydrological control

This refers to human interference in the movement of water through natural channels. The result is a non-point source pollution which can directly or indirectly affect the water quality. Examples are the construction of drainage and irrigation systems, dredging in order to create or maintain navigable stretches of water and the construction of dams and reservoirs.

Dredging can cause water pollution because pollutants that were previously held tightly within the sediment can be redissolved or resuspended in the water during the dredging operation. Dredging spoil is often dumped back into the water at a point remote from where it was taken. This can have the effect of transporting any pollution from one place to another.

Channel modifications and dam construction can irreversibly change stream flow characteristics. Apart from the immediate effects during construction, the important long term effects are mainly to do with changing patterns of sediment deposition. These effects on the water body can be similar to those caused by dredging.

13.3 Groundwater pollution

Most human activities at the land surface cause some change in the quality of water in the aquifer beneath them. The importance of the effect of a particular activity is related to the amounts and types of contaminants released. The severity of an occurrence is also related to the ability of the soil and groundwater system to degrade or dilute the contaminants, and the degree to which the contamination will interfere with uses of the water. Contamination is usually more serious in a drinking water supply than in water for other uses.

Groundwater is a good source of drinking water because of the purification properties of the soils that rainwater has to percolate through to get into the aquifer. This applies particularly to suspended matter which is effectively filtered out by the strata overlying the aquifer. Except where contaminated water is injected directly into an aquifer, essentially all groundwater pollutants enter the aquifer through recharge water from the land surface.

There are four main sources of groundwater pollution. These are:

1 industrial sources;
2 domestic sources;
3 agricultural sources;
4 environmental sources.

Industrial sources are industrial effluents, accidents such as leakage from pipes and tanks and rainwater which infiltrates and percolates through solid waste deposits. Solid wastes are deposited in landfill sites. The local Water Authority ensures that the risks to groundwater from the landfill leachate are carefully considered. Landfills that take hazardous waste are often sited in areas where there are impermeable strata, like clay, or where clay linings have been deliberately installed to prevent leachate escaping.

Domestic sources of groundwater pollution fall into more or less the same categories as the industrial sources. Leakage from septic tanks and percolation of rainwater through landfills containing domestic refuse are the main risks. Substances from septic systems include nitrogen, bacteria, viruses, synthetic organics from household cleaning products and septic tank cleaners.

Agricultural sources are potentially the most dangerous because they are non-point sources and percolation through soils into the groundwater can take place over wide areas. As a result of this, fertilizers, minerals, herbicides and pesticides can all enter aquifers. There is also a potential risk from the disposal of sewage sludge on agricultural land, which is another major disposal outlet for this material in addition to the sea.

The main environmental source of groundwater pollution is saline intrusion. In coastal locations there is an interface between the fresh groundwater flowing towards the sea and salty groundwater which tends to flow in the opposite direction. The fresh water is lighter then the salty water and therefore floats on top of it. If water is removed from wells and boreholes at too high a rate, then the lowering of the water level causes the boundary between the fresh and salt water to rise. This can result in salt water getting into the wells.

13.3.1 Behaviour of pollutants in groundwater

The most frequent and dangerous forms of pollution are those which are miscible with the water in the aquifer. The movement of these pollutants in the aquifer can be understood from the *theory of dispersion*.

Materials disperse in fluids as a result of two factors. The first is mechanical action. Water moving through the porous medium does not all do so at the same velocity. In some areas, the localized velocity will be higher than the average, and in others it may be zero. As the pollutant undergoing dispersion passes through these areas it will tend to spread out as a result of these differences in velocity.

The second factor is diffusion. It is a fact that molecules tend to move away from each other in order to create a uniformly dilute solution. They do this even in static water. This means that even in aquifers where there is little groundwater movement, pollutants entering the aquifer will slowly spread out through it. However, the rate at which this occurs may be very slow.

13.3.2 Groundwater flow

River velocities are measured in metres per second, and so it will only take a matter of days for pollution to pass downstream. Groundwater flow, however, is measured in metres or fractions of metres per year. This means that the turnover of water in an aquifer could be anything from 200 to 10,000 years.

Moreover, aquifers which are being polluted now, or which have been polluted in the past could remain so for tens or hundreds of years. Some of the pollutants which could cause problems in the future may not yet have entered or dispersed in the groundwater. They may still be percolating slowly downwards through the soil. Obviously, polluted groundwater can still be abstracted for potable supply provided that it can be treated to the appropriate standards, although this will increase costs.

13.4 Marine pollution

It is often thought that the oceans are so vast that a reasonable amount of pollution will, on dispersion, be diluted to such an extent that it is rendered harmless and will virtually disappear. Wastes are generally effectively dispersed which means that their eventual destination is unknown. In some cases however, pollution can be localized due to poor dispersion as a result of winds, tides and currents, and stratification.

Examples of polluted seas include the Irish Sea, the North Sea and the Mediterranean. Table 13.3 shows the major inputs of organic wastes and heavy metals into the North Sea.

The majority of pollution entering the North Sea comes from sources which cannot be directly controlled (i.e. rivers, and in the case of metals, atmospheric deposition). The next most significant source is direct discharges. Unlike sewage effluents discharged to inland waters, discharges to sea are often not treated or only given primary treatment.

Table 13.3 Major sources of organic wastes and heavy metal inputs to the North Sea

Source	Percentage total	
	Organic wastes	Heavy metals
Atmosphere	–	42
Rivers	56	42
Direct discharges	39	13
Sewage sludge	5	1
Other dumping	–	2

Sewage sludge is the solid residue left over after sewage treatment. In the UK, we produce about 1.2 million tonnes (as dry matter) every year and about 29% of this is disposed to sea. Major sludge dumping grounds include the Thames, Severn and Clyde estuaries, Liverpool and the Firth of Forth. These dumping grounds are broadly classified as accumulating grounds or dispersing grounds. Accumulating grounds (e.g. Firth of Forth) are characterized by fairly slow bottom water movements which allow accumulation of sludge solids at the dumping area. Dispersing grounds are characterized by strong water movements which will cause greater dispersal of pollutants over a wider area.

One other major source of marine pollution is oil, both from tanker accidents and accidental discharges from oil platforms. During the 1970s nearly 15,000 tonnes of crude oil were spilled into the North Sea as a result of tanker accidents and nearly twice as much from a single blow-out at an oil platform.

Exploitation of the sea bed for the extraction of sand and gravel can also be considered as a source of pollution. Several million tonnes of sand and gravel are extracted from the North Sea every year and this can cause significant changes in the habitat for organisms which live on the sea bottom. Also, a lot of sediment may be resuspended which may lead to the mobilization of pollutants previously contained within the sediments.

13.5 Physical pollutants

The major physical pollutants are suspended solids and temperature changes. Certain other physical effects on freshwaters can be thought of as pollution, such as reductions in flow rate caused by abstraction or changes in river bed characteristics as a result of, for example, bridge building. Certain chemical pollutants can also have direct or indirect physical effects, such as imparting colour to water.

13.5.1 Suspended solids

One direct effect of suspended solids is to cause turbidity in the receiving water. This reduces the amount of light that can penetrate and therefore will tend to reduce photosynthesis. Moreover, this could affect the reaeration of the stream. The solids may also exhibit an effect if they settle out of suspension. Deposition of solids can change the characteristics of the river bed which in turn will affect plant and animal growth and fish breeding.

The Royal Commission standard for suspended solids in sewage effluents, still widely used today, is 30 mg l^{-1}. With reasonable dilution in the receiving water, this is well within the concentration at which suspended solids begin to affect fish populations.

Suspended solids generally cause damage to fish gills, affecting their oxygen consumption and ultimately causing death at high concentrations. A summary of the main effects of suspended solids on fisheries is shown in Table 13.4. Suspended solids can have a detrimental effect on fish at concentrations below 100 mg l^{-1}.

13.5.2 Thermal pollution

The main polluting discharges which have an effect on the temperature of a water body tend to raise the temperature rather than lower it. A higher water temperature generally causes an increase in the rate of biological processes. The adverse effects of this are that an increase in growth rate causes increased primary production, one of the problems associated with eutrophication. Furthermore, the increased rate of aerobic metabolism means that the biochemical oxygen demand (BOD) is exerted more rapidly, imposing a greater requirement for reaeration of the water body. The adverse effects of this in turn are compounded by the lower solubility of oxygen in water as the temperature increases.

Course fish are more tolerant of the direct effects of temperature than game fish. Course fish can normally tolerate temperatures up to 30°C and dissolved oxygen (DO) concentrations greater than 3 mg l^{-1}; game fish have a much narrower temperature tolerance of 5–20°C and a DO greater than 5 mg l^{-1}.

In addition to the direct effects of thermal pollution, there are a number of

Table 13.4 Effects of suspended solids on fisheries

Suspended solids (mg l^{-1})	Effect
<25	No harmful effects
25–80	Some possible reduction in yield
80–400	Good fisheries unlikely
>400	Very poor or non-existent

subtle effects that may occur. These include changes in mating and other behavioural patterns of aquatic animals, selection against cold water species and the effects on the migration of fish. For these reasons, it is generally desirable that thermal discharges do not raise the final water temperature above 30°C.

13.6 Chemical pollutants

There are many potential chemical pollutants that can arise from various point and non-point sources. While any chemical pollutant presents an obvious cause for concern, some chemicals, due to their toxicological effects on the flora and fauna need to be carefully controlled in order to protect water courses. To this end, standards for water pollution control are set by the European Community (EC). As a result of this, the Dangerous Substances Directive was produced. This applies to all water bodies to which discharges of these dangerous substances are made.

13.6.1 Black and Grey List substances

In the directive, two lists of pollutants are defined. List I (the Black List) are considered to be dangerous substances because of their intrinsic toxicity and persistence in the environment. It is difficult to establish "safe" concentrations for these substances in water. Table 13.5 shows the specific substances in the Black List. It is the eventual objective of environmental policy in the EC to eliminate pollution caused by the List I substances.

List II (the Grey List) contains substances that are both less toxic and less persistent than the Black List (Table 13.6). It is possible to tolerate the presence of these in water below certain limiting concentrations, depending on water use. Ideally, programmes are required to reduce pollution by these substances.

Discharges containing Grey List substances are to be strictly controlled and require authorization from the Water Authority.

Table 13.5 List I (Black List) substances

1	Organohalogen compounds
2	Organophosphorus compounds
3	Organotin compounds
4	Substances which are carcinogenic in or via the aquatic environment
5	Mercury and its compounds
6	Cadmium and its compounds
7	Persistent mineral oils and petroleum hydrocarbons
8	Persistent synthetic substances (solid matter) which may sink or float and interfere with the use of the water

Table 13.6 List II (Grey List) substances

1	The heavy metals, metalloids and their compounds
2	Biocides and their derivatives not appearing in List I
3	Organoleptic (undesirable taste or odour) compounds
4	Toxic or persistent organosilicon compounds
5	Phosphorus and its inorganic compounds
6	Non-persistent mineral oils and petroleum hydrocarbons
7	Cyanides and fluorides
8	Substances which affect the oxygen balance, e.g. nitrates

13.6.2 Red List substances

In 1988, the UK Department of the Environment produced a "Red List" of priority pollutants (Table 13.7). This represents a limited range of the most dangerous substances which were classed as such according to clear scientific criteria. Their discharge to water courses should be minimized as far as possible.

The following sections of this chapter are intended to provide some background on some of the substances that appear in the Grey, Black and Red Lists.

13.6.3 Heavy metals

This class of elemental pollutants can be considered to be one of the most harmful as they are generally associated with toxicity effects. Elements classed as heavy metals are Cd, Cr, Cu, Pb, Hg, Ni and Zn. Some of these metals such as Cu are important as micronutrients at low concentrations, but produce toxic effects at higher concentrations.

All metals are naturally present in the aquatic environment due to their

Table 13.7 The UK Red List substances

Aldrin	Hexachlorobenzene
Atrazine	Hexachlorobutadiene
Azinphos-methyl	Gamma-hexachlorocyclohexane (Lindane)
Cadmium and its compounds	Malathion
DDT (including the metabolites DDD and DDE)	Mercury and its compounds
	Polychlorinated biphenyls (PCBs)
1,2-dichloroethane	Pentachlorophenol
Dichlorvos	Simazine
Dieldrin	Trichlorobenzene
Endosulfan	Triflualin
Endrin	Triorganotin compounds
Fenitrothion	

solublization from rocks and ore minerals together with absorption and precipitation reactions that occur at the soil–water, sediment–water interfaces.
The metal input from this source is small and results in a range of normal
background concentrations of these elements. Other sources, i.e. effluent
discharges and atmospheric deposition, are more likely to be responsible for
metal pollution in waters. Typical sources include mining, agriculture, elec-
tronics, fossil fuel combustion and waste disposal.

The extent to which metals are absorbed and precipitated depends on a
number of factors, such as properties of the metal (valency, radius, degree of
hydration, coordination with oxygen), the physico-chemical environment
(pH and redox status), the nature of the absorbent, the concentration of other
metals present and the presence of soluble ligands in the surrounding water.
For example, coprecipitation of heavy metals can occur with hydrous oxides
of Fe, Mn and Al.

In solution, metals can form a number of different chemical species with
dissolved substances. Chemical speciation is important as knowledge of this
can assist us in determining the nature of pollutant interaction with its surrounding environment. The chemical methods used to investigate speciation
include dialysis, chromatography, ion-selective electrodes and polarogra-
phy. Dissolved organic matter (i.e. humic and fulvic acids) present in
waters can interact with metals resulting in possible solubility, speciation
and toxicity changes. This is of particular interest in radionuclide studies as
humic materials may be responsible for the solubilization and transport of
these hazardous materials from disposal sites into water bodies.

13.6.4 Methylation of heavy metals

Heavy metals (particularly Hg) and metalloids have the ability to undergo
methylation reactions with micro-organisms to produce alkyl compounds
which are extremely toxic and bioaccumulative. This procedure of biomethylation occurs in sediments by reaction of the metal with anaerobic bacteria.
For example, the volatile species methylmercury (CH_3Hg^+) is hydrophilic
and lipophilic, and can accumulate in body fat of higher organisms (i.e. fish).
This bioaccumulative effect can pass further up the food chain to man.
Methylation of mercury is further discussed in Chapter 18.

13.6.5 Drinking water

The potential for high lead concentrations in drinking water exists where old
lead piping and solder is still in use. Acidic water which is standing in such
pipes can accumulate very high levels of lead as well as other metals (Cu,
Cd, Zn). The difference in heavy metal concentrations between hard and soft
waters can also be quite dramatic. This is because hard waters contain more
Ca and Mg (as carbonates) and so will not be as acidic as softer waters

due to a greater buffering capacity. According to the EC drinking water directive, the pH of domestic water supplies should be between 6.5 and 9.5.

13.6.6 Biological effects of heavy metals

Heavy metals have an affinity for sulphur and can disrupt enzyme function by reactions with —SH groups. Carboxyl and amino groups from proteins also bind with heavy metals. Some metals such as Cu, Mn, Fe and Zn can be classed as micronutrients, and are components of enzymes and proteins involved in metabolic pathways. Metals such as Cd, Pb and Hg can disrupt cell transport processes by binding to the cell membranes.

Speciation of heavy metals may also affect the toxicity and uptake of metals by organisms. For example, Cu in the free ion form (Cu^{2+}) is a biologically available species. However, Cu–fulvic acid complexes are also bioavailable, but their toxicity to aquatic organisms is known to be reduced. Also Al toxicity to fish is reduced by complexation of the metal with humic materials.

13.6.7 Organochlorine compounds

These are essentially organic compounds that have had some chlorine atoms substituted for hydrogen atoms. Some simple examples of organochlorine compounds (insecticides) are illustrated in Fig. 13.1.

Probably the most notorious of the organochlorine compounds is DDT (dichlorodiphenyltrichloroethane). Although its acute toxicity to humans is

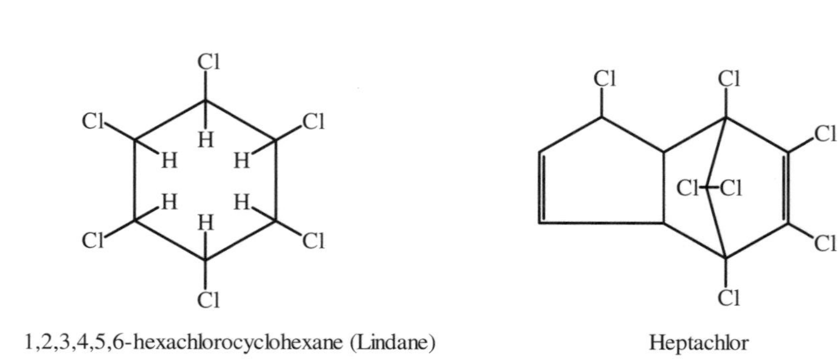

DDT

Methoxychlor

1,2,3,4,5,6-hexachlorocyclohexane (Lindane)

Heptachlor

Fig. 13.1 Organochlorine insecticides.

low, it is a persistent compound which is difficult to degrade and thus has the potential for bioaccumulation in the food chain. Use of DDT in the USA was banned in 1972 and in the UK in 1984. An alternative to DDT, methoxychlor is still used as it showed some biodegradable properties and had a low toxicity to animals as they are able to excrete the degraded products.

Other insecticides such as heptachlor and similar compounds (e.g. aldrin, chlordane) have been banned in the USA since 1983 and in the UK in seed dressings from 1986. These compounds show high persistence in the environment and are potential carcinogens.

The insecticide 1,2,3,4,5,6-hexachlorocyclohexane can exist in eight isomeric forms. However, it is the gamma isomer (Fig. 13.1), known as lindane, which actually kills the insects. A mixture of five of the isomers, including lindane, widely produced for insect control and agricultural applications is called BHC (benzene hexachloride). This insecticide has been under restricted use since the 1970s as it is considered to be carcinogenic. The gamma isomer, lindane, is still utilized in the treatment of seeds.

It is generally accepted that there is a need to reduce our dependence on organochlorine compounds and introduce the use of more environmentally friendly compounds such as organophosphorus insecticides.

13.6.8 Polychlorinated biphenyls (PCBs)

The general structure of PCBs is given in Fig. 13.2.

Taking into account all the possible isomers, there are 209 PCBs in total, all having varying toxicities. Sources of PCBs include their use as coolant fluids, plasticizers, paints and the printing industry. Atmospheric deposition can also be considered to be a potential source of PCBs in waters.

This class of compounds are chemically, thermally and biologically stable. Moreover, they are now considered to be ubiquitous in the environment and have the potential to be bioaccumulated by organisms. For example, the concentration factor of PCBs from water to mar‌‌‌‌ld be as high as 10 million times.

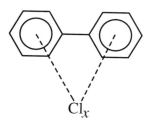

where $x = 1$ to 10

Fig. 13.2 General formula of PCBs.

In 1976 PCB manufacturing was halted in the USA and their disposal came under strict control. In the UK, an EC Directive on the disposal of PCBs has been adopted (96/59/EC). The new Directive replaces an earlier Directive (76/403/EEC) in order to reinforce the rules for the disposal of PCBs with a view towards an eventual complete ban on these substances. The disposal of PCBs can prove difficult as special incineration processes are required.

On heating PCBs in the presence of oxygen, polychlorodibenzofurans (PCDFs) are produced. The most toxic of these, 2,3,7,8-tetrachlorodibenzo-furan is illustrated below:

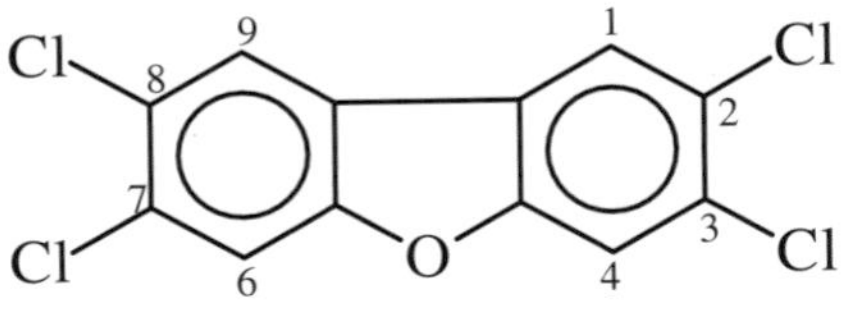

PCBs are considered to be not very toxic to humans in small quantities. However, at high doses they may cause cancer and this has been observed in test animals. It is thought that PCDF compounds are responsible for the majority of the toxicity effects associated with PCBs.

13.6.9 Dioxins

These are produced by the burning of chlorinated hydrocarbons and are a by-product of pesticide synthesis. The most toxic of these is 2,3,7,8-tetrachlorodibenzo-p-dioxin, which is shown in Fig. 13.3, along with some examples of other dioxins. The compound, TCDD, is known to be extremely toxic to some animals. In humans, it causes a skin condition called chloracne. Dioxins are both chemically and thermally stable which makes them difficult to degrade, resulting in a persistent environmental pollutant.

13.6.10 Organophosphorus compounds

Organophosphorus pesticides are considered to be an improvement over organochlorine compounds. While they are more toxic to humans, they do

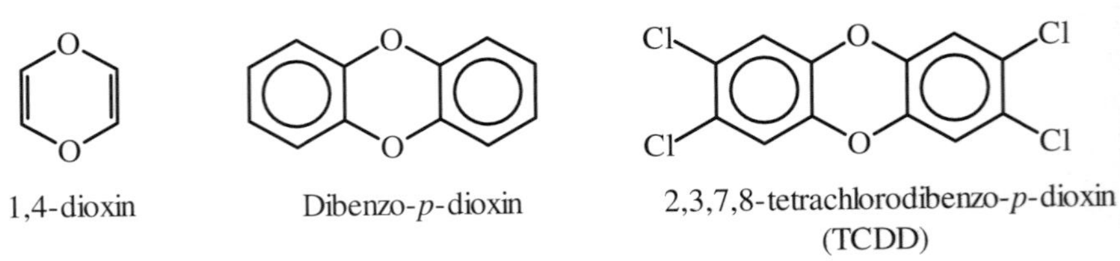

| 1,4-dioxin | Dibenzo-p-dioxin | 2,3,7,8-tetrachlorodibenzo-p-dioxin (TCDD) |

Fig. 13.3 Dioxin structures.

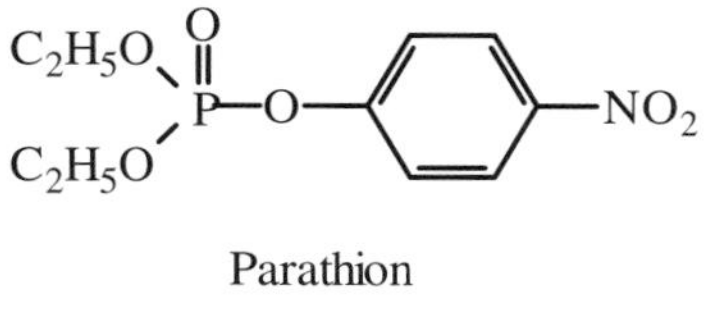

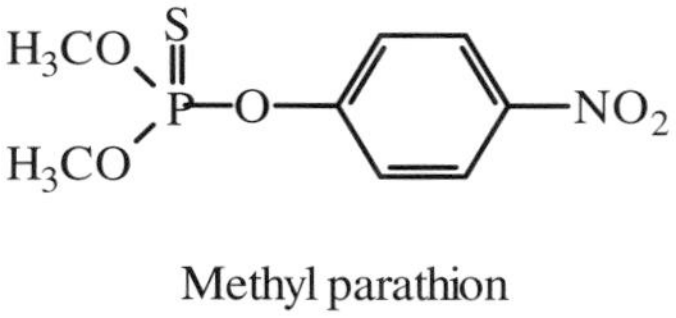

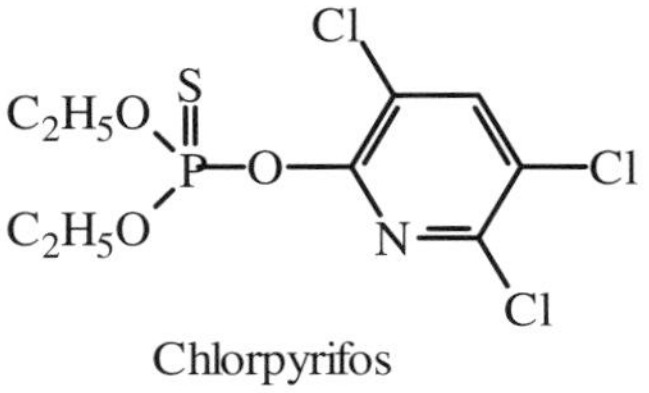

Fig. 13.4 Organophosphorus pesticides.

not generally accumulate in the food chain and can decompose in a matter of days or weeks to non-toxic compounds. Several organophosphorus pesticides are given in Fig. 13.4.

The compounds in Fig. 13.4 are types of phosphorothionates. Parathion is known to be very toxic, is non-specific to insects, and has been responsible for the deaths of several hundred people. The methyl ester was found to be less toxic to animals while still retaining the same pest activity.

Malathion is a type of phosphorodithionate. Its toxicity is very low compared with parathion (~100 times less) and it still proves fatal to insects. This is because mammals possess the enzymes to decompose malathion whereas insects do not. The compound breaks down to non-toxic products according to the reaction shown in Fig. 13.5.

As water pollutants, these compounds are of little interest due to their biodegradability.

13.6.11 Organotin compounds

These compounds have numerous commercial applications including PVC stabilizers, catalysts, fungicides and antifouling paints. The annual global production of organotin compounds is around 40,000 tonnes per year with biocides constituting about 8,000 tonnes of that total.

The most important source of organotin pollution comes from the use of antifouling paints which can be introduced into the aqueous environment due to direct contact of the ship's hull with the water. Several organotin compounds are shown in Fig. 13.6.

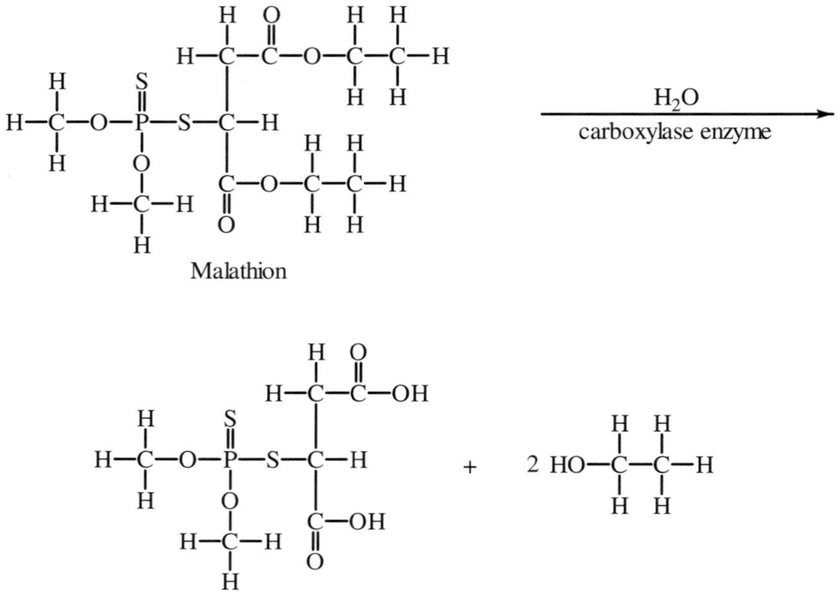

Fig. 13.5 Decomposition of malathion.

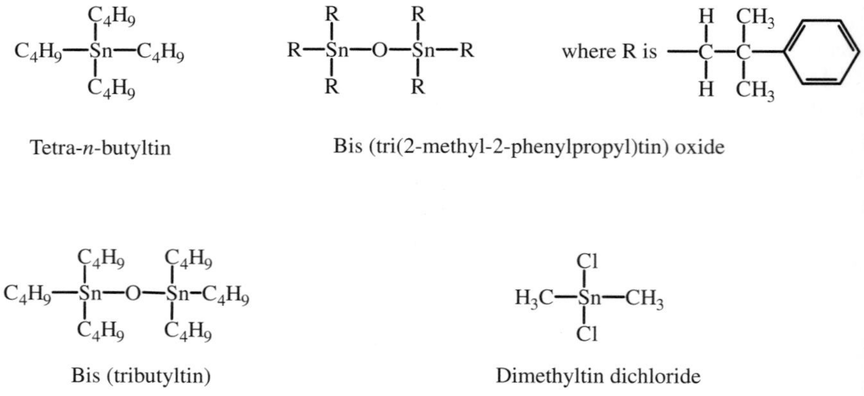

Tetra-*n*-butyltin

Bis (tri(2-methyl-2-phenylpropyl)tin) oxide

Bis (tributyltin)

Dimethyltin dichloride

Fig. 13.6 Some organotin compounds.

Toxicity of these compounds is dependent on the number and size of organic groups attached to the tin atom. The most toxic compounds are those which contain three organic groups, methyl to butyl. Thus, tributyltin (TBT) in antifouling paints is considered to be very toxic.

Oysters and other bivalve molluscs suffer shell thickening and reduced tissue growth as a result of the effects of TBT. Toxic effects are also seen in female gastropods (snails) which develop male sex characteristics. Thus, TBT has both a commercial and ecological impact. Furthermore, because of TBT's ability to bioaccumulate, there is the possibility that these compounds could be transported higher up the food chain.

Organotin compounds tend to be adsorbed on sediments which will reduce their residence time in the water column. TBT in the sediments have half-lives of several years. Their presence in the sediment could result in a possible source of methylation for tin.

The use of triorganotin antifouling products was banned in the UK in 1987 on boats under 25 metres. As a result of this ban, the use of these products on small boats was superseded by the use of products based mainly on copper oxides. In general, these contain booster biocides to improve the efficacy of the formulation in preventing organism growth such as algal slimes. These booster biocides are also of potential concern to the aquatic environment.

13.6.12 Polycyclic aromatic hydrocarbons (PAHs)

These substances are produced by incomplete fuel combustion and pyrolysis of fossil fuels. They consist of benzene rings which are fused together. PAHs can be defined as four or more benzene rings fused together. Figure 13.7 illustrates some of the more carcinogenic PAHs.

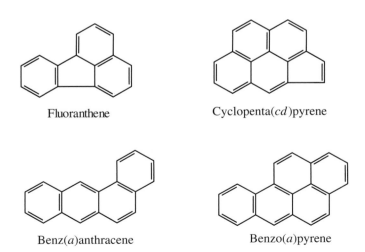

Fluoranthene Cyclopenta(*cd*)pyrene

Benz(*a*)anthracene Benzo(*a*)pyrene

Fig. 13.7 Carcinogenic PAHs.

Of the PAHs in Fig. 13.7, benzo(*a*)pyrene is the most strongly carcinogenic and is possibly linked to the development of skin and lung cancers. Larger PAHs have also shown to be bioaccumulative in some aquatic organisms and have been linked to the production of liver tumours in fish.

These compounds are serious water pollutants. PAHs enter water bodies from oil tanker spills, refineries and offshore oil rigs. They can also leach into waters from wood that has been treated with preservatives, usually creosote. Atmospheric deposition of these compounds and runoff from roadways can also provide inputs of PAHs into waters.

In drinking water, levels are typically in the order of a few $ng\,l^{-1}$, and therefore represent no significant threat to humans from this source.

13.7 Toxic pollutants

Within the context of water pollution, a toxic agent refers to any contaminant which either directly kills aquatic organisms or inhibits their metabolic activity or reproductive behaviour. This definition can be extended to cover substances which occur at low concentrations and appear to have no discernible effect on the aquatic population but, by a process of bioaccumluation, could pose a threat to the consumers of aquatic organisms, i.e. humans.

Both organic and inorganic substances can be bioaccumulated. This means that the pollutants are accumulated in the tissues of aquatic organisms, thus resulting in a contaminated food source for other organisms higher up the food chain.

The toxicity of a particular substance will clearly depend on its concentration or dose. Many substances which are essential micronutrients at low concentrations will be classed as toxic at higher concentrations. Therefore, concentration becomes a key factor in considering toxicity effects.

Persistence is a major factor to be assessed when considering the impact of a pollutant. A persistent pollutant is one that is not degraded into relatively harmless end products, i.e. it will remain toxic in its environment. These substances are considered to be candidates for bioaccumulation. One example of this is the insecticide DDT (dichlorodiphenyltrichloroethane). This compound is only slightly soluble in water, but highly soluble in fats. Aquatic micro-organisms (e.g. algae) will accumulate DDT to an extent that the concentration in the algal cells will be several times greater than in the water. Primary consumers, such as microscopic animals and shellfish, will feed on these algae, thus further accumulating the majority of the DDT. The pollutant is then passed further up the food chain (i.e. to fish, and then from fish to birds). At the end of the food chain, the concentration of DDT in biological tissue may have been magnified over twenty times. This bioaccumulation effect occurs because the DDT is not degraded or excreted by the body during its passage through the food chain.

The range of different individual elements and chemical compounds that occur in waters and wastewaters that are or can be toxic is vast. To categorize them all is not possible, but the majority will fall into one or more of the categories in Sections 13.7.1–13.7.3.

13.7.1 Synthetic organic compounds

Many of these compounds are actually in use because of their toxic properties. The majority of insecticides, pesticides and herbicides come under this category. They were designed to be fairly specific for the target organisms, although in reality they can be toxic to a wider range of organisms.

One important characteristic of these compounds is that they are not found naturally. The reason for their toxicity is that organisms find it very difficult or impossible to degrade these substances as they do not possess the metabolic capability.

Chemically, the significant groups are the organochlorine, organophosphorus and organotin compounds. These are often highly toxic. Their concentrations in UK rivers range from less than $1\,\mathrm{ng\,l^{-1}}$ to about $1\,\mathrm{\mu g\,l^{-1}}$. Chlorinated solvents may be found in certain waters at concentrations up to $100\,\mathrm{\mu g\,l^{-1}}$.

13.7.2 Heavy metals

Some of the heavy metals are of concern mainly because they could exceed safe concentrations in drinking water, have an affect on aquatic organisms and possibly accumulate in biological tissues and subsequently enter the food chain. The heavy metals are normally present in rivers at concentrations of $0.01–0.1\,\mathrm{mg\,l^{-1}}$. Of particular concern is lead in drinking water, although it can be bioaccumluated by many species. Much of the concern about lead in drinking water involves the dissolution of old lead pipes rather than the pollution of surface waters.

Mercury is a well known example of a toxic heavy metal. This metal has a greater potential to undergo methylation to produce methylmercury which is highly toxic and exhibits greater bioaccumulative effects due to its lipophilic nature.

13.7.3 Acidity/alkalinity

Natural waters generally have a pH between 6 and 8. Aquatic organisms will not normally tolerate acid or alkaline conditions outside this range with considerable damage occurring to aquatic life with decreasing pH. This is illustrated in Table 13.8. Detectable changes in the balance of species occur with a pH decrease of up to one unit. With increasing acidity to pH5 or less, the survival of fish species is impaired and aerobic degradation processes are inhibited.

Table 13.8 Effects on aquatic organisms with pH (adapted from Harrison, 1990)

pH	Effect
6.0	Crustaceans, molluscs etc. disappear
	White moss increases
5.8	Salmon, char, trout and roach die
	Sensitive insects, phytoplankton and zooplankton die
5.5	Whitefish and grayling die
5.0	Perch and pike die
4.5	Eels and brook trout die

However, some organisms are especially adapted for survival in acidic environments. Bacteria of the genus *Thiobacillus* can survive a pH as low as 1 and are abundant in acid mine drainage. They can oxidize metal sulphide ores (e.g. iron sulphide) to sulphates and sulphuric acid.

pH also has a secondary effect in that it can influence the toxicity of other substances to aquatic organisms. As an example, ammonia is more toxic to fish in alkaline waters than neutral or slightly acidic water. In contrast, heavy metals tend to be more soluble and more toxic in acidic media.

References and further reading

Alloway, B. J. and Ayres, D. C. (1997) *Chemical Principles of Environmental Pollution*, 2nd edn. Chapman & Hall, London, 395 pp.

Baird, C. (1995) *Environmental Chemistry*, W. H. Freeman, New York, 484 pp.

Harrison, R. M. (1990) *Pollution: Causes, Effects and Control*, 2nd edn, The Royal Society of Chemistry, Cambridge, p. 109.

Harrison, R. M. (ed.) (1992) *Understanding our Environment: An Introduction to Environmental Chemistry and Pollution*, The Royal Society of Chemistry, Cambridge, 318 pp.

Manahan, S. E. (1994) *Environmental Chemistry*, 6th edn, CRC Press, USA, pp. 179–222.

Meakins, N. C., Bubb, J. M. and Lester, J. N. (1994) *Int. J. Environ. Pollut.*, **4**, 27–58.

Stangroom, S. J., Collins, C. D. and Lester, J. N. (1998) *Environmental Technology*, **19**, 643–666.

Chapter 14

Aerobic wastewater treatment processes

14.1 Introduction

The development of wastewater treatment processes was originally prompted by the frequent outbreaks of waterborne diseases in densely populated areas, and to a lesser extent, the need to reduce the odours associated with putrefaction. Although the removal of pathogens remains an important aspect of treatment, the prevention of organic, and latterly nitrogen and phosphorus pollution of surface waters has become a major objective.

Aeration was initially employed as a possible means of reducing odours in wastewater holding tanks. At about the same time it was noted that during soil percolation if the flow of wastewater was stopped intermittently to allow aeration of the soil, an improved treatment efficiency and loading capacity was obtained. Later, it was realized that purification was due to biochemical oxidation and a major step forward was the development of a system which incorporated a continuous inflow of wastewater and continuous aeration. Such a system was the forerunner of the activated sludge process, while the development of soil percolation systems led to the process known as trickling or percolating filtration. These two processes, or modifications thereof, are by far the most frequently used in the UK for the treatment of domestic wastewater at municipal works today. Moreover, in the vast majority of applications they are secondary treatment processes, designed for the oxidation of soluble or dispersed colloidal materials in the wastewater. The raw wastewater, however, passes through several stages before entering the secondary treatment process. In the next section these preliminary and primary stages are described briefly, for the sake of completeness, although they are not biological processes.

14.2 Preliminary and primary treatment

14.2.1 Composition of domestic sewage

It is becoming more usual in developed countries to refer to wastewater as "used water". After all, 99.9% of it is indeed water, the remaining 0.1% (i.e.

1 g l^{-1}) being made up of approximately 70% organic and 30% inorganic solids. Approximately half of the solids content is suspended and half dissolved. The inorganic components include ammonia, chloride, metallic salts and grit, especially where combined sewerage is used. The major organic components which are important from a biological point of view are shown in Table 14.1.

14.2.2 Preliminary treatment

This is designed to remove the larger or more intractable floating and suspended materials, making the wastewater more amenable to treatment by reducing the risk of blockage or damage to pumps and valves used in subsequent stages. The large objects are often removed by bar screens, which consist of parallel bars spaced at 40–80 mm. The trapped material is raked off and typically buried or incinerated. Small stones and grit are removed in grit channels through which the sewage is passed at a constant velocity which is low enough to allow these materials to settle out but at the same time to maintain the less dense organic particles in suspension.

14.2.3 Primary sedimentation

The raw wastewater, having passed through preliminary treatment, enters the first treatment stage to have any real effect on its polluting load. The removal of particles during sedimentation is controlled by a number of factors, including size, density, surface loading and overflow rate. Sewage enters at one end (rectangular or horizontal flow tanks) or the centre of the tank (circular or radial flow tanks) and remains in the tank long enough for the solids to settle to the bottom (typically two to six hours). The settled

Table 14.1 Organic constituents of a typical domestic sewage (modified from Painter and Viney, 1959)

Class	Total organic carbon in solution (%)
Carbohydrates	30
Volatile acids	8*
Non-volatile acids	12
Free amino acids	4
Peptides and proteins	7
Surfactants	11
Others (including fats, greases and micropollutants)	28

* Acetic acid is the most common aliphatic acid.

solids are termed primary sludge and are collected by mechanical scraping into hoppers set in the base of the tank from where they are pumped to be treated further and disposed of.

Primary sedimentation removes approximately 55% of the suspended solids and about 35% of the total BOD. Little or no biological activity occurs in primary sedimentation, the BOD reduction being simply due to the fact that many of the suspended solids are organic and hence biodegradable. The settled sewage (or primary effluent) has a typical suspended solids concentration of 200 mg l^{-1} and a BOD of 250 mg l^{-1}.

14.3 Objectives of wastewater treatment

The major objectives of sewage treatment are the reduction of communicable diseases and the prevention of pollution of surface water. The first standards to be set to fulfil the latter criterion were those given by the Royal Commission on Sewage Disposal (1889–1915) which proposed that there be no more than 30 mg l^{-1} of suspended solids and 20 mg l^{-1} of BOD in the final effluent, assuming that it would be diluted at least 1 : 8 in river water itself having a BOD of less than 2 mg l^{-1}. This standard still forms the basis of many criteria in use today and has become known as the "30 : 20 standard". In certain areas a greater restriction, such as the 10 : 10 standard, may be applied, frequently incorporating a further standard for ammonia of the order of 10 mg l^{-1} or less (the 10 : 10 : 10 standard). It is apparent that the treatment efficiency of secondary biological processes must normally exceed 90% in terms of BOD and suspended solids in order to attain a suitable effluent quality.

14.4 Activated sludge process

14.4.1 Process description

The activated sludge process employs a mixed culture of aerobic organisms (the activated sludge) to oxidize the materials present in wastewater. The term activated sludge can be used in a general sense for any such process in which the biomass exists in the form of freely suspended flocs (agglomerations of numerous cells embedded in an extracellular, gelatinous matrix and typically 0.2–1 mm in diameter). The existence of the microbial cells in the form of flocs permits them to be consolidated by sedimentation under quiescent conditions in a separate secondary sedimentation tank. The supernatant thus obtained is discharged as the effluent from the process and the consolidated cells are recycled to the aeration tank where they are mixed with the incoming wastewater. The mixture of biomass and wastewater is termed the mixed liquor and is maintained in suspension by the turbulence created by mechanical aerators mounted just below the surface of the liquid or by diffusers mounted in the base of the tank which provide aeration in the form of fine bubbles of air or sometimes pure oxygen. The process can be considered to be a single-stage continuous culture system employing biomass feedback. Schematics of the entire process and of cross-sections of the aeration tank are shown in Fig. 14.1.

14.4.2 Operational parameters

(a) Biomass concentration

The concentration of biomass in the aeration tank can be expressed in simple terms as the mixed liquor suspended solids (MLSS). Since MLSS incorporates inorganic solids, a closer approximation to the biotic component may be obtained by using mixed liquor volatile suspended solids (MLVSS). Neither the MLSS nor the MLVSS are necessarily directly related to the number or mass of viable microbial cells present. By definition, these parameters comprise all suspended, or volatile suspended solids too large to pass through a filter and will therefore include both live and dead cells, extracellular components of the floc and suspended volatile matter introduced in the settled sewage.

(b) Residence times, dilution rate and growth rate

Unlike a simple chemostat the terms dilution rate (D) and growth rate (μ) are rarely used from an operational standpoint. Neither are they identical as in a simple chemostat, due to the effects of recycling a large proportion of the biomass, μ being much lower than D. The growth rate of the biomass is normally expressed in the form of the sludge age or mean cell residence time (θ_x) or sometimes as the sludge retention time (SRT). This parameter is equal to the reciprocal of the growth rate:

$$\theta_x = \frac{1}{\mu} = \frac{V_2 X_2}{Q_3 X_3 + Q_4 X_4} \tag{14.1}$$

where X_2, X_3 and X_4 are the biomass concentrations in the aerator, the effluent stream and the biomass waste stream respectively, Q_3 and Q_4 are the flow rates of the effluent and the biomass waste stream and V_2 is the volume of the aerator. The numerical subscripts refer to these quantities at different points within the system as defined in Fig. 14.2. θ_x is normally expressed in terms of days and at steady state can be considered to be the mass of biological solids in the system divided by the mass wasted daily. For the conventional process, θ_x is normally between four and ten days.

The dilution rate is normally expressed in terms of the hydraulic retention time (HRT or θ), which is the reciprocal of D. This represents the average

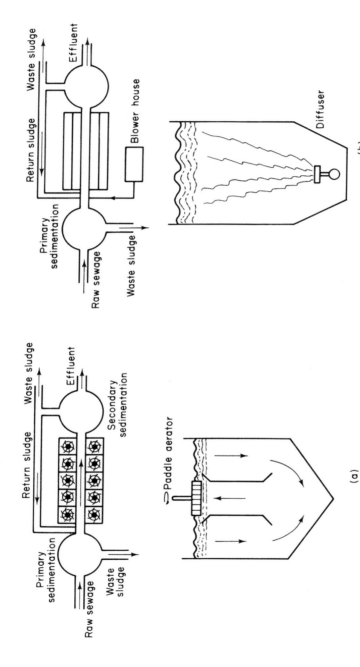

Fig. 14.1 Schematics of the activated sludge process with: (a) mechanical; and (b) diffused air aeration and cross-sections of the aeration tanks.

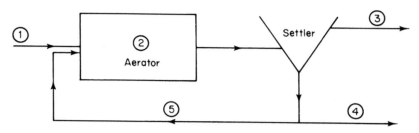

Fig. 14.2 Process diagram of the activated sludge process.

time spent by the influent liquid in the process and usually refers only to the aeration stage:

$$\theta = \frac{1}{D} = \frac{V_2}{Q_1} \qquad (14.2)$$

The term hydraulic loading is sometimes used and this is directly analagous to D, being the flow rate of influent wastewater per unit volume of the aeration tank.

(c) Substrate loading

This parameter, sometimes known as the plant loading, is the rate of substrate addition to the system. The BOD loading can be expressed simply in terms of mass of BOD per unit volume of aeration tank per day ($B_{v.\,BOD}$). Alternatively the loading can be expressed in terms of the mass of substrate (i.e. BOD) per unit mass of biomass per day. This parameter is termed the sludge loading ($B_{x.\,BOD}$) or the "food to micro-organisms" ratio (f/m) and is given by

$$B_x = \frac{Q_1 S_1}{V_2 X_2} \qquad (14.3)$$

Sometimes, the term f/m is used to describe substrate utilization rather than loading. Since in normal operation a small residual substrate concentration, S_3, is not used by the biomass for these purposes equation (14.3) is modified by the substitution of the term $(S_1 - S_3)$ for S_1.

In any event, the terminology associated with operational aspects of the process is by no means universally standardized, and if one comes across a term such as f/m, one should be careful to establish exactly what it means.

14.5 Substrate removal and growth in biomass recycle systems

The important consequence of biomass recycle is that the cell retention time (or sludge age), θ_x, can be made much greater than the hydraulic retention time of the system, θ. In other words, the specific growth rate, μ, can be made much lower than for a single-pass system such as that described earlier. The importance of recycle from the point of view of the objectives of wastewater treatment and the practical implications of these should also be considered. The objectives of treatment are that the final effluent should contain as little residual substrate as possible. According to equation (10.15) as applied to single-pass systems the residual or effluent substrate concentration S_3 will be equal to K_s if $D = \mu_{max}/2$. Values of K_s for settled sewage and activated sludge biomass are often about 100 mg l^{-1}. Therefore, in order to obtain an effluent of reasonable quality from a single-pass system a much lower value for D would have to be selected. Assuming a K_s of 100 mg l^{-1} and a μ_{max} of 0.2 h^{-1} values of D equivalent to 25%, 10% and 1% of μ_{max} would yield S_3 values of 33, 11 and 5 mg l^{-1} respectively. Given that the volume of sewage to be treated is more or less fixed, the only way to obtain improved treatment efficiency in such a system would be to utilize a larger reactor volume for the same flow rate. That reasonable effluent quality can be obtained in systems where $D > \mu_{max}$ if cell recycle is utilized will be shown by derivation from the appropriate mass balances. The same notation will be used as before, but there will be some additional terms to consider. These are the rate at which the secondary sludge is recycled to the reactor. This will be expressed as a ratio of recycle flow to influent flow, R, and the concentration of biomass in the recycle, X_5. This latter can be expressed in one of two ways: either as an independent parameter, or as a concentration factor based on the mixed liquor suspended solids, X_2. The concentration factor, c, is then X_5/X_2.

A mass balance for X in the reactor is given by the rate of change due to inflow, plus the change due to growth, minus the change due to decay and the loss in the outflow. It can be assumed that the inflow component contains only those cells in the recycle line. Hence:

$$V_2 \frac{dX_2}{dt} = Q_1 R c X_2 + V_2 \mu X_2 - V b X_2 - (1 + R) Q_1 X_2 \qquad (14.4)$$

Dividing by V and solving for μ at steady state yields

$$\mu = D(1 + R - Rc) + b \qquad (14.5)$$

A mass balance for the substrate concentration can also be calculated. The mass rate of change is due to the inflow, minus the outflow, minus utilization. In this case the inflow has two components; the first is due to the substrate concentration in the influent wastewater, and the second is the residual substrate concentration in the recycle, which is equal to S_3, assuming that no change occurs between the point at which it leaves the reactor and the point of re-entry of the recycled biomass.

The mass balance is therefore:

$$V_2 \frac{dS}{dt} = Q_1 S_1 + Q_1 R S_3 - \frac{\mu X_2 V_2}{Y} - (1 + R)Q_1 S_3 \qquad (14.6)$$

Solving for X at steady state yields:

$$X_2 = \frac{YD(S_1 - S_3)}{\mu} = \frac{YD(S_1 - S_3)}{D(1 + R - Rc) + b} \qquad (14.7)$$

A value for S_3 can also be derived by substituting the value for μ in equation (14.5) into the Monod equation, hence:

$$S_3 = \frac{K_s[D(1 + R - Rc) + b]}{\mu_{max} - [D(1 + R - Rc) + b]} \qquad (14.8)$$

In effect, then, the growth rate of the biomass is no longer equal to D, but decreases by the combined effects of the recycle flow and the cell concentration ratio. A comparison of X and S_3 for the same system as a function of dilution rate operated with and without recycle is shown in Fig. 14.3, which

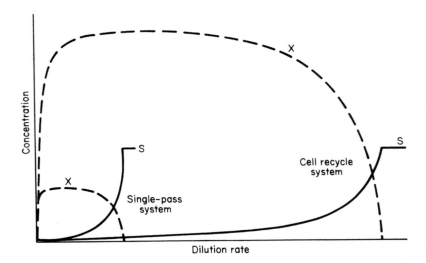

Fig. 14.3 Effect of cell recycle on substrate and biomass concentrations in continuous cultures.

clearly indicates that a much higher biomass concentration and a lower value of S_3 is maintained by the recycle system in the region of $D < \mu_{max}$. Moreover, the system remains stable at dilution rates much greater than μ_{max}.

14.5.1 Relationships between influent and effluent substrate concentration

Equation (14.8) predicts that the effluent substrate concentration is independent of the influent substrate concentration. Practical experience with activated sludge systems, however, reveals that this is not always the case since at increasingly high influent substrate concentrations there is a progressive deterioration in effluent quality. One possible explanation for this is that K_s is not a true constant. Tempest and Neijssel (1976) stated that some organisms possess two distinct "utilization systems" for certain substrates, one with a high value of K_s and one with a low value. At very high substrate concentrations the system with the higher K_s could predominate. The value of S would, according to equation (14.8), increase in response to an increase in K_s.

Another, perhaps simpler reason for the increase in S_3, is that it would be difficult to maintain the value of c constant. At increasingly high values of MLSS a point would be reached where a conventional secondary sedimentation tank would not be able to achieve the degree of consolidation required. From an operational standpoint it would seem more practicable to keep X_5 constant, and to replace c with the term X_5/X_2. With increasing values of X_2 a decrease in the value of X_5/X_2 would in turn increase the specific growth rate leading to an increase in the value of S_3.

The use of X_5/X_2 necessitates some modification of the steady state equations. Calculation of S_3 becomes more complicated due to its dependence on S_1. The relevant equations which involve the solution of a quadratic will not be given here.

14.6 Biological activities in activated sludge

In simple terms, the successful operation of the activated sludge process is dependent on four major characteristics of the biomass. These are (i) the ability of the floc to "adsorb" substrates, (ii) the assimilation and oxidation of organic matter, (iii) the oxidation of nitrogen and (iv) the maintenance of good flocculation to permit efficient sedimentation of the secondary sludge. Although throughout the course of its development and application the activated sludge process has often been treated as a dynamic engineering system it must be remembered that the basic unit of activity is the floc, which in turn is dependent for its existence on the metabolism, growth and physical properties of the microbial cells present. It seems appropriate to begin with an examination of these.

14.6.1 Microbial composition of activated sludge

With enough patience and by using a sufficiently comprehensive range of isolation media and diagnostic tests, the isolation and characterization of the different bacterial species in, say, a water sample should not be an insurmountable problem. Specific problems exist, however, when it is attempted to isolate bacteria from activated sludge. This was recognized quite early by Allen (1944) who thought it essential to adopt some means which would liberate bacteria from the interior of the sludge floc in which large numbers are embedded in a gelatinous matrix. If dilutions are prepared for the purpose of plate counts it is probable that individual colonies would not arise from individual cells, but rather from clumps of cells. Non-flocculated cells in the interstitial liquid would, however, lead to individual colonies, thus giving a distorted picture of the dominant species present in activated sludge flocs. Moreover, since some activated sludge plants operate at dilution rates greater than an attainable μ_{max} for several organisms those organisms arising from the unflocculated state may be completely unrepresentative of the stable floc community since they may be present only by virtue of chance inoculation or because they were present in the influent wastewater and would at some stage be washed out.

Allen found that homogenizing activated sludge by forcing a sample through a small orifice led to between 10- and 100-fold increases in the plate count obtained compared to unhomogenized samples. Despite these problems, several workers have attempted to characterize the bacterial composition of activated sludge using a range of dispersion techniques ranging from simple shaking to ultrasonication. However, any data on numbers and distribution of bacteria should be viewed in the light of these drawbacks.

The medium used to culture a broad range of isolates from activated sludge is also of importance. In view of what was said in Chapter 5, sewage agar might appear to be a suitable medium. Indeed Dias and Bhat (1964) found plate counts from activated sludge samples on sewage agar to be about twice as high as counts on glutamate–urea agar. A popular medium for culturing activated sludge bacteria is the casein–glycerol–yeast extract agar of Pike and Carrington (1972).

The number of viable cells in activated sludge varies according to the method used for counting but would appear to be around 10^{10} per g of suspended solids, or $10^7–10^8$ per ml. On this basis, which it must be remembered is very approximate, viable cells would constitute only about 1% by weight of the sludge which is consistent with the idea that the floc contains a significant proportion of dead cells and extracellular material.

Many different species have been reported to be present in activated sludge. At least for a while, many thought *Zoogloea ramigera* to be the only heterotrophic organism of any consequence in activated sludge since this bacterium produces a copious extracellular slime matrix and in pure culture grows in aggregates of cells embedded in the matrix forming a structure which is superficially similar to an activated sludge floc. Its assumed predominance was probably due to the fact that very few other bacteria form flocs in batch cultures in typical laboratory media. It is now generally accepted that there may be several tens of different species present as integral members of the floc community. A purely subjective attempt to list the genera commonly found is given in Table 14.2. This focuses largely on the heterotrophic members of the sludge community, but it should be remembered that autotrophs, particularly the nitrifying organisms, will frequently be present, although their isolation is rarely reported. Yeasts and algae are also rarely reported and these appear to have only a minor role, if any, in the process. Fungi (e.g. *Geotrichum*) belong to the common population of activated sludge but their role has not been studied in any great detail except for their possible significance in bulking. However, it is probable that they are not as important in this respect as the filamentous bacteria such as *Sphaerotilus*. Of the higher organisms, the protozoa are of most significance due to their contribution to floc structure and their role in reducing effluent turbidity.

In a typical activated sludge plant the mixed liquor will contain approximately 5×10^4 protozoa per ml, which represents about 5% by weight of the MLSS. Hundreds of different species may be present and these are mainly ciliates although significant numbers of amoebae and flagellates may develop under certain conditions. The dominant ciliates include *Opercularia*, *Vorticella*, *Aspidisca*, *Carchesium* and *Chilodonella*. The majority of these are attached to or crawl over the surface of the floc and feed on bacteria. Bacterial cells firmly associated with the floc are probably unavailable to the ciliates which feed mainly on freely suspended cells.

Table 14.2 Bacterial genera found in activated sludge

Major genera	Minor genera
Zoogloea	Aeromonas
Pseudomonas	Aerobacter
Comomonas	Micrococcus
Flavobacterium	Spirillum
Alcaligenes	Acinetobacter
Brevibacterium	Gluconobacter
Bacillus	Cytophaga
Achromobacter	Hyphomicrobium
Corynebacterium	
Sphaerotilus	

Table 14.3 The effect of ciliated protozoa on effluent quality parameters in laboratory-scale activated sludge systems (condensed from Curds, 1982)

Effluent parameter	Ciliates absent	Ciliates present
Total BOD (mg l^{-1})	53–70	7–24
Soluble BOD (mg l^{-1})	30–35	3–9
Suspended solids (mg l^{-1})	86–118	26–34
Viable count (10^6 ml^{-1})	160	1–9

Thus the presence of protozoa will select to some extent for floc-forming species and by scavenging free-swimming cells will reduce effluent turbidity. This important function of the protozoa has been studied extensively by Curds and his co-workers, who examined effluent quality parameters from laboratory-scale activated sludge simulations which had been established both in the presence and absence of ciliated protozoa, with the results shown in Table 14.3.

It is apparent that the presence of protozoa brings about a reduction in both effluent SS and soluble BOD. Although it is possible to culture some protozoa in a bacteria-free organic medium it seems unlikely that this could be responsible for the large reduction in BOD. Some workers have considered that the grazing activities of the ciliates could stimulate the floc in some way, but this remains unclear.

It also seems likely that protozoa can contribute directly to floc integrity. Some earlier workers suggested that ciliates were the primary promoters of floc formation but it is now apparent that floc formation can occur in their absence. Many of the species present can, however, secrete a mucous protein–polysaccharide material which may function in the same way as the extracellular polymers produced by the bacteria. Predation by protozoa is believed to be responsible for the often quite high removal of pathogens by the activated sludge processes. Removal of *E. coli* is frequently over 90% although it can be much less than this. Activated sludge also seems effective in removing viruses, with coxsackie and polio virus removals consistently in excess of 90%.

14.6.2 Flocculation

A floc is a macroscopically visible entity consisting of several million bacterial cells together with some protozoa and fungi. The factors which appear to affect flocculation are growth rate, composition of the mixed population, substrate concentration and nutrient balance. A typical floc is shown in Fig. 14.4. A low specific growth rate appears to favour flocculation. Indeed, such a low growth rate can occur in activated sludge systems *only because*

Fig. 14.4 A typical floc.

flocculation occurs, hence permitting settlement, and recycling so that $\mu < D$. In high rate systems where θ is about 0.5 d or less flocculation does not occur to any great extent and high effluent suspended solids concentrations are found.

It is sometimes stated that flocculation is a function of "endogenous respiration". However, since it seems likely that endogenous respiration occurs even if substrates are not depleted and net growth has ceased, it is probably more true to say that it is a function of low growth rate and low substrate concentration.

When carbon is in excess and other nutrients are depleted, some bacteria will accumulate storage products, in the form of polymeric molecules. One such product is poly-β-hydroxybutyrate (PHB) which has been implicated in the flocculation of some bacteria. However, some species which do not accumulate PHB will still flocculate and it is likely that a wide range of macromolecular storage products, secreted externally in the form of a capsule or slime layer, may be implicated in the same way. In activated sludge these molecules have been found to include carbohydrates and proteins as major constituents together with some lipids and nucleic acids. Also, because in large flocs organisms starve and die and because viability is relatively low in comparison to total cell numbers, much of the extracellular matrix may consist of products of cell lysis.

The precise manner in which the extracellular polymeric matrix contributes to flocculation is uncertain. It appears to act partially simply as a fairly viscous material which can physically trap or surround cells which approach it, perhaps including those organisms which are unable to flocculate themselves. In certain conditions, intertwined fibrils of cellulose-like material also appear to be responsible for the physical stability of the floc.

Extracellular polymers have been found to possess fairly large numbers of charged functional groups (e.g. carboxyl groups) which have been strongly implicated in the binding of transition metal ions thus contributing to the reduction in heavy metal pollutants which occurs in the activated sludge process. The precursors of the cell-floc appear, therefore, to behave like negatively charged colloids. If charge neutralization occurs, then the colloids will undergo some form of precipitation to form an aggregate. Such charge neutralization may be brought about by calcium and magnesium ions. A schematic of a possible structure for sludge flocs, taking these observations into account, is shown in Fig. 14.5.

In normal operating practice, the factors which affect flocculation and the maintenance of the biomass in good condition are given relatively little attention compared to that which is given when flocculation fails or is unsatisfactory for some reason.

14.6.3 Bulking and related phenomena

The failure of an activated sludge to flocculate and settle properly may be due to a number of factors, and be manifested in many different ways. There has sometimes been a tendency to place all of the phenomena relating to sludge problems in the category of bulking. However, this term is more correctly applied to a specific phenomenon; a number of different terms are used to describe distinct phenomena which result in poorly settling sludge, deflocculation and effluent turbidity. Of all common operational problems

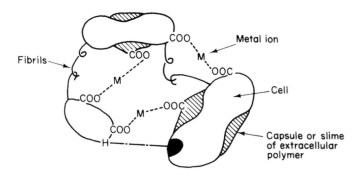

Fig. 14.5 Schematic of possible floc structure.

experienced with activated sludge, poor sludge settleability is probably the most widespread, occurring at some time or other in up to half of all the plants in the UK.

(a) Bulking

This is a condition commonly associated with an overabundance of filamentous organisms (e.g. *Sphaerotilus*) within the floc. Although aggregation of the cells still occurs, the presence of the filaments causes the floc to be bulky and loosely compacted so that its settlement is hindered and large quantities may be lost in the final effluent. This loss of sludge may cause a reduction in the MLSS and the plant may tend towards functioning more as a single-pass reactor with the result that μ will increase. In severe cases the higher effective growth rate could cause a loss of nitrification. Although there exists no precise definition of bulking, an approximate yardstick for a bulking sludge is one having an SVI of greater than 200 ml g^{-1}. However, some sludges with SVIs greater than this value may not appear to be excessively filamentous. It has been postulated that this so-called non-filamentous bulking is caused by excessive hydration of the floc matrix, the bound water reducing the mean density of the floc and therefore retarding settlement.

The types of filamentous organisms found in activated sludge include filamentous bacteria (e.g. *Sphaerotilus*, *Thiothrix* and *Beggiatoa*), actinomycetes, cyanobacteria and fungi. It is doubtful whether the actinomycetes cause true bulking since SVIs of greater than 200 ml g^{-1} are not generally associated with their presence. They can, however, form a kind of floating scum which affects effluent quality. Both the cyanobacteria and the fungi are rarely implicated in bulking, although a few cases are documented.

Since filamentous organisms appear to be present even in non-bulking sludges, some additional factors must be associated with bulking. It has been found difficult to establish a definite relationship between the total number of filaments associated with a floc and the SVI or SSVI. However, the SVI of a sludge does appear to be proportional to the total filament length per unit volume of sludge when observed under the microscope.

It has been found fairly consistently that plug-flow plants produce a better settling sludge than those operated under conditions of complete mixing, although at long hydraulic residence times the effect is slight. A plug-flow plant exposes the biomass to a high concentration of substrate at the inlet to the aeration tank which is gradually utilized as the mixed liquor flows towards the end of the tank. Thus the population behaves more as if in a batch culture in plug-flow plants than in completely mixed plants where it is continuously exposed to a uniformly low substrate concentration. It has been suggested that competition for the substrate, and hence the extent of growth, is dominated by the floc-forming bacteria in the initially high substrate concentrations in a plug-flow plant but that the reverse is true in the case of low substrate concentrations. This situation can be considered analogous to that where the floc-formers have a high μ_{max} but a low affinity for the substrate (high K_s) and the filamentous organisms have a low μ_{max} with a high affinity for the substrate (low K_s).

The response of the two types of organisms to the dissolved oxygen concentration appears to be similar. However, *Sphaerotilus* is a strict aerobe, and is intolerant of anoxic conditions for any length of time. This may explain why the inclusion of an anoxic zone (an unaerated chamber or section of the aeration tank situated at the inlet) also appears to promote good settleability. However, introduction of a separate anoxic tank also introduces a degree of plug-flow characteristics. The substrate profile with increasing numbers of such tanks is shown in Fig. 14.6.

(b) Rising sludge

The inclusion of an anoxic zone will also effect partial or complete denitrification. This process, whereby nitrate is reduced to nitrogen gas or other gaseous nitrogen compounds, is mediated by certain facultative heterotrophs which utilize the nitrate as a terminal electron acceptor in place of oxygen where the latter is absent. Evolution of nitrogen gas bubbles in the final sedimentation tank increases the buoyancy of the floc, causing it to rise, sometimes leading to high concentrations of effluent suspended solids.

(c) Pin-point floc

This condition is generally associated with plants operating at a long sludge age. It has often been stated that pin-point flocs, which are very small poorly settling particles, are the biologically recalcitrant residue which remains after the normal floc community has undergone extensive endogenous

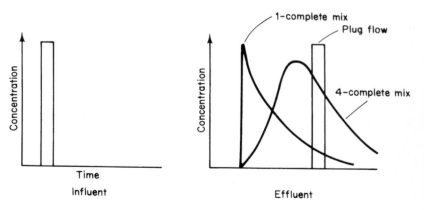

Fig. 14.6 Substrate profiles in plug-flow tanks.

respiration under conditions of low substrate availability. In its strictest sense, endogenous respiration means the utilization by cells of their own mass. If measured on a mass basis it will be difficult to distinguish between this and the utilization of storage products, the utilization of extracellular products of other species present or the relatively slow oxidation of true substrates originally very rapidly absorbed by the floc. Any or all of these phenomena could contribute to the disappearance of the extracellular binding materials of the floc, leading to pin-point flocs and high effluent turbidity.

14.6.4 Heterotrophic activity in activated sludge

Activated sludge is an extremely complex material, although it is frequently treated as a homogenous system. If a particular sludge is said to be capable of metabolizing glucose, for example, it should be remembered that probably only a fraction of the viable cells present are actively breaking down the particular substrate and converting it to biomass. Some non-viable cells will also probably be degrading it without using the breakdown products for growth, but the vast majority of the material in the floc, much of it consisting of dead and abiotic material, will be totally inactive with respect to the oxidation of the substrate. When one considers that there are probably as many as a hundred distinct substrates and a similar number of different microbial species present it can be seen that however subtle the approach to studying the microbiological and biochemical aspects of activated sludge, it is likely to be inadequate to some degree. This is probably the reason why much of the data on activated sludge are inconsistent and even contradictory. Some general principles, however, have been established.

Domestic sewage has a BOD : N : P ratio of approximately $100 : 5 : 1$ and this typically provides the biomass with sufficient nitrogen and phosphorus to remove most of the BOD. In general, fungi appear to have a lower nitrogen requirement than bacteria and if the BOD : N ratio increases to $100 : 4$ or greater their increased predominance may cause bulking. A BOD : P ratio of $100 : 0.5$ or more may have the same effect. Normally, supplementation of the nitrogen requirements of the process by nitrogen fixation does not occur to any great extent, but in very nitrogen-deficient wastes nitrogen fixing *Azotobacter* spp. may become predominant.

Nutritional factors in general receive relatively little attention when a plant is operating normally, tending to become important only if problems arise. However, the limited available data indicate that heterotrophs responsible for carbonaceous BOD removal have μ_{max} values in the region of $0.7\,h^{-1}$, typically fifty times higher than the nitrifiers.

14.6.5 Nitrification

Ammonia is an important pollutant for two major reasons. Firstly, it is toxic to aquatic life in concentrations in the parts per million range and secondly it exerts a high oxygen demand. Although the concentration of ammonia in raw sewage is only in the region of 30 mg l^{-1} it can contribute up to 40% of the total BOD exerted. Moreover, much of the organic nitrogen present is in a form which can readily be converted to ammonia by heterotrophic bacteria. Its removal during wastewater treatment processes is therefore important. Effluent standards imposed are frequently <10 mg l^{-1} ammonia (as N) and sometimes <5 mg l^{-1}.

Some of the ammonia present will be removed and assimilated by heterotrophs to satisfy their nitrogen requirement, but in domestic sewage, which is typically a carbon-limited medium, this is not a significant route. A major biological transformation is that due to certain aerobic autotrophs which use ammonia as a source of energy, oxidizing it to nitrate via nitrite in a process termed nitrification. Organisms of the genus *Nitrosomonas* (and possibly some other related genera) perform the first stage of this process:

$$2NH_4^+ + 3O_2 \rightleftharpoons 2NO_2^- + 2H_2O + 4H^+$$

In the second stage the nitrate is further oxidized by a different organism, *Nitrobacter*:

$$2NO_2^- + O_2 \rightleftharpoons 2NO_3^-$$

An important point to note is that the process generates hydrogen ions which can cause significant reductions in pH. If the pH becomes sufficiently low, it may completely inhibit nitrification. Indirect effects on treatment performance could include induced changes in the balance of the heterotrophic population resulting in bulking, for example. Although nitrification effectively neutralizes the oxygen demand exerted by ammonia, it does not constitute effective nitrogen removal since the ammonia is converted stoichiometrically to nitrate.

Nitrifying bacteria are slow growing in comparison to the heterotrophs in activated sludge, and are inhibited by a wide range of toxic organics and inorganics as well as by low temperatures. For this reason it is important to maintain a sufficiently high sludge age to avoid washout of these organisms. Kinetic data indicate values of μ_{max} for both *Nitrosomonas* and *Nitrobacter* in the region of 0.5 d^{-1} at 20°C falling to perhaps 0.2 d^{-1} at 15°C. K_s values for ammonia appear to be about 1 mg l^{-1} but perhaps as high as 5 mg l^{-1}. Based on an influent ammonia concentration of 35 mg l^{-1} and using optimum estimates for μ_{max} and K_s a minimum sludge age required to prevent washout would be just over 2 d. However, nitrification would be desirable in many instances at much lower influent concentrations where strict effluent quality legislation is in force. With an influent ammonia concentration of only 10 mg l^{-1}, a K_s value of 5 mg l^{-1} and a μ_{max} of 0.2 d^{-1} a sludge age of nearly 8 d would be necessary to prevent washout of the nitrifiers.

14.6.6 Denitrification

Although nitrification reduces effluent BOD and prevents the toxic effects of ammonia in the receiving stream, nitrite and nitrate in the effluent are also undesirable for a number of reasons. The former, if present in drinking water, can react with amines in the diet to form nitrosamines, which are highly carcinogenic, and the latter can cause methaemoglobinaemia in young children (Blue Baby Disease). Hence, complete removal of soluble nitrogen compounds may be desirable.

Denitrification is the process whereby facultatively aerobic heterotrophs, notably *Pseudomonas* spp., can use nitrate as a terminal electron acceptor in the absence of dissolved oxygen. The adventitious denitrification which occurs in secondary sedimentation tanks is inconsistent, inefficient and replaces the nitrate problem with that of a turbid effluent. The major problem in effecting complete denitrification of a well nitrified effluent with low BOD is the scarcity of respiratory substrates available to donate electrons to nitrate. Thus, the use of a second, post-aeration anoxic state to effect denitrification will almost invariably require the addition of an exogenous carbon source as a substrate unless complex recycle systems are used. Methanol is probably the most economically efficient substrate, but it has the advantage of contributing more to the overall sludge production.

Various modifications to the activated sludge process designed to effect biological denitrification without the need for additional substrate input have been considered. These range from the inclusion of an anoxic zone between the aeration and final sedimentation stages, to a four-stage system incorporating alternate anaerobic and aerobic reactors with multiple cell and effluent recycle systems. Such systems are discussed in greater detail by Barnard (1978). The most successful systems would appear to incorporate a high rate of effluent recycle to the beginning of the process where the incoming wastewater provides respiratory substrates for denitrification in an anoxic zone followed by an aerobic zone for nitrification of the ammonia in the incoming wastewater. It appears that the nitrifiers are not inhibited to any significant extent by exposure to anoxic conditions for a few hours.

14.7 Operational modifications

The two major operational parameters which can be controlled to provide the most appropriate conditions are the sludge loading and the sludge age. These parameters can be varied over approximately 100-fold ranges. There are, within these ranges, three major operational modes, and typical values

Table 14.4 A comparison of typical operational parameters for conventional, high rate and extended aeration activated sludge

	High rate	Conventional	Extended aeration
Sludge loading	1–5	0.2–0.6	0.03–0.1
Hydraulic loading	1.5–3.5	0.5–1.5	0.25
MLSS (mg l^{-1})	1,500	2,500	5,000
Sludge production	1.0	0.5	0.2–0.3
Sludge age (d)	<0.5	4–10	24
Hydraulic retention time (h)	>2	6–10	
O$_2$ requirement		1.0	1.25
BOD removal (%)	60–70	90–95	95

for the associated operational parameters and performance characteristics are given in Table 14.4.

The conventional process is normally applied to readily degradable organic wastes and to domestic sewage. The high rate process provides rapid partial treatment, with sludge loadings typically ten times higher than the conventional mode. The process is frequently used for high strength wastes and produces an effluent which can be fully treated by a secondary, conventional process. Although the higher loadings are maintained at the expense of a high effluent BOD, the rate of removal per unit tank volume is greater than in the conventional process so that the overall size of plant necessary for complete treatment in a conventional plant may be significantly greater than that of a two-stage high rate–conventional plant linked in series. At very high rates with sludge ages of less than 0.5 d rapidly growing organisms are encouraged and flocculation is inhibited, so as well as containing a high BOD the effluent also contains high suspended solids.

The term extended aeration is applied to plants having sludge ages of 20–30, even 40 d and operating under conditions of low loading compared to the conventional process. However, the volume of aeration tank required is only fractionally higher since higher MLSS concentrations can be maintained. Its major advantage is that the net sludge production is much lower than in the conventional process. This is because the high sludge age is such that b becomes a significant parameter in comparison with μ. What sludge is produced is relatively stable and inert and easier to handle, treat and dispose of. It was originally hypothesized that such a system may be operated such that $\mu = b$ and no net growth of biomass would occur at steady state. Indeed there have been some reports that extended aeration plants have been operated for long periods without the necessity to waste sludge. On the other hand, it appears unlikely that the process could generate an effluent totally free of suspended solids, so that the equality of μ and b would not be a necessary condition for stable operation without sludge wastage.

Some extended aeration plants (oxidation ditches) are operated without primary sedimentation. In this case some solids accumulation, representing the inert fraction of the raw sewage, would be expected although it may be quite low since these plants are frequently used in the treatment of purely domestic sewage from small rural communities. In any event, sludge production is very low and a higher degree of mineralization is attained which is reflected in the oxygen requirement which is typically about 25% higher than in the conventional process.

14.8 Design modifications

A fundamental design parameter is the nature of mixing in the aeration tank. In theory, plants can be designed to operate in a complete mixing mode, where the influent wastewater is distributed throughout the aeration tank and there is a resultant uniformly low substrate concentration throughout, or in a plug-flow mode where the wastewater enters at one end of the tank and the mixed liquor flows to the outlet end in a "plug" with no longitudinal mixing. In this mode the substrate concentration is highest at the inlet end of the tank and lowest at the outlet end. Initially, therefore, the oxygen demand is higher than the uniform demand encountered in the equivalent completely mixed process, while at the end of the tank oxygen consumption by the biomass is much lower. This situation led to the development of the tapered aeration system, where greater aeration capacity is provided at the inlet end of the tank and the degree of aeration is matched to the BOD remaining at intervals along the tank. At the outlet end of a plug-flow tank the rate of BOD removal per unit volume is quite low, and therefore inefficient use is being made of this tank volume. A design known as stepped loading, which is essentially intermediate between a plug-flow and a completely mixed regime, introduces the influent wastewater at several points along the length of the tank so that a more uniform BOD is created throughout.

Effective use of the aeration tank volume is the principle behind the contact stabilization process. Based on the observation that much of the initial BOD is removed from solution or colloidal suspension in a very short time following contact with the biomass, the recycled sludge is accordingly aerated for a very short time in contact with the influent wastewater, passed into a sedimentation tank, and the recycled consolidated sludge aerated for a longer period in the stabilization tank. It is presumed that the initial BOD removal in the contact tank is simply due to the adsorptive properties of the sludge and that biochemical oxidation occurs only in the stabilization tank. However, it has been noted that some oxygen uptake does occur in the contact tank, so the initial adsorption stage might be more properly described as "oxidative assimilation". This concept is discussed at greater length by Gaudy and Gaudy (1980). Since most of the oxygen demand is exerted in the stabilization tank where the sludge is present in concentrated form, the total aeration tank capacity is much lower than for the equivalent conventional process.

A number of processes are designed to enhance the rate of oxygen transfer into solution so that higher BOD loadings can be treated. The enhancement is achieved by increasing the partial pressure of oxygen in the gas phase either by using pure oxygen or by increasing the total pressure of the system. The best-known process involving the latter method is the ICI Deep Shaft which utilizes a deep liquid column to increase the hydrostatic pressure. The column is divided into downflow and upflow sections. The wastewater enters at the top of the downflow section, passes to the bottom of the column and returns to the top via the upflow section. Air is injected into the downflow section of the system which cycles at a velocity sufficiently high to prevent the bubbles rising. As the depth increases the air bubbles dissolve due to the increase in pressure. The bacterial population in the system appears to be physically unaffected by the high pressures obtained.

14.9 Percolating filters

14.9.1 Process description

A percolating filter typically consists of a circular or rectangular bed of broken rock, gravel, clinker or slag, with a particle size of 5–10 cm. The filling is termed the medium ("Medium" is actually almost always referred to as "media" (the plural form)) and is usually 1.5–2.0 m in depth and supported on a layer of larger stones over underdrains. The bed is surmounted by a distribution system which trickles settled sewage over the surface; on circular filters this usually consists of four radial sparge pipes whose movement is maintained by the impetus of the settled sewage flow, although on larger circular and rectangular beds the distributor may be powered. A cross-section of a typical circular filter is shown in Fig. 14.7.

The biomass develops as a thin film coating the surfaces of the medium and oxidizes the settled sewage as it trickles slowly through the interstices whose volume comprises typically 50% of the total bed volume. As in the activated sludge process there is a net production of biomass which is manifested in the development of the film to such a degree that its tenacity is reduced and portions are sloughed off by the shear forces created by the flow of wastewater. The treated wastewater containing the sloughed off biomass leaves the bed via the underdrain at the base and passes into a sedimentation tank, known as the humus tank, the consolidated sludge being termed the humus sludge.

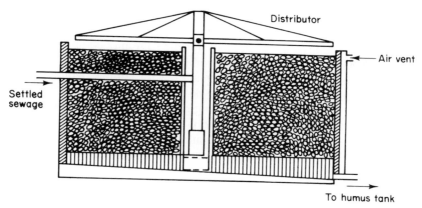

Fig. 14.7 Cross-section of a typical percolating filter.

14.9.2 Operational parameters

Unlike the activated sludge process it is very difficult to measure or control the concentration of biomass within a percolating filter. Hence there is no equivalent to MLSS in normal use. Moreover it is difficult to estimate μ, θ_x or θ with as much ease as in the activated sludge process. Because of stratification in the biomass component of the filter and temporal (seasonal) variations in humus solids the meaning of μ, θ_x and related terms has much less significance than in suspended growth systems. Hence the application of operational parameters is normally limited to BOD loadings per unit volume of filter bed. Sometimes the hydraulic loading (flow rate per unit volume) is specified.

In general, filters can be operated as low or as high rate systems, the former being characterized by a BOD loading of $0.5\ \text{kg m}^{-3}\ \text{d}^{-1}$. Often the term "rate" is more commonly expressed in terms of hydraulic loading. In the UK a high rate process has a hydraulic loading of $>3\ \text{m}^3\ \text{m}^{-3}\ \text{d}^{-1}$, whereas in the USA the term "standard rate" is used for hydraulic loadings of $<2\ \text{m}^3\ \text{m}^{-3}\ \text{d}^{-1}$.

The Royal Commission "20 : 30" standard is normally attainable in conventional filters only by operating at a low rate. Indeed, in a conventional filter without the added sophistication of effluent recycle BOD loadings as low as $0.1\ \text{kg m}^{-3}\ \text{d}^{-1}$ may be necessary to obtain such an effluent quality. High rate treatment is sometimes employed in "roughing" filters, which are designed to effect only partial treatment prior to treatment of the weaker effluent in a conventional process.

Although the biomass in a filter is difficult to determine, it can be estimated on the basis of a few assumptions. Typical film thickness can vary between <0.1 mm to 0.3 mm for a compact, relatively nonfilamentous biomass. For a conventional rock packing with a diameter of 50 mm the specific surface area for colonization will be about $100 \, \text{m}^2$ per m^3 of bed volume. Assuming the biomass to have broadly similar properties to activated sludge, perhaps as much as 3% of the film may comprise dry solids. Thus, the upper limit for a typical biomass concentration in a conventional filter, expressed in terms of mass per unit reactor volume, would be approximately $1000 \, \text{mg} \, \text{l}^{-1}$. Thus a BOD loading of $0.5 \, \text{kg} \, \text{m}^{-3} \, \text{d}^{-1}$ would be roughly equivalent to a sludge loading of $0.5 \, \text{kg} \, \text{kg}^{-1} \, \text{d}^{-1}$ for optimum film development. Thus, low rate filtration is roughly equivalent to conventional activated sludge in terms of sludge loading.

14.9.3 Substrate removal and growth

Percolating filters operate under partial plug-flow characteristics. Hence, the substrate concentration decreases as the wastewater descends through the bed. Unlike the activated sludge process, the biomass does not travel with it, but remains spatially fixed. The rate of substrate removal is therefore not so much a function of time as of depth. A substrate mass balance can therefore be made for a differential element (i.e. a horizontal "slice") of the bed. Hence:

$$\text{influent} - \text{substrate used} = \text{effluent} + \text{accumulation}$$

The substrate concentration in the influent is $(S + \mathrm{d}S)$ and in the effluent, S (see Fig. 14.8). Hence for a steady state condition, the accumulation term being zero, we get:

$$Q_1(S + \mathrm{d}S) - \frac{\mu M_x}{Y} = Q_1 S \tag{14.9}$$

Here M_x is the mass of active biomass in the differential element and is given by

$$M_x = (Aafd_m)\mathrm{d}H \tag{14.10}$$

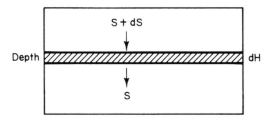

Fig. 14.8 A differential element of a percolating filter.

where A is the cross-sectional area of the bed, a the specific surface area of the medium, f the depth of microbial film, d_m the microbial density within the film and H the filter depth. Replacing μ by the Monod relationship yields:

$$\frac{dS}{dH} = \frac{\mu_{max}}{Q_1 Y} aAfd_m\left(\frac{S}{K_s + S}\right) \tag{4.11}$$

For a given filter a number of these terms will be effectively constant at steady state. The biomass potential constant, P_B, can be expressed as

$$P_B = \frac{\mu_{max}d_m f}{Y} \tag{4.12}$$

so that

$$\frac{dS}{dH} = -\frac{P_B aA}{Q_1}\left(\frac{S}{K_s + S}\right) \tag{14.13}$$

From an operational point of view, assuming the biological parameters (P_B) and the influent substrate concentration to be constant, the rate of substrate utilization can be increased by reducing the flow, increasing the specific surface area of the medium, or increasing the cross-sectional area of the filter. Solution of equation (14.13) permits the definition of a particular bed height required to achieve a given effluent quality. The generalized relationship between effluent substrate concentration and bed height is shown graphically in Fig. 14.9.

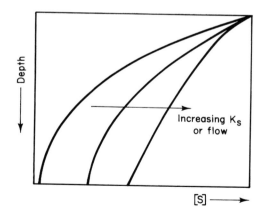

Fig. 14.9 Effect of bed height on effluent substrate concentration in a percolating filter.

A biomass balance for a trickling filter is difficult if not impossible to construct because the rate of biomass removal is not amenable to operational control, being a purely adventitious process dependent on a number of complex factors.

14.9.4 Biological activities in percolating filters

The ability of the biomass to adhere to the surface of the packing medium is in a way analogous to its ability to flocculate in the activated sludge process and is equally important to the successful operation of the process. It is probable that the mechanisms of film cohesion and the adhesion of the film itself to the medium are similar to the mechanisms of flocculation, not least because many of the bacterial species present are those which also predominate in activated sludge. These include *Flavobacterium, Pseudomonas, Achromobacter*, and some filamentous organisms, notably *Sphaerotilus*.

In the lower levels of the filter, nitrifying organisms will be present. It would appear that *Nitrobacter* cannot successfully compete with the heterotrophs in the upper part of the filter. In heavily loaded filters and especially at low temperatures the heterotrophic zone may extend to the base of the filter, causing extensive inhibition of the nitrifiers.

The removal of pathogens during percolating filtration is quite variable. Bacterial pathogen removal similar to that in activated sludge may be expected, but virus removal is generally inferior.

Although filters can be modelled under steady state conditions, the rate of biomass growth and its loss by sloughing cannot be considered a balanced process. Factors beyond the control of the operator will come into play. For significant periods the biomass concentrations may not be in steady state. Although film growth is controlled by substrate utilization its rate of accumulation is limited by two major factors. These are the flushing and scouring effects of the percolating wastewater and removal by grazing. A percolating filter is altogether a more complex habitat than activated sludge, consisting as it does of the solid support, biological film, wetted surfaces due to the flow of wastewater, air spaces, and relatively dry regions. A filter will support a variety of air-breathing macrofauna, including insect larvae and nematode worms. These organisms feed directly on the film, reducing its mass and reducing overall secondary sludge production by their respiratory activity. Their activity is highly seasonal, so that the amount of film present in a filter is highest in the winter months and lowest during the summer. The seasonal variation in filter biomass is shown in Fig. 14.10.

The greater complexity of the filter habitat is also more suited to the growth of fungi. There may be a greater proportion of fungal growth in a healthy filter than in activated sludge, but, like the latter, filter operation can be compromised by excessive fungal growth. In conditions conducive to fungal development the hyphae may proliferate to such an extent that parts

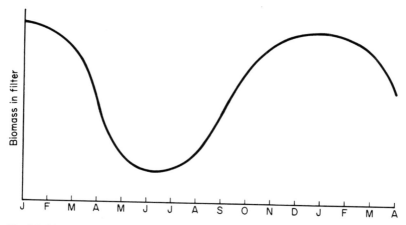

Fig. 14.10 Seasonal variation in filter biomass.

of the filter become blocked and a condition known as ponding develops. Pools of wastewater lie on the surface if the growth is sufficiently excessive to impede its movement through the bed. Under such conditions anaerobiosis can occur, leading to a decrease in effluent quality.

The vertical stratification and immobility of the bed permits the development of algae on the surface of the bed. These do not contribute significantly to waste stabilization and the oxygen requirement of the filter; however, in the presence of excessive algal growths the surface of the filter can become blocked, causing a decrease in process efficiency and odour problems. The other major group of organisms normally present in filters are the protozoa. These are predominantly ciliates and include *Opercularia*, *Vorticella* and *Epistylis*. Their role is similar to that performed in the activated sludge process. Thus, bulk film accumulation in the filter is controlled to a large extent by grazing macroinvertebrates while the protozoa feed on detached and free-swimming bacteria.

14.9.5 Design modifications

At the basal level, filter operation can be termed high rate or low rate, terms which are normally defined with respect to hydraulic loadings, but which in many cases will also be related to BOD loadings where domestic sewage is being treated. There are, however, a number of modifications which permit independent control of hydraulic and BOD loadings through the use of effluent recycle.

The minimum and maximum limits of hydraulic loadings are generally constrained respectively by the liquid flow rate necessary to keep the biofilm wetted throughout the bed and the flow rate which would exceed the rate of percolation resulting in flooding. With respect to the former, recycle may be employed during conditions of low flow in order to maintain wetting of the film. It may also be used to reduce effectively the BOD loading on the filter, for example in order to maintain nitrification. It should be pointed out that the recycle normally consists of settled effluent from the humus tank, rather than the liquid flow from the base of the filter, thus avoiding recycle of the humus solids which have been sloughed off. Thus, the recycle of active biomass and the reseeding of the mixed liquor which occur in the activated sludge process have no operational parallel in the percolating filter. While recycling of detached biomass may be hypothetically advantageous in some instances it is likely that it would only result in clogging of the filter bed.

Certain high rate filters are termed "roughing filters". Whereas low rate filters are designed to achieve an effluent quality of "20 : 30" standard, roughing filters are used for partial reduction of the BOD loading prior to further treatment of the effluent.

Double filtration makes use of a roughing filter connected in series to a polishing filter which brings the final effluent to a high quality. The consistently high loading on the first filter tends to generate a high degree of film accumulation. In order to prevent clogging, the roughing filter is packed with a coarser medium than the polishing filter.

In alternating double filtration the same principle is applied but in this case the two filters will contain the same type of medium. Clogging of the roughing filter is prevented by periodically reversing the flow of wastewater through the system, so that the roughing filter becomes the polishing filter and vice versa. Dosing of the previously heavily loaded filter with a more dilute effluent may also encourage endogenous respiration, thus reducing sludge production.

Some interest has been shown in a filtration process which is analogous to the extended aeration process. The raw wastewater does not undergo primary sedimentation but is applied directly to the filter. The effluent from the filter is recirculated without settling and so there is a considerable solids flux throughout the system which is designed to enhance oxidation of the solids but which inevitably increases the risk of clogging. In order to prevent this a high voidage medium may be used. Some plastic media can provide a void space within the bed of up to 95%, compared with about 50% for conventional media. This property may also be advantageous for high rate systems where biomass production is expected to be high. Some synthetic media can have a specific surface area greater than twice that of conventional media.

In contrast, where very low biomass production is expected a finer grade of medium may be used. An example of such an application is in nitrifying filters. Unlike a nitrifying activated sludge plant these filters receive effluent from an aerobic process which has already effected the removal of carbonaceous BOD. Hence, the biomass consists almost exclusively of nitrifiers. Film growth is minimal since in the case of *Nitrosomonas* only about 50 mg of biomass is produced for each g of nitrogen oxidized.

References and further reading

Allen, L. A. (1944) The bacteriology of activated sludge. *J. Hygiene, Camb.*, **43**, 424–431.

Andrews, J. F. (1983) Kinetics and mathematical modelling, in *Ecological Aspects of Used Water Treatment* (eds C. R. Curds and H. A. Hawkes), Vol. 3, Academic Press, London, pp. 113–172.

Atkinson, B. and Mavituna, F. (1990) *Biochemical Engineering and Biotechnology Handbook*, 2nd edn, Macmillan Publishers, Basingstoke.

Barnard, J. L. (1978) The Bardenpho process, in *Advances in Water and Wastewater Treatment, Biological Nutrient Removal* (eds M. P. Wanielista and W. W. Eckenfelder), Ann Arbor Science, pp. 79–114.

Bruce, A. M. and Hawkes, H. A. (1983) Biological filters, in *Ecological Aspects of Used Water Treatment* (eds C. R. Curds and H. A. Hawkes), Vol. 3, Academic Press, London, pp. 2–111.

Curds, C. R. (1982) The ecology and role of activated sludge. *Ann. Rev. Microbiol.*, **36**, 27–46.

Dias, F. F. and Bhat, J. V. (1964) Microbial ecology of activated sludge. I. Dominant bacteria. *Appl. Microbiol.*, **12**, 412–417.

Eckenfelder, W. W. and Musterman, J. L. (1995) *Activated Sludge Treatment of Industrial Wastewater*, Technomic Publishing.

Gaudy, A. F. and Gaudy, E. T. (1980) *Microbiology for Environmental Scientists and Engineers*, McGraw-Hill Book Company, New York.

Hawkes, H. A. (1983) Activated sludge, in *Ecological Aspects of Used Water Treatment* (eds C. R. Curds and H. A. Hawkes), Vol. 2, Academic Press, London, pp. 77–162.

Hawkes, H. A. (1983) The applied significance of ecological studies of aerobic processes, in *Ecological Aspects of Used Water Treatment* (eds C. R. Curds and H. A. Hawkes), Vol. 3, Academic Press, London, pp. 173–333.

Horan, N. J. (1990) *Biological Wastewater Treatment Systems: Theory and Operation*, John Wiley & Sons, Chichester.

Mudrack, K. and Kunst, S. (1986) *Biology of Sewage Treatment and Water Pollution Control*, Ellis Horwood Ltd, Chichester.

Painter, H. A. and Viney, M. (1959) Composition of a domestic sewage, *J. Biochem. Microbiol. Technol.*, **1**, 143–162.

Pike, E. B. (1975) Aerobic bacteria, in *Ecological Aspects of Used Water Treatment* (eds C. R. Curds and H. A. Hawkes), Vol. 1, Academic Press, London, pp. 1–63.

Pike, E. B. and Carrington, E. G. (1972) Recent developments in the study of bacteria in the activated sludge process. *Water Pollut. Control*, **71**, 583–605.

Tempest, D. W. and Neijssel, O. M. (1976) Microbial adaption to low nutrient environments, in *Continuous Culture 6: Applications and New Fields*, (eds A. C. R. Dean, D. C. Ellwood, C. G. T. Evans and J. Melling), Ellis Horwood Ltd, Chichester, pp. 283–296.

White, J. B. (1987) *Wastewater Engineering*, 3rd edn, Edward Arnold, London.

Chapter 15

Anaerobic wastewater treatment

15.1 Introduction

In the treatment of domestic wastewaters and other wastes of similar composition containing a significant proportion of the organic load in the soluble and colloidal phases, anaerobic processes have not traditionally been widely used. Anaerobic treatment is fairly slow by comparison to aerobic treatment and is more suitable for wastes with a comparatively high BOD. The recent development of high rate anaerobic processes (see Chapter 17) has led to their use for the treatment of largely soluble high strength industrial wastes. However, the anaerobic digestion process has been favoured in developed countries for many years for the treatment and stabilization of the waste sludges from conventional sewage treatment prior to their disposal. Approximately 50% of all the sludge produced in the UK is anaerobically digested.

The major objectives of sludge digestion are the neutralization of odours and other offensive characteristics, the reduction of its tendency to putresce and a decrease in the number of pathogens and the microbial activity in general. During sludge digestion the content of organic matter is reduced through the conversion of 30–40% of the solids to gas, so that the stabilized product is more amenable to handling and disposal.

The major end-product of anaerobic metabolism is methane, which, since it is a gas, does not contribute to the BOD of the final product. In this respect therefore, a BOD reduction can be achieved without the need for aeration. This is the major advantage of anaerobic treatment since aeration is costly and aerobic processes become limited by the rate of oxygen transfer as the BOD increases. However, this advantage is offset by the lower rates of treatment attainable in anaerobic systems and the resultant longer retention times, the greater sensitivity to toxic materials and a general tendency to process instability.

15.2 Anaerobic sludge digestion

15.2.1 Process description

Anaerobic digestion employs a complex mixed culture of anaerobic micro-organisms to hydrolyse, ferment and ultimately convert to methane the organic solids present in sludge in the form of fats, proteins and polysaccharides. Commonly, the organic solids content of the sludge is reduced by up to 50% by these means.

The majority of conventional digesters function as single-pass reactors without biomass recycle. In many, but not all, cases the gas is collected and the entire process is enclosed in an airtight tank with a floating gas holder cover to facilitate this. Alternatively, digestion may occur in a reactor with a fixed volume and the gas stored in a separate holder.

The degree of mixing required is much less than in aerobic processes. Low rate digestion rarely employs mixing at all, while in higher rate systems either intermittent mechanical mixing or sparging with the gas produced is used. Most conventional digesters are operated in a semibatch manner, with the addition of raw sludge and the withdrawal of the stabilized product occurring intermittently, sometimes at intervals of up to one week.

Frequently, two-stage systems are used in which the secondary digester functions as a sludge storage tank with a low residual level of anaerobic activity or as a sludge thickener. Effective settlement of the sludge solids does not occur if significant methanogenesis is still proceeding, hence the need for the second reactor. The supernatant liquors from anaerobic digestion are normally recycled to the head of the treatment works to undergo further aerobic treatment. A schematic diagram of a typical digestion system is shown in Fig. 15.1.

15.2.2 Operational parameters

(a) Temperature

Since digestion is such a slow process, it is frequently necessary to apply heat to accelerate the biochemical reactions involved. Unheated systems are more frequently termed sludge lagoons than digesters. However, "cold", or psychrophilic digestion at temperatures below 20°C is sometimes employed. The majority of conventional digesters are operated in the mesophilic range, between 25 and 40°C. Both psychrophilic and mesophilic anaerobic bacterial populations are found in nature, the former in the bottom sediments of lakes and marshland and the latter in the rumens of herbivorous animals. Thermophilic populations are not so common in the natural environment, although anaerobic digestion can proceed in the thermophilic range of 50–70°C. However, thermophilic digestion is employed

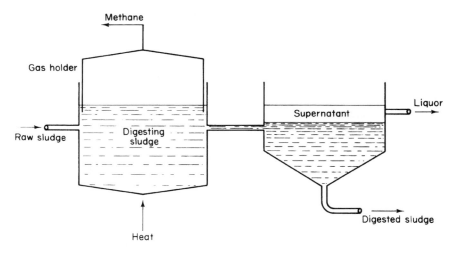

Fig. 15.1 A typical anaerobic digester.

comparatively infrequently compared to mesophilic digestion. Since it prob-ably depends on a bacterial population largely different from the more common mesophiles the benefits may be marginal.

(b) Retention times

For cold digestion mean sludge residence times are often in excess of 100 days. Thus, if a significant volume of sludge is produced, a large land area may be needed to provide sufficient stabilization capacity for it. Mesophilic and thermophilic digesters can operate at mean sludge retention times typically in the range 25–35 days, and sometimes as low as 12–15 days.

(c) Substrate loading

Rather than using COD or BOD as parameters to calculate loading, volatile solids are used as a measure of the organic content of the sludge. Hence loadings are normally expressed in terms of kg of volatile solids per m^3 per day. Conventional sludge digesters typically operate in the volatile solids loading range of $1-2 \text{ kg m}^{-3} \text{ d}^{-1}$. However, since the sludge and the sub-strate are one and the same thing, if a feed containing a lower concentration of biodegradable organics were to be added at a rate sufficient to maintain the normal organic loading the higher volumetric loading required would reduce the retention time of the biomass in the system. Thus, the operational parameters described here are relevant only in the case of wastes having properties similar to primary or mixed primary sewage sludge.

15.3 Biological activities in anaerobic digestion

15.3.1 Microbial composition of digesting sludge

Anaerobic digestion is a microbiologically complex process. Although fungi and protozoa have been found to be present their numbers are low, and the digestion process itself is believed to be almost entirely due to bacterial activity. The types of bacteria found in the process can be broadly classified into three groups: hydrolytic, fermentative and methanogenic bacteria. There is, however, considerable overlap between members of the first two groups. The most significant source of anaerobes in raw sewage is faeces. While this material is a carrier of faecal coliforms and other organisms used as indicators of faecal pollution (e.g. clostridia, streptococci) the most significant bacteria numerically are probably members of the genus *Bacteroides*. These are typically present in faeces at densities of about 10^{10} g^{-1}, outnumbering other obligate anaerobes by a factor of 10^2–10^4. Accurate counts of obligate anaerobes in wastewater and sludge treatment are few and far between. However, the limited available data suggest that 10^4–10^5 lipolytic bacteria per ml, up to 10^6 proteolytic bacteria per ml and 10^5–10^6 cellulolytic bacteria per ml are typical counts for anaerobic digestion. Many of the isolates are clostridia and *Bacteroides* spp., together with *Pseudomonas* and *Bacillus* spp. Among the fermentative organisms can be found some facultative bacteria but the majority of the bacterial population of anaerobic digesters are obligate anaerobes. *Clostridium* spp. are also prominent as fermentative organisms, together with *Butyribacterium*, *Propionibacterium* and *Megasphera* spp. Streptococci and lactobacilli can also be found. The reducing conditions in anaerobic processes give rise to the characteristic hydrogen sulphide odour. Although some of the sulphide originally derives from sulphur-containing amino acids, the majority is probably formed from the reduction of sulphate present in the raw sewage. The anaerobic bacterium *Desulfovibrio desulfuricans* can perform this reduction. If this occurs due to anaerobiosis in the sewer prior to arrival of the wastewater at the treatment works the sulphide evolved can be reoxidized aerobically in the ventilated airspace by *Thiobacillus* spp. to form sulphuric acid which attacks the concrete lining of the sewer.

The methanogenic population is also strictly anaerobic. These organisms are difficult to grow in pure culture, hence little is known about them. At one time the group was thought to be nutritionally versatile, converting a wide range of fermentation end-products to methane. However, it is now known that most of the group are limited to one- or possibly two-carbon compounds provided by another group of organisms which perform a secondary fermentation of the major fermentative end-products. These organisms are also obligate anaerobes and difficult to study in detail because of their close syntropic relationship with the methanogens.

15.3.2 Survival of pathogens

Anaerobic digestion of sewage sludge is now frequently practised prior to disposal of the stabilized sludge to agricultural land. In this respect, the organisms most often cited as being of concern are *Salmonella* spp. and eggs of *Taenia saginata*. The survival of these organisms in anaerobic digestion depends on a variety of factors including temperature, retention time and type of process involved. Results from different plants cannot easily be compared.

Mesophilic digestion seems moderately successful in destroying salmonellae, with a median reduction of about 90%. Although higher removals than this are often reported, there have been few recent studies in this area. Salmonellae are frequently detectable in digested sludge.

The destruction of *T. saginata* infectivity is quite effective in mesophilic and thermophilic digestion, with removals approaching 100% in several reported cases, although at lower temperatures survival for several months seems possible.

The removal of *Vibrio cholerae* and *V. El Tor.* appears to be very good, whereas *Mycobacterium* spp. appear to survive to some extent. *Entamoeba* and *Giardia* do not survive very well in anaerobic digestion. The removal of viral pathogens shows no consistent pattern, although it appears to improve with increasing temperature. Viruses can frequently be recovered from anaerobically digested sludge.

15.3.3 Biochemistry of anaerobic digestion

Early studies of anaerobic digestion led to the postulation of three distinct stages in the breakdown of high molecular weight organic molecules to methane. These were the hydrolysis of insoluble polymers, the fermentation of the monomeric breakdown products and the generation of methane using fermentative end-products as substrates. The later observation that methanogenic bacteria can use only a limited range of substrates led to the discovery of a fourth stage, intermediate between fermentation and methanogenesis, involving the formation of acetate and hydrogen from higher volatile alkanoic acids. The entire process, showing all four phases is illustrated in Fig. 15.2.

(a) Hydrolysis

The initial phases of the hydrolysis of a substrate such as sewage sludge occur extracellularly since the large, insoluble polymers cannot be trans-

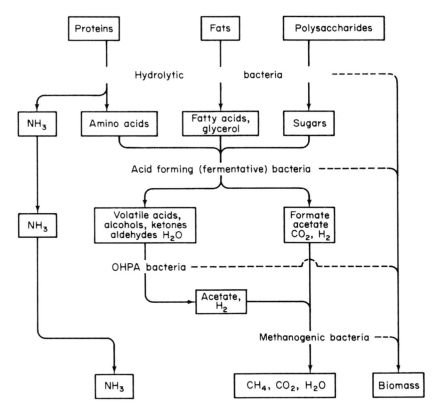

Fig. 15.2 Biochemistry of anaerobic digestion.

ported into the cell. In general, only a few bacteria are able to produce the enzymes necessary for the degradation of the very large molecules, with the number of species able to degrade the breakdown products increasing with decreasing molecular size.

Little is known about lipid hydrolysis in anaerobic digestion. Clostridia and micrococci appear to produce most of the extracellular lipases in anaerobic digesters. These enzymes attack triglycerides to produce fatty acids and glycerol as reaction products.

A wide variety of extracellular proteases are found in anaerobic digestion. Some of these enzymes are very specific, while others will degrade a range of proteins and peptides. The proteases also operate in a fairly broad pH range 5–11. The particular enzymes which are active in the pH range 7–8 are often zinc-containing metalloproteins which are inhibited by powerful chelating agents such as EDTA. Certain of the other proteases are particularly sensitive to organophosphorus compounds.

Among the polysaccharides in anaerobic digestion cellulose (poly-1-4β-glucose), hemicelluloses (complex mixtures of poly-1-4β-xylose and pectins) and starch (poly-1-4- and 1-6-glucose) are important examples. All are degraded initially by extracellular hydrolases produced by a variety of different bacterial genera. Pectin and dextran can also be degraded in anaerobic digestion.

(b) Fermentation

Amino acids and sugars are the most readily fermentable substrates, but some anaerobes can ferment alkanoic acids, purines and pyrimidines. The sugars are usually fermented to alcohols. Pyruvate is an important intermediate in the fermentation of amino acids, from which lactate, propionate, butyrate, formate and acetate are formed. Acidogenic fermentations are quantitatively the most important in anaerobic digestion, with acetate as the main end-product. Acetate, propionate, butyrate, caproate, caprylate, valerate and heptanoate constitute the group of compounds traditionally known as volatile fatty acids (VFA) which are important intermediates in the anaerobic digestion process.

Acetate is also generated from the breakdown of long chain alkanoic acids by β-oxidation, a cyclical process which removes two-carbon fragments from the end of the molecule during each turn of the cycle. Anaerobic β-oxidation of alkanoic acids results in the formation of considerable quantities of $NADH_2$.

Two additional fermentative end-products which have an important role in anaerobic digestion are carbon dioxide and molecular hydrogen. These are generated from the degradation of pyruvate via formate and at intermediate stages of several of the other fermentative pathways.

(c) Acetogenesis

For many years it was believed that the methanogens could convert most, if not all of the fermentative end-products directly to methane. For example, a presumed pure culture of "*Methanobacillus omelianskii*" could perform the following reaction:

$$2CH_3CH_2OH + CO_2 \rightleftharpoons 2CH_3COOH + CH_4$$

However, "*M. omelianskii*" has more recently been shown to consist of a very close symbiotic association of two strict anaerobes. The first member of this association generates acetate and hydrogen:

$$CH_3CH_2OH + H_2O \rightleftharpoons CH_3COOH + 2H_2$$

Since acetate appears to be the most complex organic substrate which can be directly converted to methane, the activity of this and related organisms, collectively known as the obligatory hydrogen-producing acetogenic (OHPA) bacteria is obviously a crucial aspect of the treatment process.

(d) Methanogenesis

Although acetate is an important substrate, only a few and perhaps only one species can generate methane from it:

$$CH_3COOH \rightleftharpoons CH_4 + CO_2$$

The acetoclastic organism *Methanosarcina barkeri* can also use methanol (CH_3OH) as a substrate. Some methanogens can utilize formaldehyde (HCHO) and all, so far as is known, obtain energy from the oxidation of molecular hydrogen using carbon dioxide as a terminal electron acceptor:

$$4H_2 + CO_2 \rightleftharpoons CH_4 + 2H_2O$$

It is this reaction which forms the basis of the unity of an otherwise fairly diverse group of bacteria. However, about twice as much methane is generated from acetate than from the reduction of carbon dioxide in anaerobic digestion.

15.3.4 Stability of anaerobic digestion

Since anaerobic digestion depends on the interactions of several microbial subpopulations, maintaining the ecological balance is important in preventing failure of the process. Although hydrolysis and fermentation may proceed slowly, the organisms responsible have a robustness typical of heterotrophs in general. During stable operation of the process the important intermediates acetate and hydrogen are present in low concentrations, typically <10 mg l^{-1} and $<0.1\%$ by volume of the gas phase, respectively. Since the latter stages of the process are most susceptible to inhibition, impending failure is most frequently characterized by the accumulation of fermentative intermediates.

A mild overload of carbohydrate, for example, will result in elevated concentrations of acetate and hydrogen:

$$C_6H_{12}O_6 + 2H_2O \rightleftharpoons 2CH_3COOH + 4H_2 + 2CO_2$$

If the rate of production of acetate were to exceed its rate of utilization by the methanogens for a significant length of time its accumulation would lead to a fall in pH with a resultant inhibitory effect on the whole system. However, the OHPA bacteria are inhibited by their own metabolic product, hydrogen, normally depending on the activity of the methanogens for its removal. Acetate production slows down, and the accumulated hydrogen diverts acidogenesis to butyrate instead. With the acetate production rate temporarily retarded, the methanogens have an opportunity to remove the moderate accumulations of acetate and hydrogen, thus restoring the original balance of the system, often without the need for external pH control.

Under more severe overload conditions the retarding effect of the hydrogen on acetogenesis is lifted to some extent because at higher concentrations it is utilized in the formation of propionic acid.

$$C_6H_{12}O_6 + 2H_2 \rightleftharpoons 2CH_3CH_2COOH + 2H_2O$$

This permits the OHPA bacteria to continue to produce acetate, thus lowering the pH of the system further and eventually resulting in total failure of the system.

Despite the susceptibility of anaerobic digestion to failure due to volatile fatty acid accumulation, the concentrations of these acids under normal conditions and at the point of failure probably differ by a factor of 100 or more. Acetic acid is a weak acid (i.e., it is only partially ionized in aqueous solution) and this, combined with the natural buffering capacity of the system provided by carbon dioxide in the form of bicarbonate alkalinity, means that a significant increase in the concentration of VFAs will occur before any significant decrease in pH is observed. Thus, VFA determination can act as an early warning indicator of impending failure which can be countered by external pH control in the form of calcium hydroxide addition. The precise concentration of VFAs which will induce failure depends on a number of factors, including the initial pH, the alkalinity and the organic loading, but would typically be 2,000 mg l^{-1} (as acetate) and possibly >5,000 mg l^{-1}.

Apart from acid inhibition, the methanogenic phase is inhibited by several toxic substances. These include the heavy metals, cadmium, copper, chromium, nickel, lead and zinc. Although these will be present largely as insoluble sulphides, which have a low toxic potential, the heavy metals are probably the most frequent cause of toxic inhibition in the UK. Amongst the toxic organics, chloromethane and other chlorinated compounds and some detergents can have an inhibitory effect.

15.3.5 Microbial growth parameters

It has often been stated that the low growth rate of the methanogenic phase is the rate limiting factor in anaerobic digestion. However, there are indications that the OHPA bacteria grow more slowly, having μ_{max} values of the order of 0.1–0.5 d^{-1} compared with 0.5–5.0 d^{-1} for the methanogens. To prevent failure of the system, therefore, retention times should be able to accommodate the slowest-growing subpopulation. Thus typical μ_{max} values for the entire process appear to be about 0.14–0.30 d^{-1}. This indicates an absolute minimum retention time of just over 3 d. However, mean K_s values for the whole process, based on COD, appear to be 10–50 times larger than the equivalent values for aerobic wastewater treatment, so that at all but the highest organic loadings microbial growth will proceed at submaximal rates.

The obvious and principal advantage of anaerobic treatment is that much of the organic carbon is converted to gaseous end-products. This means that only a small proportion of it is assimilated by the biomass. This is reflected in the observed yield coefficients for anaerobic processes. Mean yield coefficients are typically 0.2 kg per kg of BOD removed, being only 50% of equivalent values for aerobic processes.

15.4 Waste stabilization ponds

The term "pond" is a somewhat imprecise term, but in general implies a biological treatment system of lesser complexity from an engineering standpoint (but not biologically) than the more technologically advanced activated sludge and percolating filter processes. The greater simplicity of pond designs means that they are cheaper to construct and require less intensive maintenance and control. These important advantages are, however, obtained at the expense of long retention times and a correspondingly large land area. Thus, stabilization ponds are employed most frequently in areas where financial resources and skilled labour are at a premium, but land is not.

15.4.1 Process description

A pond consists of a large shallow basin, normally of artificial construction, but sometimes consisting of a natural body of water, which is designed for low cost oxidation of organic wastewaters. Although high rate aerobic processes of simple design and incorporating mechanical mixing and aeration are sometimes classified as ponds they will not be discussed in detail here, being biologically related to the conventional processes described in Chapter 14.

Three major types of pond are in existence. These are described as facultative, anaerobic and maturation. In many cases, the pond achieves the entire treatment process within one reactor. For example, a facultative pond can be used without pre- and post-treatment stages and achieves sedimentation, waste oxidation and sludge stabilization simultaneously. Ponds can, however, be connected in series to give multistage treatment, and as its name suggests, a maturation pond is normally used as a secondary or tertiary treatment stage for effluent polishing.

15.4.2 Facultative ponds

(a) Operational parameters

A facultative pond achieves treatment by a combination of aerobic oxidation, photosynthesis and anaerobic digestion. It is typically 1–1.5 m in depth and contains two major zones of microbial activity. The upper zone is aerobic, the oxygen being supplied by the photosynthetic algae, and the lower zone is anaerobic. (Maturation ponds are typically quite shallow, consisting entirely of the aerobic layer.) No mixing or mechanical aeration is employed, mixing of the contents being dependent on wind action and thermal gradients within the pond. The relatively quiescent conditions permit the suspended solids to settle out to the anaerobic layer at the bottom of the pond, from where desludging is done on a very infrequent basis.

BOD removals of up to 90% can be obtained in facultative ponds, although attainable loadings vary widely, being dependent mainly on climatic conditions. The climate affects the process in two major ways. Since daylight is the primary driving force of the process, providing the energy input to generate oxygen, the intensity and length of the daylight hours is an important factor. It is difficult to separate light and temperature, although the latter is of equal importance in determining the biochemical reaction rates. Although the aerobic phase is fairly tolerant of temperature variations, the anaerobic phase is very sensitive, and activity almost ceases below 17°C.

One of the bases for design is the light availability, which is a function of surface area, rather than volume, of the pond. Thus loadings are normally expressed in terms of this surface area. BOD loadings can vary between 10 and 350 kg ha^{-1} d^{-1} depending on the climate, and very long retention times of >100 d may sometimes be required.

(b) Microbial composition of facultative ponds

The dominant members of the aerobic heterotrophic bacterial population are not dissimilar to the representative members of activated sludge biomass and include genera such as *Pseudomonas*, *Achromobacter* and *Flavobacterium*. The fungi are not a particularly significant group, and amongst the grazing organisms the protozoa have not been credited with the same level of activity as in activated sludge, but higher organisms such as rotifers and copepods contribute significantly to grazing activity, thus controlling effluent turbidity.

The algae are an important group in facultative ponds, and numerous genera may be represented. Those frequently found include *Chlorella*, *Chlamydomonas*, *Scenedesmus*, *Euglena* and *Oscillatoria*. The relative abundance of these organisms is controlled by complex factors.

(c) Microbial activity in facultative ponds

Approximately 20–30% of the organic load on a facultative pond may be broken down initially by anaerobic metabolism. Much of the remaining soluble organic matter will be aerobically oxidized in the upper layers of the pond, the oxygen required being supplied by the photosynthetic activity of the algae. The algae are also important with respect to their inorganic nutrient requirements. Most algae prefer nitrogen in the form of ammonia, which is the major end-product of anaerobic breakdown of nitrogenous organic compounds. They may also have a high phosphorus requirement, and biological phosphorus removal is sometimes described as "luxury uptake". Thus ponds with significant algal activity are often effective in controlling eutrophication.

The algae also have a beneficial effect in assimilating simple organic molecules arising as a result of bacterial activity. The increase in pH resulting from algal activity also tends to stabilize the process against the effects of anaerobic acidogenesis and is probably also instrumental in controlling the removal of pathogens.

Removal efficiencies of up to 99.99% have been observed for faecal indicator organisms. Virus removal is also considerable and is probably enhanced by the light intensity. Ponds are also comparatively effective in removing parasites. The efficiency of pathogen removal is probably due, at least in part, to the long retention times afforded by ponds. The complex microbial activities which occur in facultative ponds are shown schematically in Fig. 15.3.

15.4.3 Anaerobic ponds

In anaerobic ponds a considerable BOD reduction may be achieved, but if a reasonable effluent standard is required further treatment may be necessary. Such ponds are therefore often used as pretreatment processes.

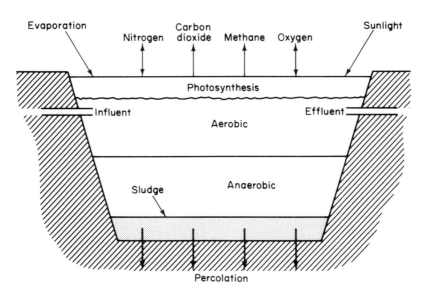

Fig. 15.3 Microbial activities in facultative ponds.

The performance of an anaerobic pond depends upon the development of an active sludge layer. Biochemically, the pond will behave in a similar fashion to an anaerobic digester. Like a digester, activity is dependent on temperature. Below 10°C little biological activity occurs in anaerobic ponds, but they continue to function as sedimentation tanks. Since there is no aerobic layer, an anaerobic pond can be much deeper than the facultative equivalent. Organic loadings can be as high as 0.1–0.2 kg m^{-3} d^{-1}.

References and further reading

Crowther, R. F. and Harkness, N. (1975) Anaerobic bacteria, in *Ecological Aspects of Used Water Treatment* (eds C. R. Curds and H. A. Hawkes), Vol. 1, Academic Press, London, pp. 65–91.

Hawkes, H. A. (1983) Stabilisation ponds, in *Ecological Aspects of Used Water Treatment* (eds C. R. Curds and H. A. Hawkes), Vol. 2, Academic Press, London, pp. 163–217.

Horan, N. J. (1990) *Biological Wastewater Treatment Systems: Theory and Operation*, John Wiley & Sons, Chichester.

Mosey, F. E. (1983) Anaerobic processes, in *Ecological Aspects of Used Water Treatment* (eds C. R. Curds and H. A. Hawkes), Vol. 2, Academic Press, London, pp. 219–260.

Stronach, S. M., Rudd, T. and Lester, J. N. (1986) *Anaerobic Digestion Processes in Industrial Wastewater Treatment*, Springer-Verlag, Berlin.

Chapter 16

Composting

16.1 Introduction

Composting can generally be described as a thermophilic aerobic decomposition process. Solid substrates are degraded over a period of weeks by a succession of microbial populations to form a dark brown, granular, humus-like end-product sometimes described as "loamy". This compost can be used beneficially as a soil conditioner, improving the characteristics of both excessively clayey and excessively sandy soils.

The advantages of composting do not lie simply in the production of a soil conditioner. It can be a viable method of domestic refuse disposal. The composting of domestic refuse in Europe has received significant attention since the 1920s. It is less popular in the USA because there is less demand for soil conditioner, the typical refuse composition is not as suitable, and alternative means of disposal appear more economically attractive.

A range of materials, from municipal refuse and paper to sewage sludge and mixtures of these, can be composted. The primary objective of composting in most cases is to convert an unstable, potentially offensive material into a stable end-product. In contrast to the anaerobic digestion of sewage sludge, which is also used mainly to stabilize degradable, potentially offensive material, composting is an aerobic process.

16.2 Process description

From a mechanical standpoint composting involves five major stages. These are salvage, degradation, drying and curing and finishing. The entire process can take up to two to three months.

16.2.1 Salvage

A typical domestic refuse in the UK contains 27% paper, 22% organic material, 16% glass, plastic and ceramic, 7% metal and 5% textiles. About 23% consists of fines and unclassified material. The metal, glass and ceramic, together with much of the plastic and textiles, often amounting to up to 28% of the total, will not be compostable. It is therefore desirable that this is removed.

Ferrous metal can be removed magnetically. Other methods available for salvage include manual picking and screening for very large, non-compostable objects.

16.2.2 Grinding, pulping and homogenization

For effective composting the material to be treated should be a finely divided, granular material, which permits aeration, mixing and invasion by the degradative micro-organisms. The normal particle size aimed for is about 4–7 cm. This can be achieved by grinding, shredding or wet pulping. The last method produces a slurry containing about 5% solids, which requires drying to about 50% solids to be suitable for composting.

Sometimes, minimal pretreatment is required because breakdown of the fibres in the material to be composted occurs as a result of agitation in the initial stages of composting. At this stage the moisture content can be adjusted. The ideal moisture content of about 50% often has to be attained by adding water. The use of sewage sludge in admixture achieves this end while also enhancing the final product. Various raw, dewatered, digested, or combined sludges can be composted to a different degree of success. Raw primary or secondary sludges are generally preferred to digested sludge.

When combined refuse–sludge composting is practised, refuse to sludge ratios of roughly 2 : 1 by weight may be employed. A moisture content of between 45 and 65% by weight is desirable. Either raw or digested sludge can be used, although raw sludge is preferred since it can be dewatered more readily and has a higher nutrient content.

Only a small quantity of sewage is composted in the UK. In the rest of Europe typically less than 5% of each nation's output of sludge is composted, although in Austria about 20% is treated in this fashion. Sewage sludge destined for agricultural land is preferred by UK farmers in liquid or cake form, rather than as compost, because of the greater availability of nitrogen.

16.2.3 Biological degradation

There are two fundamental methods of composting. The simplest is to pile the material into windrows, which are heaps resembling pyramids, about 3 m high with a base about 9 m in width. Mechanical shovels are used to turn the piles every few days in order to promote aeration. Stabilization of the material usually occurs within four to six weeks. There are a number of disadvantages associated with windrow composting. Since it is done in the open air it is significantly affected by the local climate. If the compost generates odours these may be difficult to control and may give rise to complaints by local residents. Windrow composting requires a fairly large land area and may therefore be unsuitable for urban areas.

An alternative method of composting involves mechanical mixing and aeration inside closed systems. These do not suffer the same problems as windrow composting. A number of processes are used in mechanical composting, of which perhaps the most well known is the Dano process. This employs drums which are up to 30 m long and 5 m in diameter which rotate slowly to facilitate mixing and aeration. With better control over mixing, aeration and temperature, the stabilization process may be complete in a shorter time than that required for windrow composting.

16.2.4 Curing and drying

If complete stabilization of the compost has not occurred prior to its addition to the soil, then it may actually remove plant-available nitrogen from the soil in the final stages of its degradation. A curing period of about two weeks in shallow piles is normally used following windrow composting to ensure that complete stabilization has occurred. In mechanical composting, curing periods can range up to three weeks or can be absent altogether.

16.2.5 Finishing

The amount of finishing required is largely dependent on the method of marketing or disposal of the compost. Often, the compost must be dried mechanically to a moisture content of less than 30% for agricultural or horticultural use. Its consistency is also important; for this reason regrinding and screening of the compost to remove large pieces of debris may be undertaken. For sale in small quantities, the compost may be pelletized before bagging. In contrast, if the compost is to be disposed of by landfilling, for example, then finishing may not be required at all.

16.3 Microbiology of composting

Biodegradable material is decomposed naturally by mesophilic microorganisms which initially utilize the most readily degradable carbohydrates and proteins. When such material is gathered up into heaps, the insulating effect leads to a conservation of heat and a rise in temperature. The maximum temperature achieved and the time taken to achieve it depend on many process parameters including composition of the organic wastes, availability of nutrients, moisture content, size of heap, particle size and degree of aeration and agitation. The process may be divided into four stages: mesophilic, thermophilic, cooling and maturation or curing.

At the start of the process the mass is at ambient temperature and is usually slightly acidic. As the indigenous mesophilic organisms multiply, the temperature rises rapidly. Among the products of this initial stage of degradation are simple organic acids produced by acidogenic bacteria and these cause a drop in pH. At temperatures above 40°C the activity of the mesophiles is reduced and degradation is taken over by the thermophilic fungi. The pH turns alkaline and ammonia may be liberated if excess readily available nitrogen is present. At 60°C the thermophilic fungi die off and the reaction is sustained by the spore-forming bacteria and the actinomycetes. At temperatures of over 60°C waxes, proteins and hemicelluloses are readily degraded, although cellulose and lignin fractions are scarcely attacked. As the rapidly degradable material is depleted, the reaction rate slows down, until eventually the rate of heat generation becomes less than the rate of heat loss from the surface of the heap and the mass starts to cool down. The bacterial and fungal biomass generated in the initial phases of composting can become food for a succession of higher organisms which may be associated with the process, including protozoa, rotifers and nematodes.

Once the temperature falls below 60°C the thermophilic fungi from the cooler outside of the mass can re-invade the heap centre and commence their major attack on the cellulose. The hydrolysis and subsequent assimilation of the polymeric material is a relatively slow process; hence the rate of heat generation decreases still further and the temperature falls toward ambient. At about 40°C the mesophilic organisms recommence activity, either from the germination of heat resistant spores or by re-invasion from outside. The pH drops again slightly but usually remains slightly alkaline. The nutrient levels are an important factor affecting the time course of composting. If the C : N ratio is greater than 80, composting will generally not occur at all. At just under 80 composting will occur, but slowly. The rate of composting will increase as the C : N ratio falls. This is because at low nitrogen levels, the composting process can only be maintained if the nitrogen is recycled through successive generations of degradative organisms while carbonaceous degradation takes place. This is a slower process than would occur if the nitrogen were in a plentiful, more readily available form. Because the nitrogen is conserved within the process the C : N ratio changes as the composting process proceeds. The changes in pH, temperature and C : N ratios during composting are shown in Fig. 16.1.

Aerobic thermophilic composting is a dynamic process which is brought about by the combined activities of a rapid succession of mixed microbial populations, each suited to an environment of relatively limited duration and each being active in the decomposition of one particular type or group of organic materials. Some idea of the type and numbers of organisms involved in composting is given in Table 16.1.

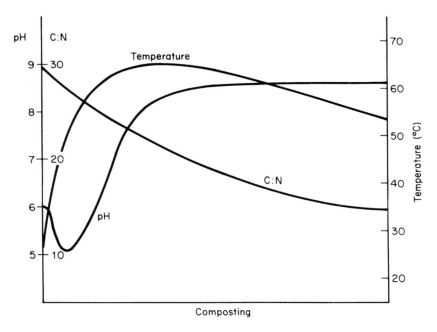

Fig. 16.1 Changes in pH, temperature and C : N ratio during composting.

Table 16.1 Organisms involved in composting

Organisms	Number per g of compost
Micro-organisms	
Bacteria	$10^8–10^9$
Actinomycetes	$10^5–10^8$
Fungi	$10^4–10^6$
Protozoa	$10^4–10^5$
Higher organisms	
Nematodes	
Ants	Various
Springtails	
Millipedes	

Despite the numerical supremacy of bacteria, they represent less than half the microbial biomass due to their relative size. Moist conditions are normally required for active growth but some bacteria form endospores which can withstand heat and desiccation and then germinate and grow when environmental conditions are more favourable.

The levels of temperature and available food supply appear to exert the greatest influence in determining the species comprising the population at any one time. Low molecular weight, water soluble materials can pass readily through the cell wall and hence can be metabolized by a wide range of non-specialized organisms. High molecular weight polymeric materials cannot pass into the cell and are useless in their natural state as substrates for many organisms. With these materials, notably hemicellulose and cellulose, the cell produces an extracellular enzyme which hydrolyses the long polymeric chains into short mono- or oligosaccharides. Only a limited number of specialized organisms can achieve this hydrolysis but almost all organisms can assimilate the resulting fragments.

The first three stages of the composting cycle take place reasonably quickly, being over in a matter of days or weeks. The final stage, maturation or curing, normally requires a period of months. This takes place at ambient temperature with mesophilic organisms predominating and macrofauna appearing. Heat evolution and weight loss are small. During this period complex secondary reactions of condensation and polymerization are taking place which give rise to the final end-product, humus, and, more particularly, the stable and complex humic acids.

16.4 Fate of pathogens during composting

The thermophilic nature of the composting process is widely claimed to be effective in reducing the number of viable organisms present. The temperature is probably the most critical factor. A temperature in excess of 56°C will usually be sufficient to inactivate 99% or more of *E. coli*, faecal streptococci and *Salmonella* spp. Polio virus is also destroyed at similar temperatures in as short a time as thirty minutes. At higher temperatures even better pathogen removal is obtained. For example, *Mycobacterium tuberculosis* will be killed by a temperature of 65°C after fifteen days exposure.

Some of the thermophilic fungi are classified as secondary pathogens. This means that they may infect individuals who are already infected by other pathogens or who are particularly susceptible to respiratory complaints. Airborne spores of *Aspergillus fumigatus* are frequently found in the vicinity of composting windrows. A number of antigenic substances having a microbiological origin may also be found associated with dust particles. *Aspergillus flavus*, which produces aflatoxin, a potent human carcinogen, may be found in composting processes.

Any adverse effects caused by the composting organisms themselves are likely to be highly localized, so that there will be no threat to residential areas. Although a number of infections of workers on composting plants have been reported, this particular facet has not been widely studied.

References and further reading

Anderson, J. G. and Smith, J. E. (1987) Composting, in *Biotechnology of Waste Treatment and Exploitation* (eds J. M. Sidwick and R. S. Holdom), Ellis Horwood, Chichester, pp. 301–321.

Hogan, J. A. (1998) Composting, in *Biological Treatment of Hazardous Wastes* (eds G. A. Lewandowski and L. J. DeFilippi), John Wiley & Sons, New York, pp. 357–393.

Hughes, E. G. (1980) in *Handbook of Organic Waste Conversion* (ed. M. W. W. Bewick), Van Nostrand Reinhold, New York, p. 108.

Pavoni, J. L., Heer, J. E. and Hagerty, D. J. (1975) *Handbook of Solid Waste Disposal: Materials and Energy Recovery*, Van Nostrand Reinhold, New York, p. 7.

Chapter 17

Industrial waste treatment

17.1 Introduction

The range of different materials found in industrial wastes (frequently termed trade effluents) is vast. Some industrial wastewaters resemble domestic sewage to a remarkable degree, whereas others may contain high concentrations of toxic or non-biodegradable materials. Clearly, biological processes would be of considerable potential in the treatment of wastes of the former type, but would be of little use in the case of wastes of the latter type.

Historically, the options available for the treatment of industrial wastewaters have been broadly similar to those used for domestic wastewaters. These are direct discharge to surface water, direct discharge to sea through long outfalls or treatment prior to discharge of an effluent of acceptable quality. Treatment can be conducted at the industrial site, in which case the plant will often (but not always!) have been designed specifically to cope with a certain waste composition. Alternatively industrial wastewaters can be discharged to sewers with or without on-site pretreatment and treated in admixtures with domestic sewage at a municipal treatment works employing conventional processes.

Discharge to sewers has been practised for as long as sewers have been in existence. Clearly, not all wastewaters are suitable for discharge. This was recognized at an early stage, and led to the imposition of trade effluent controls by the authorities responsible for wastewater treatment. The development of industrial wastewater treatment processes has been due largely to changes in trade effluent control policy which have imposed economic pressure upon the dischargers.

17.2 Trade effluent control

Trade effluent control in the UK extends back to the turn of the century in areas like West Yorkshire where the chemical industry was at that time already well established. In the earliest application of what is now known as the "Polluter Pays Principle" (PPP) a private Act of Parliament provided for the removal of all solid matter, fibre and grease in trade effluents discharged to sewer in Huddersfield, or for payment to the Corporation in lieu of that control. The charge in 1906 was fixed at 1 penny per thousand gallons (equivalent to £0.009 m^{-3}), irrespective of the actual composition of the wastewater.

At present, the cost of discharge to sewer is most commonly assessed by the Regional Water Authorities by application of the Mogden formula. This is:

$$C = R + V + B\left(\frac{O_t}{O_s}\right) + S\left(\frac{S_t}{S_s}\right)$$

(17.1)

where:

C = unit charge in p m^{-3} for the discharge.
R = unit cost in p m^{-3} for conveyance and reception of sewage.
V = unit cost in p m^{-3} of the volumetric treatment of sewage.
B = unit cost in p m^{-3} of the biological treatment of sewage.
S = unit cost in p m^{-3} of the treatment and disposal of primary sludge.
O_t = the oxygen demand in mg l^{-1} of the trade effluent after 1 hour quiescent settlement.
S_t = the total suspended solids concentration in mg l^{-1} in the trade effluent.
O_s = the standard strength of settled sewage in terms of its oxygen demand in mg l^{-1}.
S_s = the standard strength of settled sewage in terms of its suspended solids concentration in mg l^{-1}.

Since some industrial wastewaters may contain only partially biodegradable material, both COD and BOD are often taken into account when calculating the factor $(O_t/O_s)B$. As an example, 80% of the charge might result from the BOD and 20% from the COD.

The unit cost elements vary from region to region and generally increase with time. By the early 1980s, however, the cost of discharging an industrial effluent of similar composition to standard strength sewage, depending on the region, had risen to between £0.06 and £0.20 per m^3.

In addition to the charges levied for the discharge, there are normally limits attached to the consent which is granted by the Water Authority. These will specify maximum acceptable values for toxic metals, cyanide, sulphide, sulphate and a range of other potentially harmful materials as well as temperature and pH. Some substances, particularly the anthropogenic, persistent organic compounds included in the European Economic Community's List 1 ("Black List") should be excluded altogether from industrial effluents, which in practice means that they should not be detectable at concentrations above a few parts in 10^9.

Both the increasing costs of discharging effluents to sewer and the restrictions on the concentrations of specific substances in such effluents have made industry give more serious consideration to the on-site treatment of industrial wastestreams. Much of the treatment technology involves physical or physicochemical processes. However, for the treatment of wastewaters with a high oxygen demand or for the removal of certain specific substances, biological processes are often the most appropriate.

17.3 Biological aspects of industrial waste treatment

It is a widely held view that biological processes are, in a great many situations, the most cost effective techniques for treating aqueous wastestreams containing organic constituents. However, the composition of the wastes will be highly variable from site to site, so that the scope for a general discussion of the principles involved is quite limited. The potential impact of industrial wastewaters on biological treatment processes can be placed into one of four categories: (i) inhibition of biological degradation, (ii) incidental removal of the pollutant by absorption, adsorption or precipitation and a consequential concentration into the sludge produced, (iii) passage through the process unaffected and (iv) biological degradation.

17.3.1 Inhibition

Probably the most well-known group of toxic substances which can inhibit biological processes are the heavy metals. This group includes many transition elements (e.g. cadmium, chromium, copper, mercury, nickel, zinc), some non-transition metals (e.g. lead) and the metalloids (arsenic and selenium). Many organic compounds are also toxic, particularly chlorinated derivatives such as trichloromethane (chloroform). Examples of some groups of organic compounds which can have inhibitory effects on biological processes are shown in Table 17.1.

Anaerobic processes can be particularly susceptible to inhibition by excess of sulphide as hydrogen sulphide. High sulphide concentrations can occur in effluents from molasses fermentation, tanneries, petroleum refineries and paper mills. Between 50 and 100 mg l^{-1} of soluble sulphides appears to be the tolerable limit although much larger quantities of insoluble sulphides can be tolerated without ill effect. Ferrous sulphide is insoluble; hence iron can be used to good effect in controlling sulphide inhibition.

Anaerobic processes are inhibited at pH values outside the range 6.5–8.2. Volatile fatty acid concentrations in excess of 2,000 mg l^{-1} and ammonium ion concentrations in excess of 3,000 mg l^{-1} are also toxic. Free ammonia at a concentration of 150 mg l^{-1} is severely toxic.

Table 17.1 Types of inhibitor of biological treatment processes

Substances	Industrial uses/ source of effluents	Examples of effects on biological processes
Alcohols	Pharmaceuticals, antifreeze	Inhibitory concentrations of 10–1,000 mg l^{-1}
Phenols	Industrial syntheses, plastics, dyes, pharmaceuticals, petroleum refining	Diverse reactions – both toxic and degradable depending on concentration and whether biomass is acclimated
Chlorinated compounds	Solvents, agricultural chemicals, preservatives, propellants, cleaning fluids	Anaerobic digestion inhibited by 10 mg l^{-1} CCl$_4$, 16 mg l^{-1} CHCl$_3$ and 1 mg l^{-1} CH$_2$Cl$_2$
Agricultural chemicals	Insecticides, pesticides, herbicides	Various – many inhibit nitrification
Organic nitrogen compounds	Various	For example, <1 mg l^{-1} thiourea inhibits nitrification
Surfactants	Domestic and industrial cleaning	Reduce efficiency of aeration

17.3.2 Incidental removal

The heavy metals also fall into this class. Metals like lead, which tend to form fairly insoluble salts, may precipitate out directly, while others, such as copper and cadmium will become associated with the settling sludge by a process of adsorption or complexation by the extracellular polymeric constituents of the flocs. Activated sludge has a considerable capacity to bind heavy metals, although competitive binding by soluble ligands usually results in widely variable removal efficiencies for heavy metals in the region of 50–60% and often much lower for certain elements. The metal binding potential of activated sludge is, however, very high. This is very largely due to the extracellular polymeric components of the cell-floc matrix. These possess metal binding sites which produce complexes with equilibrium constants of 10^5–10^7 at neutral pH. *Zoogloea ramigera* is a prolific polymer producer which in pure culture is capable of immobilizing up to 40% of its own mass in heavy metals. Viable commercial processes based upon this have not, however, been developed.

In anaerobic digestion, the heavy metals form sparingly soluble sulphides which promotes their removal. However, anaerobic bacteria appear to be more sensitive to heavy metals than aerobic heterotrophs. Mosey (1976) derived a formula to predict the inhibitory effects of heavy metals in anaerobic digesters. The total metal load may be expressed in the form of K, where:

$$K = \frac{\dfrac{[Fe]}{27.9} + \dfrac{[Zn]}{32.7} + \dfrac{[Ni]}{29.4} + \dfrac{[Pb]}{103.6} + \dfrac{[Cd]}{56.2} + 0.67\dfrac{[Cu]}{31.8}}{\text{solids concentration in kg l}^{-1}}$$

K is expressed in terms of meq kg^{-1}. The factor 0.67 is introduced to account for the partial reduction of copper II to copper I. Anaerobic digestion failure is considered likely if K exceeds 400 meq kg^{-1} and almost certain if it exceeds 800 meq kg^{-1}. To permit a margin of error, a safe working value of K would be 200 meq kg^{-1}.

Some organic compounds may be removed incidentally by association with settleable solids or floatable scum or grease. This may occur if the compounds are insoluble or slightly soluble in water or are hydrophobic.

17.3.3 Biodegradation of specific pollutants

In the treatment of municipal wastewater it is likely that each of the individual organic compounds present is being degraded by only one, or a very few bacterial species. The heterogeneity of the wastewater determines the heterogeneity of the microbial biomass to a considerable extent. Many industrial effluents are much less complex than municipal wastewater, and some consist principally of only one organic compound. In these cases, a complex biomass resembling, say, the basis of the activated sludge floc, is not necessary for treatment.

(a) The microbial infallibility principle

In its simplest form, the microbial infallibility principle states that whatever man or nature can make, micro-organisms can degrade. This might be expected in the case of naturally occurring compounds, since it would involve essentially a reversal of existing biosynthetic pathways, but less so in the case of xenobiotic compounds (i.e. those which do not occur naturally) since micro-organisms would not be expected to possess the enzyme systems necessary for degradation. The principle appears to be true to a considerable extent; certainly it is difficult to disprove. Even a compound such as 2,3,7,8-tetrachlorodibenzo-p-dioxin, long considered to be one of the most persistent compounds known to man, can in fact be degraded very slowly by a few microbial strains. Mechanisms of degradation are considered further in Chapter 18.

(b) Selection of appropriate strains

The most likely sources of micro-organisms capable of degrading toxic or recalcitrant compounds are sites which have historically been exposed to such compounds. Contaminated soils in the region of industrial operations are a good example of such sources. The samples taken can be used as inocula in enrichment cultures in the laboratory designed to select organisms capable of degrading the compounds of concern. Further selection of desirable characteristics in the degradative organisms can be made by exposing them to irradiation or mutagenic chemicals which may enhance their capabilities as a result of genetic change. Such techniques are on the verge of what is currently termed genetic engineering. It is now possible to construct organisms which carry genetic material derived from a variety of microbial sources so that they are capable of degrading compounds which no naturally-occurring organism could degrade. *Pseudomonas* spp. capable of degrading oil and 2,4,5-trichlorophenoxyacetic acid have been engineered in this way. The consequences of releasing genetically engineered organisms into the environment are largely unknown, however, and this has mitigated against their use for pollution control. However, there are a large number of micro-organisms which have not been genetically engineered and which are capable of degrading a broad range of recalcitrant compounds. Some examples are shown in Table 17.2.

(c) Structure–biodegradability relationships

The addition of chemical substituents to organic molecules can transform them from biodegradable compounds to persistent compounds. Such substituents include amino-, methoxy-, sulphonyl-, nitro- and chloro- groups, a range of *meta-* substituents on the benzene ring, ether linkages and carbon chain branching. Only very generalized guidelines can be derived from these, however.

Chlorinated compounds are of particular interest partly because some representatives are extremely toxic and persistent (e.g. DDT) and partly because organochlorine compounds are rare in nature. The biodegradation of many of these compounds is dependent on an initial reductive dechlorination which occurs anaerobically. This is particularly important in the case of chlorinated aliphatic compounds. In contrast, chlorinated benzene derivatives and polychlorinated biphenyls (PCBs) appear to be degraded only aerobically.

Although aerobic processes are still very much in favour for industrial waste treatment, anaerobic conditions appear to be more favourable for reactions involving nitrosamine degradation, epoxide and nitro-reduction and degradation of certain aromatic compounds. Since much of the information available on structure–degradability relationships is based upon

Table 17.2 Examples of xenobiotic compounds which can be degraded by micro-organisms

Compound	Organism
Halogenated aliphatic compounds	
A range of short chain di-, tri- and tetra-halo compounds	Aerobic wastewater treatment
Aromatic compounds	
Benzene, nitrobenzene, 2,4- and 2,6-dinitrotoluene	Aerobic wastewater treatment
Cresol, phenol, toluene	*Bacillus* and *Pseudomonas* spp.
Halogenated aromatic compounds	
Chlorinated benzenes and benzoates, 4-chlorophenol	Aerobic wastewater treatment, *Pseudomonas* spp.
Polyaromatic hydrocarbons	A few degraded in aerobic wastewater treatment
Polychlorinated biphenyls and related compounds	*Pseudomonas, Bacillus, Chromobacter, Achromobacter, Flavobacterium*, some fungi
Pesticides	
Lindane	*Chlorella, Chlamydomonas*. Attacked anaerobically by clostridia and pseudomonads
Dieldrin	Partial degradation anaerobically by some cyanobacteria
DDT	Reductive dechlorination by several bacteria under anaerobic conditions

generalities, it is usual when considering potential treatment options to conduct biodegradability tests on the wastestream in question.

(d) Cyanide degradation

Cyanide (CN^-) is highly toxic to a wide range of organisms. Hydrogen cyanide concentrations in industrial effluents will be limited typically to $1 \, \text{mg} \, l^{-1}$ or less, while in drinking water, CN^- concentrations in excess of $0.05 \, \text{mg} \, l^{-1}$ are not permitted in the EEC.

Cyanide occurs naturally in certain plants in the form of cyanogenic glycosides. An example is shown in Fig. 17.1. When the plants containing this material are attacked by pathogenic fungi the cyanide is released and kills the fungi. Some fungi have evolved so that they can detoxify enzymatically the cyanide to formamide ($HCONH_2$) or ammonia and carbon dioxide. ICI

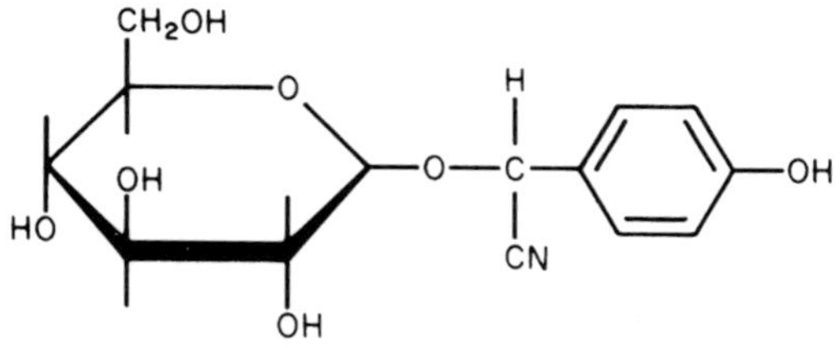

Fig. 17.1 A cyanogenic glycoside.

have evaluated the ability of a number of fungi, notably *Fusarium monili-forme* to detoxify effluents containing several thousand $mg\,l^{-1}$ of cyanide.

(e) Biodegradability tests

Relative biodegradability of organic compounds falls into three categories. Ready biodegradability is applied to compounds which will rapidly and reliably degrade in the environment. Inherently biodegradable compounds will degrade under favourable test conditions, but it cannot be assumed that rapid and reliable degradation in less favourable environmental conditions will occur. These two categories are considered in a little more detail in Chapter 18. The third category involves simulation tests. A procedure is recommended by the OECD Expert Group on Degradation/Accumulation as a test for the determination of the ultimate biodegradability of test materials under conditions which simulate activated sludge treatment. Two bench-scale activated sludge systems of the type shown in Fig. 17.2 are operated in parallel. Transinoculation is used to ensure that the nature of the biomass in each system and the operating efficiencies are the same. Hence, the test is known as the coupled units test. The systems are maintained using a synthetic sewage containing peptone, meat extract and urea as the sources of carbon substrates together with adequate nitrogen, phosphorus and trace nutrients. A composite inoculum is made up of secondary sewage effluent, soil suspension and mesosaprobic surface water.

To one unit is added the test substance at a concentration equivalent to a COD of 40 $mg\,l^{-1}$ or more. The other unit is maintained on synthetic sewage only. Every day 1.5 l of mixed liquor from each aerator is transferred to the other aerator to ensure that the two simulations are operating in as similar a fashion as possible. The main operating parameters are shown in Table 17.3.

The units are allowed to stabilize and the biomass acclimate to the substrate over a period of six weeks and then the evaluation of the test substance is made over the next three weeks. The extent of removal of the test substance is detected from differences in the COD or dissolved organic carbon (DOC) concentrations in the effluent from each unit. A minimum of twenty

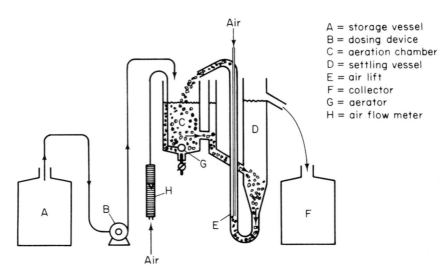

Air

A = storage vessel
B = dosing device
C = aeration chamber
D = settling vessel
E = air lift
F = collector
G = aerator
H = air flow meter

Air

Fig. 17.2 A bench-scale activated sludge system.

Table 17.3 Main operating parameters for the OECD "coupled units" test

Parameter	Approximate value
Aeration tank volume (l)	3
Mean hydraulic detention time (h)	3
Sludge age	Not controlled*
MLSS (mg l^{-1})	2,500
Dissolved oxygen (mg l^{-1})	> 2
Influent COD (mg l^{-1})	300†
COD loading (kg m^{-3} d^{-1})	2.4†
Sludge loading, $B_{x, COD}$ (d^{-1})	1.0†

* Sludge wasted in order to control MLSS concentration.
† Multiply by 0.5 to obtain approximate equivalent values based on BOD.

observations is recommended and differences between effluent COD or DOC are evaluated statistically.

17.4 Aerobic treatment processes

Both percolating filters and activated sludge processes have been applied to a range of industrial wastes, including pharmaceutical, chemical and petrochemical, coal conversion, food and dairy processing effluent. These wastes usually differ from those normally treated by these processes at municipal works in one or more of the following respects. They may have a much higher oxygen demand, they may contain substances which are toxic to the microbial population in the treatment process (this category includes extremes of pH) or they may be deficient in one or more essential nutrients (e.g. nitrogen or phosphorus). Toxicity can sometimes be overcome by dilution; if the diluent is domestic wastewater then nutrient deficiencies may be relieved at the same time. The oxygen demand may also be decreased by dilution, but in cases where it is very high, anaerobic processes may be a more suitable treatment alternative.

The applicability of percolating filters to industrial wastewater treatment has been enhanced by the use of synthetic media rather than the traditional stone or clinker. Plastic media can be packed into towers of much greater height than traditional media because they are much lighter. Designs involving extensive corrugation or perforation can produce specific surface areas of over $200 \text{ m}^2 \text{ m}^{-3}$, which are equal to or greater than the upper end of the range for granite or slag media, while having a voidage of over 90% compared with 40–50% in traditional filters. Thus, high biomass concentrations within the filter can be maintained without excessive risk of blocking the voids. The high voidage is also of advantage where forced ventilation is necessary to deal with high loadings.

Both low and high rate filters have been used in the treatment of meat processing, poultry and dairy effluents. Plastic media filters are probably the most popular option for fruit and vegetable processing effluents with BODs of up to $1,000 \text{ mg l}^{-1}$. Sometimes roughing filters upstream of an activated sludge installation may be used, while in other applications double filtration or alternating double filtration may be appropriate. Although filters are reasonably resilient, especially if operated with effluent recycle, they are not suited to the treatment of wastewaters containing high fat concentrations, since these can lead to blockage or ponding.

The range of modifications designed to improve activated sludge systems was discussed in Chapter 14. These modifications are often used to permit the higher organic loadings of industrial effluents to be tolerated, for example by increasing the solubility of oxygen in the mixed liquor. Higher biomass concentrations are also advantageous in dealing with high strength wastes, and these can be achieved in a number of ways, including the use of carrier-assisted systems.

Where specific types of wastewaters are concerned, existing treatment systems can be modified or "retrofitted" to improve performance. An example of this is the use of powdered activated carbon (PAC) as an additive in the activated sludge process. This absorbs organic compounds which cannot be degraded. It also reduces the level of colour in effluents and can enhance heavy metal removal. Another advantage of PAC is its effect in reducing the effective levels of inhibitors.

The range of applications of activated sludge processes in treating industrial wastewaters is very broad and cannot be considered here. In most cases it represents the most cost effective technology (although frequently cost evaluations have ignored the high rate anaerobic processes described in Section 17.5). The most common problems involve wastewaters which are deficient in nutrients, have a pH outside the range 6–8.5, or a temperature outside the mesophilic range. Poor flocculation of the biomass seems to be associated with wastewaters containing only a very few different degradable substrates, and is probably due to the presence of only a few species of bacteria and consequently insufficient diversity in the components of the floc.

An example of a wastewater containing both recalcitrant and toxic constituents which can be effectively treated by the activated sludge process if applied in the correct manner is the effluent from coal processing. Coal gasification liquor typically contains high concentrations of phenols, ammonia, thiocyamates, cyanides, sulphates, sulphides and chlorides. Where phenol and ammonia concentrations are particularly high, they may be recovered from the wastestream. However, 1,000 mg l^{-1} phenol can be treated by activated sludge if suitably acclimated. Dilution may be required to attain this concentration. However, even after dilution the toxicity of phenols and thiocyanates can be such that starting up new plants depends upon the use of heavy inocula of acclimated seed sludges. Nowadays, these are tankered in or used in the form of freeze-dried cultures (see Section 17.4.1).

Although nitrification is notoriously sensitive to toxicity at low levels, it has been attained in an activated sludge plant treating coking plant wastewater by the use of up to 50 mg l^{-1} PAC to alleviate the inhibition.

17.4.1 Bioaugmentation

Bioaugmentation is a relatively new technique. It involves the addition of bacterial formulations to wastewater treatment plants which do not perform as required for reasons related to their biological conditions. Problems due to bulking, deflocculation, poor BOD removal due to the presence of compounds like lipids in the wastewater, lack of nitrification, and poor removal of specific compounds such as phenols have been addressed using bioaugmentation.

The bacterial formulations consist of freeze-dried bacterial suspensions generated from pure cultures and then blended to produce a mixture of several different species together with wetting agents, nutrients, enzyme preparations and other ingredients. Dosages of the preparations appear to be in the tens of kg range, but frequently smaller follow-up doses are required because the organisms are gradually lost from the system. The exact composition of these formulations has not been publicized for commercial reasons. A very wide range of applications is advertised, with a good level of success in alleviating problems due to biological factors.

17.5 Anaerobic treatment processes

In recent years, the development of newer, higher rate anaerobic processes has shown a considerable potential for industrial wastewater treatment. Although aerobic processes are probably still in the majority for many readily biodegradable wastewaters the emphasis has shifted somewhat towards anaerobic technology for several reasons. Included among these are the demonstrated ability of anaerobic systems to tolerate much higher loadings than were previously thought possible, a better understanding of the reasons for process failure and ways of avoiding it, lower operational costs due to the avoidance of aeration and the potential of methane as a source of energy.

Although there has been a proliferation of different reactor designs, the processes used in industrial wastewater treatment can be divided into two groups: suspended biomass systems and attached film systems. These are very roughly analogous to activated sludge and percolating filters respectively in terms of the form of the biomass, although their biochemistry is very similar to that of conventional anaerobic digestion. Unlike systems for the anaerobic digestion of sewage sludge, however, in many industrial wastewater treatment processes the substrate is in a soluble form. This permits separation of the biomass from it and allows operation under conditions where $\theta \ll \theta_x$. Thus the success of anaerobic processes is very largely due to the development of suspended solids separation. Enhancement of their potential has been achieved by techniques designed to ensure the retention of as large a concentration of biomass within the system as possible. The range of reactor types which have been developed from this approach is shown in Fig. 17.3.

17.5.1 Suspended biomass systems

(a) Contact process

This process is the anaerobic equivalent of the activated sludge process. It consists of a completely mixed reactor with gas take off followed by a settling tank. The settled sludge is recycled to the reactor. Settlement of the anaerobic biomass is critical because the sludge tends not to form such cohesive flocs as its aerobic counterpart and the continued production of methane in the settler can cause problems due to the resultant buoyancy of the biomass. The latter effect can be avoided to some extent by degassing the mixed liquor en route to the settler or by rapidly cooling it from ambient operating temperature (e.g. 35°C) down to 15°C, thus arresting gasification. Chemical coagulants may also be used to assist flocculation and sedimentation. Maximum biomass concentrations developed in the reactor rarely exceed 10 g l^{-1}.

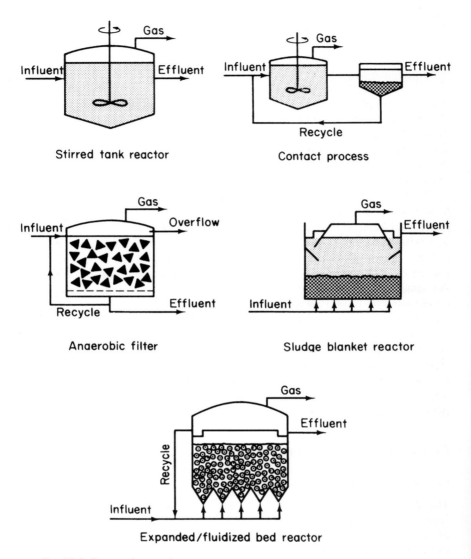

Fig. 17.3 Range of anaerobic reactor types.

COD loading rates of up to 10 kg m^{-3} d^{-1} are attainable, although rates of 2–5 kg m^{-3} d^{-1} are more typical. The process appears to operate well with high wastewater CODs of typically 2–10 g l^{-1}. Hydraulic retention times as low as 0.5 d with moderate strength wastes and <6 h with dilute wastes can be achieved, although hydraulic retention times of a few days are more typical. Depending on the type of wastewater being treated, typical operational regimes can ensure COD removals of >70% and often as much as 90%.

(b) Upflow anaerobic sludge blanket reactor

Wastewater enters a sludge blanket reactor at the base and flows upwards. Within the base of the reactor there forms a suspended blanket of anaerobic biomass which is retained there by its own density and is agitated gently by the formation of gas. At the top of the reactor is usually a baffle system which assists in retention of the less dense fraction of the biomass and disentrains the gas.

The biomass exists as compact grains or granules of up to 3–4 mm in diameter. The biological mechanism of granule formation is as yet unknown, but it would appear that granulation is encouraged by no or minimal mechanical agitation of the blanket, with only a gentle rolling motion occurring. The suspended solids concentration within the lower levels of the blanket or bed can be as high as $70 \, g \, l^{-1}$, decreasing in the upper layers. A high settling velocity of up to $50 \, m \, h^{-1}$ permits a significant upflow velocity to be used. A combination of these two factors permits effective treatment of high strength wastewaters at high organic loadings. Indeed, it is thought that high organic loadings are necessary for granule formation to occur. The presence of inert suspended solids entering the process with the wastewater can interfere with sludge blanket development and stability. Hence the process is probably suited to wastewaters with low suspended solids concentration. Depending upon the type of wastewater being treated, efficient COD removals may be obtained at loadings of up to $15 \, kg \, m^{-3} \, d^{-1}$.

17.5.2 Fixed film processes

(a) Anaerobic filter

This process is the anaerobic counterpart of the percolating filter, but differs from it in more than just its anaerobicity. Filters can be operated in upflow or downflow mode, with the filter bed being completely submerged. In order to avoid blockage media of >2 cm in diameter are frequently used. Although fairly thick microbial films adhere to the media surfaces, a substantial proportion of the biomass in an anaerobic filter will exist in the form of flocs trapped in the interstices between media particles. Biological solids concentrations of up to $20 \, g \, l^{-1}$ can be achieved. A variety of media have been used ranging from crushed stone to random plastic media. Film attachment is affected by porosity, pore size and surface roughness.

Anaerobic filters have been used for both strong and weak wastewaters.

COD loadings in the range $1-10 \text{ kg m}^{-3} \text{d}^{-1}$ appear to be typical. COD removal efficiencies in excess of 90% can be achieved.

(b) Expanded and fluidized bed reactors

The anaerobic degradation of industrial waste in suspended biomass systems is limited to a significant degree by the maximum attainable biomass concentration. This can be overcome to an extent by stationary fixed film processes like the anaerobic filter, although new limitations are introduced due to restrictions on the specific surface area of the media in order to avoid clogging and diffusional limitations in thick films. Expanded and fluidized bed systems employ very small diameter media with high specific surface areas which permit the growth of thin biological films maintaining high biomass concentrations within the reactor. Clogging is avoided by introducing the influent wastewater through a distribution system at the base of the reactor at a sufficient upflow velocity to impart an upthrust to the bed particles. In an expanded bed the particles remain in fairly close contact, but the bed is expanded by about 5%, whereas in a fluidized bed the particles are continuously moving and bed expansion is much higher. In expanded beds biomass concentrations of up to 60 g l^{-1} supported on media of typically less than 3 mm in diameter have been observed. Biomass concentrations of 100 g l^{-1} appear theoretically possible. In fluidized beds slightly smaller particles may be used. The effects of shear will limit the attainable biomass concentrations, but 40 g l^{-1} has been achieved. High recycle ratios are necessary to maintain fluidization and ensure a reasonable hydraulic retention time.

COD loadings of about $20 \text{ kg m}^{-3} \text{d}^{-1}$ can be tolerated with COD removal efficiencies of greater than 90%. Even at loadings of 40 or $50 \text{ kg m}^{-3} \text{d}^{-1}$, 70% removal of COD is possible in the mesophilic range. The process is particularly suited to very high strength wastewaters.

17.5.3 Multistage phase-separated systems

In a single reactor, the operating conditions will not permit acidification and methanogenesis both to proceed at optimum rates. Since the reactions involved in anaerobic degradation occur sequentially rather than in parallel, the volatile fatty acids representing the important intermediates in the reaction sequence, physical separation of the acidogenic phase from the methanogenic phase is feasible. Each phase can then be controlled separately for independent optimization. Phase separation can be achieved by linking a primary acidification reactor in series with a secondary methanogenic reactor. Significant methanogenic activity in the primary reactor can be prevented by kinetic control (i.e. high dilution rates), chemical control, by the addition of inhibitors of methanogenesis, or physical retention of the acidogenic bacteria, for example, by means of a dialysis membrane.

Several advantages of two-phase operation are claimed. A sedimentation and sludge recycle step can be introduced immediately after the acidification reactor with improved settleability due to the absence of gasification. Hydraulic residence times in the primary reactor need to be only 5–10% of those in the secondary reactor; hence a smaller reactor can be used for acidification. The acidification reactor acts as a buffer, protecting the more sensitive methanogens from shock loads and improving the stability of the process generally. Higher loading rates, greater conversion of substrates to methane and higher COD removals are all possible with two-phase systems.

17.5.4 Range of applications of anaerobic treatment

The anaerobic processes described above are applicable to a wide range of high strength organic wastewaters including those from agriculture, meat processing, soft drink production, alcoholic drink production, fruit and vegetable processing, dairy products and some chemical and pharmaceutical manufacturing operations. Much of what is known about the performance of the anaerobic processes has been gleaned from laboratory- and pilot-scale reactors. By 1982, over 500 full-scale anaerobic installations (excluding those used in sewage sludge digestion) had been brought into use. The majority were treating agricultural wastes, although 16% were being used for industrial waste treatment. In the UK only thirty-five plants were in operation, five of which (14%) were treating industrial wastes. However, the number of systems being used in this way is increasing, and there is no doubt that anaerobic technology will have an increasingly important role in industrial waste treatment in the future.

References and further reading

Dyer, J. C., Vernick, A. S. and Feiler, H. D. (1981) *Handbook of Industrial Wastes Pretreatment*, Garland STPM Press, New York.

Eckenfelder, W. W., Patoczka, J. and Watkin, A. T. (1985) *Chem. Eng.*, **2**, 60–74.

Edwards, J. D. (1995) *Industrial Wastewater Treatment: A Guidebook*, CRC Press, New York.

Kobayashi, H. and Rittman, B. E. (1982) *Environ. Sci. Technol.*, **16**, 170A–183A.

Lester, J. N., Sterritt, R. M., Rudd, T. and Brown, M. J. (1984) in *Microbiological Methods for Environmental Biotechnology* (eds J. M. Grainger and J. M. Lynch), Society for Applied Bacteriology Technical Series No. 19, Academic Press, London.

Mosey, F. E. (1976) Assessment of the maximum concentration of heavy metals in crude sewage which will not inhibit the anaerobic digestion of sludge. *Water Pollut. Control*, **75**, 10–18.

Organization for Economic Co-operation and Development (1981) *Guidelines for Testing of Chemicals*, OECD, Paris.

Richardson, D. J. and Shieh, W. K. (1986) *Water Res.*, **20**, 1077–1090.

Schuckrow, A. J., Pajak, A. P. and Touhill, C. J. (1982) *Hazardous Waste Leachate Management Manual*, Noyes Data Corporation, New York.

Stronach, S. M., Rudd, T. and Lester, J. N. (1986) *Anaerobic Digestion Processes in Industrial Wastewater Treatment*, Springer-Verlag, Berlin.

Chapter 18

Environmental microbiology

18.1 Introduction

The environmental theme runs throughout this book, although for the most part this is within the context of waste, wastewater and water treatment. These areas are covered specifically in Chapters 14–17 and 20. In this chapter, the importance of microbial activity in the aquatic environment, as opposed to engineered treatment processes, is considered. This subject is given separate consideration because of the significant differences in key features between the two types of ecosystem involved, especially in terms of organism density (biomass concentration) and substrate concentration.

In the aquatic environment, the role of micro-organisms in degrading organic matter mirrors the same process occurring in wastewater treatment plants, but in addition to this, micro-organisms in the aquatic environment can be important as indicators of pollution or even act as pollutants themselves. An important example of the latter is pathogenicity. This is covered in detail in Chapter 19.

18.2 Biodegradation in the aquatic environment

18.2.1 Bacterial metabolism at low substrate concentrations

Unlike wastewater treatment plants, many natural waters contain very little organic matter and consequently have low microbial densities. If biodegradation of specific compounds does occur, it frequently proceeds under conditions of severe nutrient limitation. Low limiting nutrient concentrations will not permit growth except at the slowest rates. Consequently, many pollutants are only degraded slowly. This may mean that the energy flux in such systems is insufficient for the sustenance of the microbial population. The concept of a minimum threshold value for substrate concentration below which a viable population cannot be maintained has been put forward. Experimentally, under aerobic conditions, such minimum threshold values typically fall in the range $0.1–1$ mg l^{-1}. Another potential consequence of very low substrate concentrations is a failure to induce synthesis of the appropriate enzymes.

Organisms competing at very low substrate concentrations could do this more effectively if their metabolism could be switched towards a high-affinity (i.e. very low K_s) system for substrate utilization; there is evidence that some organisms do in fact possess both high and low K_s systems. Nevertheless, some compounds may not be present in sufficient concentration to support growth. However, they may be degraded in other ways.

18.2.2 Cometabolism

Cometabolism or co-oxidation is a mechanism whereby micro-organisms can chemically alter a particular compound without deriving assimilable carbon or energy for growth from it. This is thought to occur because some of the enzymes involved in degrading the organisms' main carbon source are not totally specific for their substrates and can act on other compounds, albeit quite slowly. Cometabolism appears to be an important route for the degradation of hydrocarbons, especially the more recalcitrant alicyclic compounds, in the environment. It may also be an important pathway for pesticide degradation. 2,4,5-Trichlorophenoxyacetic acid and 2,3,6-trichlorobenzoic acid can both be cometabolized. Sometimes, cometabolism by mixed cultures can be demonstrated, suggesting that fairly extensive degradation of compounds previously thought to be recalcitrant could occur. It is, however, much easier to demonstrate cometabolism of some of these compounds in laboratory cultures than it is in the environment.

It is clearly a feature of cometabolized compounds that they cannot act as sole carbon sources and so would give negative results in several of the tests currently used to assess biodegradability (Section 18.2.3). Whether cometabolism occurs will also depend upon the main substrate being used; if the enzymes responsible for the cometabolic transformation are inducible, rather than constitutive, then the inducer must also be present for cometabolism to occur.

18.2.3 Biodegradability testing

The impact of a chemical pollutant released into the aquatic environment depends to a large extent on its concentration, which in turn is determined by the rate at which it degrades. Biodegradation in water is an important route, so much so that a variety of tests have been developed to quantify it for the ever increasing number of chemical substances which are manufactured. To be used on a routine basis, these tests must be rapid and simple. The Organization for Economic Co-operation and Development (OECD) has recommended a number of tests for biodegradability in the environment falling into the two main categories of ready biodegradability and inherent biodegradability tests. A third test, involving simulation of wastewater treatment conditions is considered in detail in Chapter 17.

(a) Ready biodegradability

Readily biodegradable substances are those which can be used as sole sources of carbon and energy by test inocula and which degrade extensively within twenty-eight days of the commencement of the test. Some of the tests resemble the BOD test (Chapter 5) in that a decrease of dissolved oxygen concentration is monitored in closed bottles containing a dilute solution of the test compound in a nutrient and mineral salts formulation and inoculated with an appropriate source of bacteria. Other tests involve measurement of the carbon dioxide generated from aerated test mixtures. The OECD gives details of five ready biodegradability tests. The features common to all of these are that the pass level should be reached within twenty-eight days of commencing the test, although once degradation has started the pass level should be reached within the next ten days.

The modified AFNOR test involves the measurement of dissolved organic carbon (DOC) reduction in a test solution concentration of 40 mg l^{-1} (as C). The pass level is 70% DOC reduction. An inoculum is made up from three samples of polluted surface water containing at least 10^5 bacterial cells per ml. The cells are collected on a 0.22 μm filter and resuspended to a final concentration of $5 \pm 3 \times 10^7$ per ml. Suitable activity of the inoculum is checked by the degradation of glucose, 80% of which must degrade within seven days. Glucose degradation can also be used to examine whether the test compound is inhibitory. A sterile control is also included.

The modified Sturm test is based upon the measurement of carbon dioxide evolution from 10 and 20 mg l^{-1} solutions of the test compound aerated with carbon dioxide-free air. The pass level is a yield of 60% of the theoretical carbon dioxide production. The carbon dioxide is trapped as insoluble barium carbonate in barium hydroxide solution and the remaining barium hydroxide determined by acid titration. The inoculum used is the supernatant following settling of blended activated sludge taken from a works receiving little or no industrial effluent.

The modified MITI test originates from Japan, and involves measurement of BOD reduction in a closed system fitted with a DO probe. The inoculum used is coarse-filtered activated sludge supernatant obtained from a mixture of samples taken from widely differing locations. The closed bottle test, in which a 60% BOD reduction is the pass level, is very similar to the BOD test. Between 2 and 10 mg l^{-1} of the test material is used. The inoculum used is a mixture of equal volumes of coarse-filtered garden soil suspension, sewage effluent and effluent from a laboratory-scale activated sludge simulation. The modified OECD test uses a similar inoculum to measure the ultimate biodegradability of between 5 and 40 mg l^{-1} (as C) of the test substance during aeration in the dark. The pass level is 70% reduction in DOC.

Standardization of these techniques is difficult because there are bound to be variations in the quality and activity of the inoculum. Although reference substances are not recommended, controls using aniline, sodium acetate or sodium benzoate may be used to check inoculum activity.

(b) Inherent biodegradability

Tests for inherent biodegradability are designed to assess whether a substance has any potential for biodegradation. A negative result in an inherent biodegradability test often means that the substance can be assumed persistent in the aquatic environment. A level of greater than 20% degradation can be regarded as evidence for inherent primary degradability and greater than 70% mineralization as evidence for ultimate degradation.

The modified SCAS test involves a fill and draw activated sludge unit which is aerated for 23 h, allowed to settle and the supernatant replaced with settled domestic sewage. The test is run in parallel, one unit acting as control and the other containing the test material at a concentration of 20 mg l^{-1} as DOC. It is recommended that the test is run for least twelve weeks. In the modified Zahn–Wellens test, activated sludge (0.2–1.0 g l^{-1}) is aerated for twenty-eight days with a solution of nutrients and trace elements containing the test material at a concentration of 50–400 mg l^{-1} as DOC. The modified MITI (II) test resembles the equivalent test for ready biodegradability, which is very similar to a traditional BOD test, but which uses an activated sludge seed at a concentration of 100 mg l^{-1}.

18.3 Degradation in heterogeneous environments

The biodegradability tests detailed in the previous section are effectively batch cultures in that they commence with a relatively high substrate concentration and low biomass concentration and end when the substrate runs out. This situation rarely occurs in the environment, and it was not until the advent of continuous culture techniques that microbiologists were able to simulate the completely homogeneous conditions at the other extreme of the spectrum of microbial ecosystems, where the steady state concentrations of biomass and substrate are uniformly high and low, respectively. There are more environmental situations which resemble continuous cultures than those which resemble batch cultures. However, there are also situations which resemble neither, and these can be broadly described as heterogeneous environments. They may involve discrete interfaces, such as air/water or water/solid substrata, or perhaps the strata in a stratified lake, or gradually changing conditions, such as solute gradients. A good example of the latter is an estuarine situation in which there will be a chloride gradient.

Laboratory culture techniques are not well developed for the recreation of heterogeneous environments. However, a recent development is the multistage bidirectional chemostat. This consists of several (e.g. five or six) stirred reactors forming a cascade so that reactor one overflows into reactor two, which overflows into reactor three, and so on. In addition the culture from each reactor is pumped back into the reactor immediately before it in the cascade sequence. Fresh culture medium can be pumped into the system at each end or both ends of the array and spent culture overflows or is pumped out at each end or both ends. If the system is operated with the same flow rate in both directions with a solvent added at one end of the cascade a stable solute gradient is established according to the following relationship:

$$X_n = \frac{[(N+2)a - b]}{N+1} + \frac{(n+1)(b-a)}{N+1} \tag{18.1}$$

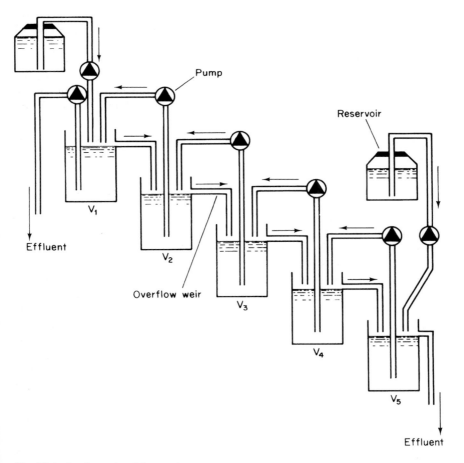

Fig. 18.1 A schematic of the gradostat.

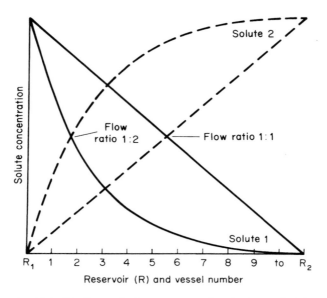

Fig. 18.2 Gradients of solute concentrations in the gradostat.

where X_n is the solute concentration in the nth vessel, N is the number of vessels and a and b are the concentrations of different solute in each reservoir. A schematic of the system is shown in Fig. 18.1 and a graph of the gradient of solute concentration in Fig. 18.2.

18.4 Microbial activity in trace heavy metal cycling

Microbial processes feature significantly in the cycling of the heavy metals. Many metals in aquatic systems are associated with dissolved and particulate organic matter. This material gradually becomes incorporated into sediments. Micro-organisms can act on these sediments in a variety of ways. They can destroy organic matter, thus affecting the complexation of metal ions, they can change the physico-chemical conditions (e.g. pH and redox potential), thus affecting the solubility of metal ions, for example, and they can convert certain inorganic metallic species into organometallic compounds.

Certain organic compounds, both naturally occurring (e.g. fulvic and humic acids) and man-made (e.g. nitrilotriacetic acid (NTA)) bind heavy metals quite strongly by complexation or chelation reactions. In the case of NTA, which is added to detergents as a "builder", designed to chelate the hardness ions (see Section 18.5.2), the chelation process results in enhanced mobility of the metal ions in aquatic systems. NTA prevents the metals from binding to the settleable sludge fraction in wastewater treatment processes and can even resolubilize metals already bound to wastewater sludges and river and lake sediments. The impact of NTA on heavy metal solubility and mobilization would clearly be reduced if it were to be degraded fairly rapidly in wastewater treatment and receiving waters. Although NTA has been found to degrade in the activated sludge process, not all of it is removed by this process. Moreover, the introduction of NTA to such systems often results in acclimatization times of thirty days or more before the onset of degradation. Biodegradability tests (Section 18.2.3) are also inconsistent.

An example of physico-chemical changes mediated by micro-organisms contributing to metal cycling is bacterial leaching of metals. Bacteria of the genus *Thiobacillus* are capable of oxidizing elemental sulphur and sulphide to sulphate as a means of obtaining energy for autotrophic growth. These organisms can exist in a highly acidic environment, some of the acid being generated by the bacteria themselves. A simplistic way of representing this is

$$2S + 2H_2O + 3O_2 \rightleftharpoons 2H_2SO_4$$

T. ferrooxidans can also oxidize ferrous ions in acid conditions:

$$4Fe^{2+} + O_2 + 4H^+ \rightleftharpoons 4Fe^{3+} + 2H_2O$$

Heavy metals (e.g. Zn) can be mobilized from insoluble sulphide deposits by reactions of the following type:

$$ZnS + 8Fe^{3+} + 4H_2O \rightleftharpoons Zn^{2+} + SO_4^{2-} + 8H^+ + 8Fe^{2+}$$

The third way in which micro-organisms can influence metal cycling is exemplified by mercury transformations. Mercury is one of the more toxic heavy metals, and the ability of some bacteria to methylate it is thought to have arisen as a detoxification mechanism. Methylmercury can be excreted by micro-organisms which form it. Unfortunately, it is highly toxic to higher organisms. Many bacteria can also mediate transformations of inorganic mercury.

Hg^{2+} ions can be reduced to metallic mercury in reducing conditions by enzymic reactions mediated by organisms such as *Pseudomonas*. This also can be considered a detoxification mechanism because Hg^0 can be lost from an aqueous system as a result of its volatility. Mercury may be methylated by the transfer of methyl groups from methylcobalamin (vitamin B_{12}—CH_3) in extracts of methanogenic bacteria. Microbial methylation of mercury can, in fact, occur both aerobically and anaerobically. The predominant product is monomethylmercury (CH_3Hg^+), but some dimethylmercury (($CH_3)_2Hg$) is also formed. Some organisms can also demethylate monomethylmercury, generating elemental mercury in the process. The methylmercury compounds are of particular concern because they accumulate in the tissues of fish and shellfish due to their lipophilicity. The cycles of inorganic and

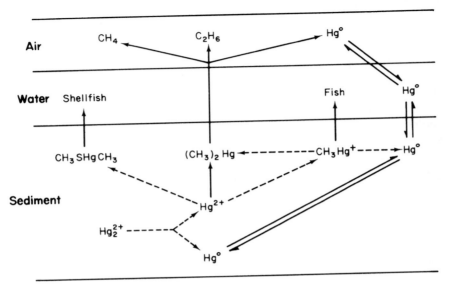

Fig. 18.3 Mercury cycling in the environment.

organic mercury species in the environment are shown in Fig. 18.3, which highlights the transformations mediated by bacteria.

Biomethylation processes are not restricted to mercury. Volatile methyl compounds of arsenic, selenium, lead and tin can also be formed. Again strains of *Pseudomonas* have been implicated in these processes.

18.5 Eutrophication

In a healthy, natural aquatic ecosystem such as a lake or a river, there are three main biotic components existing together in a balanced condition. These are the producers, consumers and decomposers. The producer organisms include the green plants, the algae and the photosynthetic bacteria and cyanobacteria. The consumer organisms include herbivorous aquatic animals and the carnivores which in turn feed on these. When the consumer oganisms die and decay they constitute a source of food for the decomposer organisms, which by a process of mineralization make the nutrients available to the primary producers once again. The entire cyclical process is driven by the energy of sunlight and is regulated by the availability of nutrients.

18.5.1 Limiting nutrients

The limiting nutrient in any system is the one which imposes a restriction on the extent of sustainable growth of the biomass. This concept was discussed earlier in Chapter 10. In wastewater treatment processes (and, usually by design, in laboratory bacterial cultures) carbon is the limiting nutrient. The C : N : P ratio of domestic sewage, for example, is about 20 : 19 : 1. Contrast this with the ratio of these elements in algal protoplasm, which is about 101 : 16 : 1. In natural inland waters, the nutrient limiting algal growth is most frequently phosphorus. Carbon dioxide is in plentiful supply from the atmosphere, as is nitrogen, which many algae can fix directly. As much as 50% of the nitrogen in the algal biomass in some lakes may have been fixed directly from the atmosphere. In the marine environment, however, nitrogen can be the limiting nutrient.

18.5.2 Causes of eutrophication

Eutrophication can be defined as the enrichment of waters by nutrients and the consequent deterioration of quality due to the luxuriant growth of plant life and its resultant effect on the ecological balance of the affected waters.

Eutrophication is caused most frequently in freshwaters by an increase in the orthophosphate (PO_4^{3-}) concentration. This can occur for a variety of reasons. Phosphate can arise from both point and non-point sources. The most notable point source is effluent from wastewater treatment plants. Efficient phosphorus removal in wastewater treatment rarely occurs because these systems are carbon-limited, although processes which make use of "luxury uptake" of phosphates have been designed (see Chapter 14). Raw sewage often contains condensed phosphates, such as sodium tripolyphosphate (STPP). Many synthetic laundry detergents contain a significant concentration of STPP which acts as a "builder", enhancing the efficiency of the surfactant by sequestering the hardness ions calcium and magnesium and so preventing the deposition of insoluble scums on the washed articles. STPP hydrolyses fairly rapidly to orthophosphate in wastewater treatment. Where its use in detergents has been uncontrolled, it has been estimated that more than 30% of the phosphate in wastewater treatment effluents originates from the use of detergents. The implication of this in causing eutrophication has led to the development of legislative provision for the control of STPP in detergents in many parts of the world. However, detergents are not the sole cause of eutrophication. Other major sources of phosphorus in freshwaters include agricultural runoff and phosphate mining.

Relatively small increases in available phosphorus concentration can have very large effects on algal productivity. An algal protoplasm with C : N : P ratio of 101 : 16 : 1 implies an increase of over 100 mg l^{-1} in algal dry matter for each 1 mg l^{-1} in available phosphorus.

The concentration of phosphate which will initiate eutrophication is uncertain. Problems due to algal growth have been associated with phosphorus levels as low as 0.01 mg l^{-1} in surface waters. Where eutrophication is already a problem, such as in the Great Lakes of North America, chemical treatment of wastewater effluents is needed to maintain such low phosphorus levels.

18.5.3 Effects of eutrophication

(a) Lake stratification

The effects of eutrophication in deep lakes are closely connected with the phenomenon of thermal stratification. This phenomenon is due to the fact that water is at its densest while still liquid at 4°C, which also explains why ice floats and ponds freeze from the surface downwards. In the summer, the water at the surface of the lake becomes heated, and as it warms up it becomes less dense than the colder water at 4°C which forms a layer at the bottom of the lake. The surface layer, usually only a few metres deep, is called the epilimnion and the lower, colder layer the hypolimnion. Between them is a zone of transition called the thermocline or metalimnion in which the temperature changes significantly with quite small changes in depth. This acts as a kind of barrier which prevents the exchange of material between the epilimnion and the hypolimnion. In the autumn, the epilimnion cools. When it reaches 4°C the lake is isothermal throughout and mixing can then occur; this is called the autumn turnover. As the surface of the lake cools below 4°C the lake again becomes stratified, but the less dense water near the surface is colder than the denser bottom layer at 4°C. In the spring, the surface layers become warmer and another turnover occurs as the lake becomes isothermal. If particularly cold winters are not experienced, then only one annual turnover rather than two may occur.

The thermocline acts as a barrier to the movement of dissolved oxygen from the upper epilimnion, where it may be abundant due to photosynthesis and wave action, to the hypolimnion. Hence, when productivity in the epilimnion is high, the hypolimnion, representing the majority of the total volume of the lake, may be completely anaerobic.

(b) Biological effects

One of the earliest noticeable effects in a polluted ecosystem is a progressive reduction in the number of different species present, but a significant increase in the number of individuals of certain species. As a result of this food chains become shorter and less complex and the community is generally destabilized.

The micro-organisms which come to dominate in eutrophic lakes and rivers differ. In rivers the filamentous blanket weed *Cladophora* may grow rapidly and cause problems. In lakes, the normal dominance of diatoms or green algae in oligotrophic conditions is shifted in favour of the cyanobacteria when the water becomes eutrophic. These organisms possess a gas vacuole which allows them to float and to form dense surface layers of biomass called "blooms". The ability of these organisms to float reinforces any stratification which may already exist in the lake. The dissolved oxygen levels depend on the balance between respiration and photosynthesis. In the upper layers photosynthetic activity during the day results in a positive oxygen balance, but at night respiration can deplete this so that anoxic conditions prevail temporarily. Below a certain depth, where the light intensity falls to less than 1% of its value at the surface, respiration always exceeds photosynthesis and if the lake is thermally stratified, the hypolimnion will become anaerobic and remain so until the turnover.

In the epilimnion, intense photosynthesis can result in quite significant pH increases due to the removal of dissolved carbon dioxide and the formation of insoluble carbonates which are in turn hydrated to form hydroxides:

$$2HCO_3^- \rightleftharpoons CO_3^{2-} + CO_2 + H_2O$$
$$CO_3^{2-} + H_2O \rightleftharpoons CO_2 + 2OH^-$$

The hydroxide ion concentration can reach $10^{-3}M$ (pH 11) at which level it can be lethal to fish populations.

In the presence of blooms, the processes of death and decay continue, with algal biomass sinking through the thermocline into the hypolimnion and exerting an excessive oxygen demand upon it. As the hypolimnion becomes anaerobic reduced inorganics, sulphides, ammonia and methane are formed. The aerobic flora and fauna disappear, leaving a much less diverse population behind. In the long term, the accumulation of sedimentary material from the degradation of algal biomass can be so great that the lake fills up and shallow lakes may disappear altogether. The sediment can also act as a store for phosphate because under aerobic conditions in the hypolimnion any iron present is oxidized to the ferric ion which forms sparingly soluble salts with the phosphate which in turn becomes associated with the aerobic upper layers of the sediment. Oxygen depletion in the hypolimnion as a result of algal biomass input from the epilimnion reduces the iron to ferrous ion, releasing soluble phosphate which becomes available to the algae when the next turnover occurs.

(c) Socio-economic effects

The amenity value of eutropic waters is reduced because they are aesthetically unpleasant, plant growth and silt hinder navigation, the angling stock is reduced and there may be health hazards to swimmers.

Severe economic effects may also occur. If the water is abstracted for potable supply the costs of treatment may be increased. Algal biomass may physically clog filters so that they have to be cleaned more often. The water may be unpalatable because of organoleptic compounds excreted by the algae. Chemical treatment systems may require higher dosing rates to achieve effective removal of undesirable materials.

Many algae and cyanobacteria excrete toxins which can have a direct effect on fish or on livestock drinking the water directly. Algal toxins have been implicated in both human and animal diseases in several countries. Symptoms of gastroenteritis and allergic reactions have been observed following the development of *Microcystis* and *Anabaena* spp. in sources of abstraction.

18.6 Microbiological indicators of pollution

The most important use made of indicator organisms is in the detection of faecal pollution of water (see Chapters 5 and 20). However, they can also be useful as indicators of pollution by organic and inorganic substances.

An overall biological indication of water quality can be obtained from an examination of saprobity. If an effluent containing biodegradable material is discharged to a surface water the reaches downstream of the discharge can be divided into four zones. The polysaprobic zone corresponds to the reach immediately below the point of discharge. Degradation is very rapid, and may cause anaerobiosis to develop with reduced end-products being formed. The water is typically very turbid. The next zone is termed α-mesosaprobic. Degradation is slowing down and free oxygen may be present. In the β-mesosaprobic zone, free oxygen is abundant, especially during the day in eutrophic waters. There are only traces of polymer degradation products left. Photosynthesis is an important contributor to reaeration. The final zone is termed oligosaprobic. Typically the dissolved oxygen concentration has reached saturation and mineralization is complete.

Each of these zones can be characterized by indicator species which exist exclusively in particular zones. Each of the indicator species is assigned a number from one to four corresponding to the zone in which it is typically found, where the oligosaprobic zone is assigned the value one and the polysaprobic zone the value four. Each of these numbers is multiplied by a frequency rating from one (random occurrence) to five (massive development). The products are summed for all species examined and divided by the sum of frequency ratings observed to give the saprobity index, a number between one and four. The saprobity index is considered reliable only if a sufficiently large number (ten or more) of indicator organisms is used. The types of organisms which can be used as indicators in saprobity indices are not confined to micro-organisms, but cover a broad spectrum of animal and plant life.

There are some advantages to using microbial indicators only, however, because they tend to respond more quickly to changes in environmental conditions than the macro-flora and fauna. If they are sessile (i.e. anchored to a substratum) they remain fixed at the sampling point. In the polysaprobic zone are found *Euglena*, *Thiobacillus* spp. and filamentous bacteria such as *Thiothrix* and *Beggiatoa* which can use hydrogen sulphide as an energy source, and *Sphaerotilus*, which is also implicated in activated sludge bulking. In the α-mesosaprobic zone filamentous green algae and filamentous blue-green bacteria are dominant, with green algae continuing to dominate the less polluted β-mesosaprobic and oligosaprobic zones.

A particularly unsightly indicator of organic pollution is a growth called "sewage fungus". This is a mixed population of filamentous organisms, including *Sphaerotilus natans*, *Zoogloea* and *Beggiatoa*, together with fungal and algal forms such as *Geotrichum* and *Stigeoclonium*. The filaments confer a certain cohesivity on the community so that it forms a dense unsightly growth attached to submerged substrata. The growth attracts other higher organisms which use it as a source of food. "Sewage fungus" can be a particular nuisance where discharges contain predominantly carbohydrates. Pulp and paper industry effluents may also cause its development.

References and further reading

Dart, R. K. and Stretton, R. J. (1980) *Microbiological Aspects of Pollution Control*, 2nd edn, Elsevier Scientific Publishing Company, Amsterdam.

Higgins, I. J. and Burns, R. G. (1975) *The Chemistry and Microbiology of Pollution*, Academic Press, London.

Horan, N. J. (1990) *Biological Wastewater Treatment Systems: Theory and Operation*, John Wiley & Sons, Chichester.

IHD–WHO Working Group on Quality of Water (1978) *Water Quality Surveys*, Unesco/WHO.

Moat, A. G. and Foster, J. W. (1995) *Microbial Physiology*, 3rd edn, John Wiley & Sons, Chichester.

Wimpenny, J. W. T. and Lovitt, R. W. (1984) in *Microbiological Methods for Environmental Biotechnology* (eds J. M. Grainger and J. M. Lynch), Academic Press, London, pp. 295–312.

Chapter 19

Pathogens

19.1 Introduction

Pathogenicity is the ability of a micro-organism to cause disease. Different species of bacteria are generally pathogenic to different organisms, i.e. plant pathogens are unlikely to affect animals or man and vice versa. Certain pathogens called zoonoses, however, have low host specificity and can be transmitted from animals to man. The individual strains of the same species of bacteria may cause varying degrees of infection of the host organism; the relative ability of the different strains is termed virulence. As an example, cells of *Streptococcus pneumoniae* which have extracellular polymer capsules are more virulent in causing pneumonia than non-capsulated strains. Microbial agents of disease generally originate from outside the body, although under conditions of lowered resistance, the body's natural microbiota may become harmful. Susceptibility to pathogens varies with a number of factors, but is likely to be increased when the host is very young or very old, suffering from malnutrition or environmental stress, or when the organism is wounded or already infected with another disease.

The major mechanisms by which micro-organisms affect the host are invasion and toxin production. Some bacteria, e.g. *Bacillus anthracis*, must proliferate throughout the entire body in order to kill the host, whereas others remain localized but produce sufficiently large quantities of toxin to affect all the body tissue. An example of the latter is *Corynebacterium diphtheriae*, which grows in the throat. Two types of bacterial toxin are formed, called exotoxins and endotoxins. *C. diphtheriae* produces an exotoxin, which is excreted from living cells and is subsequently circulated through the host via the bloodstream. Diseases caused by exotoxins include botulism, cholera, whooping cough, dysentery and gas gangrene. Endotoxins are retained within viable cells and are only liberated on death or lysis of the cell. In contrast to exotoxins, endotoxins are less toxic, more heat stable and cannot be chemically modified to reduce their toxicity. They are frequently responsible for the secondary effects of infection such as fever and shock. The symptoms caused by bacterial invasion and toxin production are normally manifested in full, but in some cases the host acts only as a carrier of disease, displaying subclinical or no symptoms. The control of a disease such as typhoid is hindered by its long-term persistence in carriers, as the bacterial cells are excreted only sporadically, making detection difficult.

19.2   Modes of entry and transmission of pathogens

In order to cause an infection, pathogens must gain entrance to the body of the host. Micro-organisms cannot penetrate normal, undamaged skin but can pass through mucous membranes, which thus form the most common portals of entry. These occur in the alimentary, respiratory and genito-urinary tracts. Additional routes of entry are via abrasions or small openings in the skin where local infections such as boils may occur, via wounds from which infection may spread throughout the body by means of the circulatory system, or via animal or insect bites. Examples of pathogenic micro-organisms which use each method of entry into the host are given in Table 19.1, together with the diseases that they cause. The degree of harm caused by bacterial toxins varies with the mode of entry into the body. Tetanus toxins, for example, cause disease when introduced into skin or muscle but are relatively harmless when passing through the digestive tract.

Once having established an infection, the continued survival of a pathogen necessitates its transmission to other susceptible organisms. An infected organism cannot sustain the pathogen indefinitely, but either dies or develops an immune resistance such that the invading micro-organisms can no longer proliferate. Pathogens must therefore be dispersed from a current host by means of a portal of exit and transmitted to a new host. In cases where the mode of transmission involves a period outside a suitable host, the micro-organism must be able to withstand the unfavourable conditions.

The mode of transmission is largely dependent upon the location of the relevant infection in the host. Thus bacteria causing respiratory diseases such as pneumonia and tuberculosis are transmitted by droplet exhalation when an infected individual sneezes, coughs and even speaks loudly. In urban areas of high population density, a highly infectious pathogen such as the influenza virus can spread from one to 10^6 people within two months. Enteric pathogens, i.e. those that occur in the intestinal tract, are transmitted through faecal contamination of food and water. They leave the host in the faeces or urine and must enter the next host via the mouth to relocate in the intestines. In underdeveloped countries, poor sanitation frequently results in water supplies and associated food sources being directly contaminated with untreated sewage, thus transmission rates of enteric pathogens are high. Where adequate sanitation is maintained, however, other transmission routes are more significant. These include infection by flies which are in frequent contact with both faeces and food and by humans involved in food

Table 19.1 Examples of some diseases caused by pathogenic micro-organisms

Mode of transmission	Portal of entry	Pathogenic micro-organism	Disease
Waterborne	Gastro-intestinal tract	Salmonella typhi	Typhoid fever
		Vibrio cholerae	Cholera
		Shigella dysenteriae	Bacterial dysentery
		Legionella pneumophila	Legionellosis
		Salmonella spp.	Gastro-enteritis
		Campylobacter spp.	
		Rotavirus	
		Hepatitis A virus	Hepatitis
		Entamoeba histolytica	Amoebic dysentery
Airborne	Respiratory tract	Corynebacterium diptheriae	Diptheria
		Mycobacterium tuberculosis	Tuberculosis
		Streptococcus pneumoniae	Pneumonia
		Neisseria meningitidis	Meningitis
		Rhinovirus	Common cold
		Varicella-zoster virus	Chicken pox
Direct contact	Genito-urinary tract	Treponema pallidum	Syphilis
		Neisseria gonorrhoeae	Gonorrhoea
	Minor abrasions	Bacillus anthracis	Anthrax
		Leptospira icterohaemorrhagiae	Leptospirosis
	Wounds	Clostridium tetani	Tetanus
		Clostridium perfringens	Gas gangrene
Animal/insect vector	Flea bite	Yersinia pestis	Plague
	Tick bite	Francisella tularensis	Tularemia
	Louse bite	Rickettsia prowazekii	Typhus fever
	Mosquito bite	Yellow fever virus	Yellow fever
	Dog, fox bite	Rabdovirus	Rabies
	Mosquito bite	Plasmodium falciparum	Malaria
		P. vivax, P. ovale, P. malarie	
	Tsetse fly bite	Trypanosoma gambiense	Sleeping sickness
		T. rhodesiense	

handling and processing, who although healthy may be carriers of a number of enteric organisms.

In contrast to those organisms transmitted by droplet exhalation and faecal contamination, a number of pathogens are incapable of surviving outside the host for any length of time, and must therefore be transmitted by direct contact. Examples of these include the pathogens which cause the sexually transmitted diseases syphilis, *Treponema pallidum*, and gonorrhea, *Neisseria gonorrhoeae*. Other diseases spread by direct contact are anthrax, brucellosis and tularemia. These are contracted by infection of small abrasions in the skin following handling of infected animals, carcasses or skins. The fourth major mode of transmission is through an intermediate host which carries the pathogen and subsequently introduces it into another host, i.e. man, through its bite. Intermediates are frequently insects, such as the anopheles mosquito which transmits malaria, the flea which carries plague and the tick which harbours tularemia bacteria. Vertebrate carriers of pathogenic organisms include dogs, cats, foxes, bats, skunks and others which transmit the rabies virus through their bite. In some cases, a pathogen can grow in more than one host, giving a selective advantage over those which are restricted. The plague bacterium *Yersinia pestis* multiplies not only in man, but also in its vector, the flea, and in its natural reservoir, the rat. As a result both its survival and transmission rates are high, accounting for the degree of devastation caused by plague epidemics. An example is that of the fourteenth century known as the "Black Death" which is reported to have caused the death of approximately one third of the world's population.

With respect to the environmental engineer, those pathogens associated with water are of most significance. These include both the enteric microorganisms which are transmitted by faecal contamination of water supplies and those pathogens carried by insect vectors which live and breed on or in water, such as the malarial mosquito. The following sections will thus be focussed on these aspects in particular.

19.3 Diseases caused by waterborne pathogens

The majority of waterborne diseases arise as a result of the contamination of water used for drinking with human or animal faeces. The pathogens contained in the faeces of a diseased human or animal or in those of a symptomless carrier are distributed within the water body and may be ingested by others, either directly by drinking or by consuming crops irrigated or washed with the same water. The types of infection transmitted via this route are called enteric infections and include cholera, typhoid fever, dysentery, shigellosis, diarrhoea and *Salmonella* infections. On a global scale, infections attributable to inadequate water or sanitation have been estimated by the World Health Organisation to comprise 80% of all diseases and that at any one time, half the hospital beds in the world are occupied by people suffering from water-related diseases.

The major source of the aetologic agents of such diseases is human sewage. Monitoring the pathogenic content of sewage has proved to give a good indication of the types of illness likely to prevail in a community at any time. Pathogenic organisms which occur in sewage are listed in order of significance in Table 19.2. In addition to sewage, however, pathogens which infect humans have been identified in livestock, fowl, wildlife, pets and even fish in contaminated water. This highlights the range of mobility which exists for enteric pathogens and explains the level of concern over faecal pollution. Increased intercontinental travel further permits the spread of pathogenic organisms on a large scale, and the speed of travel means that symptoms of an enteric infection acquired abroad may not become evident until some time after the host's return. A case of typhoid fever reported in the UK in 1983 is an example of this.

Of the twenty to thirty recognized water-related diseases, not all are caused by the consumption of contaminated water. In the less developed

Table 19.2 Pathogenic organisms that can occur in sewage, listed in order of importance

Bacteria	Viruses	Intestinal parasites
Salmonella typhi	Enteroviruses	Schistosoma spp.
S. paratyphi	Poliovirus	Ascaris lumbricoides
other spp.	Echovirus	Trichuris trichuria
Shigella spp.	Coxsackieviruses	Taenia spp.
Vibrio cholerae	New enteroviruses	Diphyllobothrium latum
Mycobacterium turberculosis	Hepatitis type A	Ancylostoma duodenale
Leptospira icterohaemorrhagiae	Norwalk virus	Necator americanus
Campylobacter spp.	Rotavirus	Entamoeba histolyica
Listeria monocytogenes	Reovirus	Giardia lamblia
Candida albicans	Adenovirus	Naegleria spp.
Yersinia enterocolitica	Parvovirus	Acanthamoeba spp.
Enteropathogenic Esherichia coli		Cryptosporidia
Pseudomonas aeruginosa		
Klebsiella spp.		
Staphylococcus aureus		
Aeromonas hydrophila		
Mycobacterium paratuberculosis		
Erysipellothrix rhusopathiae		
Bacillus anthracis		
Clostridium spp.		
Yersinia pestis		
Brucella spp.		

countries other routes of infection are commonly by direct contact with pathogenic parasites in water, or by bites from infected arthropod vectors which undergo a stage of their life cycle in the aquatic environment. In developed countries where sewage and water treatment are widely practised, there are fewer cases of direct contamination of potable water supplies, other than those involving accidental leakage. More concern is directed to the health risks of marine sewage outfalls contaminating bathing beaches and shellfish beds and the problems arising from pathogens present in sewage sludge disposed of to land.

To measure faecal pollution, as for instance on a bathing beach, the presence of a group of bacteria known as faecal coliforms is determined, the most common of which is *Escherichia coli*. Since enteric bacterial and viral pathogens are frequently present in low numbers and are difficult to detect, the occurrence of *E. coli* is taken as an indication that the waterbody in question has been contaminated with sewage and may contain harmful organisms. *E. coli* is thus an indicator organism for the assumed presence of enteric pathogens. Methods of enumerating faecal coliforms in water include the multiple tube test and membrane filtration method, both of which are described in Chapter 5.

The following sections of the current chapter present descriptions of some of the major bacterial, viral and parasitic diseases caused by waterborne pathogens. Although not strictly microbial in nature, helminth and worm infections which may occur as a result of faecal contamination of water have also been included, since these are relevant to the environmental engineer.

19.3.1 Bacterial infections

(a) Cholera

Cholera is endemic in Asia, notably in India and Bangladesh. At various times large epidemics (pandemics) have spread from Asia to Europe, Russia, Africa and even America. One pandemic reaching Europe in 1854–62 caused 20,000 deaths in England alone. During this pandemic a London doctor, John Snow, correlated an outbreak of cholera at Broad Street in London with one particular water pump and removed its handle to control the spread of the disease. It was later found to have originated from sewage contaminating the well. Thirty years later, the causative agent *Vibrio cholerae* was identified, and the role of pathogens in disease began to be recognized.

V. cholerae is visible in Gram-stained preparations of the excreta from infected hosts as a motile, Gram-negative rod with a single polar flagellum. Two biotypes are recognized, the classical and the El Tor, and these biotypes are further subdivided by their serological characteristics into two serotypes, the Inaba and Ogawa. *V. El Tor* initially caused a much less severe form of cholera than *V. cholerae*, but during the twentieth century has apparently increased in virulence and, due to its more robust nature, has recently spread more rapidly than *V. cholerae*.

The cholera bacteria colonize the small bowel and after an incubation period of one to three days produce an enterotoxin (choleragen) which causes the disease cholera. The major effect is extremely rapid loss of fluid, giving symptoms of severe diarrhoea and vomiting. The resultant dehydration, accumulation of acid metabolites in the blood and shock can cause death within a few hours. If, however, fluid and electrolytes, commonly a solution of salt and sugar, are replaced promptly, recovery can be equally as rapid. Cholera is a self-limiting disease provided that dehydration and shock are prevented. Treatment methods include the use of tetracycline.

Excretion of cholera vibrios by convalescent carriers may occur for up to fifteen months, while symptomless carriers excrete the pathogen intermittently for periods of six to fifteen days. Identification of *V. cholerae* from excreta can be made by culture on selective media for 18 h. Definitive identification requires biochemical testing and agglutination testing with specific antisera.

Cholera outbreaks most frequently originate from polluted water and are typically explosive in nature. Infection can also be spread by faecal contamination of food, either directly or by flies, or through contact with contaminated clothing. Vaccines composed of heat-killed mixtures of *V. cholerae* and *V. El Tor* are available but provide limited protection for three to six months only. Prevention thus necessitates the purification of water supplies, proper disposal of sewage and careful personal hygiene.

(b) Typhoid fever

Typhoid fever occurs globally but most frequently in areas with poor sanitation. As late as 1909 there were up to 500,000 annual incidents of typhoid fever in the USA, with over 40,000 deaths. The disease was gradually controlled by filtration and chlorination of water supplies, but remains endemic in many other areas of the world, including the Mediterranean.

The aetiologic agent *Salmonella typhi* is transmitted through water or food contaminated directly or indirectly with human faeces. Typhoid fever is specific to humans; although it does not occur in animals and is thus not transmitted by other species, it is problematical due to its persistent survival in a variety of habitats including water, ice, dust and dried sewage.

S. typhi is a motile, flagellated, Gram-negative, non-sporeforming rod. It is facultatively anaerobic, fermenting glucose but not sucrose or lactose. The possession of capsular (Vi), somatic (O) and flagellar (H) antigens allows its identification by serological means and the reaction to bacteriophage typing indicates over eighty different varieties of the bacterium, each of which may become the source of an outbreak.

On entering the gastro-intestinal tract, *S. typhi* locate intracellularly, for instance within phagocytes, and multiply rapidly during an incubation period of approximately ten to fourteen days. Subsequently the bacteria enter the bloodstream and are widely disseminated throughout the body. Heavy infection of the bile duct occurs, from where millions of bacteria are released into the intestine. Loci of infection also become established in the lungs, bone marrow and spleen and reinfection can occur from these sites.

The symptoms caused by *S. typhi* endotoxins include fever, abdominal swelling, rash, headache, nausea, vomiting and diarrhoea. If untreated the illness can be fatal within weeks. *S. typhi* can be isolated from the faeces optimally one month after infection, when the bacteria are excreted in large numbers. Excretion may continue for two to three months and in some cases for up to a year. Treatment of typhoid fever is by chloramphenicol, while ampicillin is used to eradicate the bacteria in healthy carriers. Preventative vaccines are available and are effective in 70–90% of cases but, in common with other waterborne diseases, the most appropriate prevention method is the provision of adequate sanitation and the prevention of carriers from handling food.

(c) Salmonellosis

The majority of infections associated with the water cycle are those causing gastro-enteritis. This can be caused by a variety of micro-organisms, including the recently recognized *Campylobacter jejuni*, which has been causing a rapidly increasing number of outbreaks. However, one of the most significant aetiologic agents of gastro-enteritis remains the *Salmonella* bacterium. The total number of *Salmonella* serotypes known to be pathogenic to man exceeds several hundred and their frequency of isolation fluctuates both temporally and spatially. It has been estimated that 1% of the human population may excrete salmonellae at any one time. Not surprisingly, therefore, salmonellae are commonly detected in polluted waters including sewage, stabilization ponds, irrigation water, stormwater, streams and tidal water. In addition, they are carried by a wide range of animals. Percentages of farm animal populations which are healthy but carry *Salmonella* have been evaluated as 13–14% cattle, 4–15% sheep and 7–22% pigs. Contamination from animals which share common water supplies with humans, or from agriculturally-based industrial effluents such as abattoirs and meat processing plants can thus occur on a large scale. It is also recognized that animals can become affected by grazing pasture upon which sewage sludge has been spread. Salmonellae which have survived sewage treatment may remain viable for periods of up to several weeks on grassland, therefore grazing restrictions for three weeks after sludge application have been set by the Commission of the European Communities to minimize potential infections.

Salmonella are facultatively anaerobic Gram-negative rods, motile by means of peritrichous flagella. Infection is contracted by ingesting large numbers of viable bacteria in faecally contaminated food or water. The nature of the pathogenicity of *Salmonella* is to cause either enteric disease, i.e. related to infection of the gastro-intestinal tract, or systemic infection, where the bacteria are disseminated throughout the body. Enteric diseases have incubation periods of one to two days; the most common form is gastro-enteritis which is caused by species such as *S. typhimurium* or *S. enteritidis*. Symptoms are severe diarrhoea and abdominal cramps with nausea and vomiting. Systemic infections have longer incubation periods of seven to fourteen days and are characterized by enteric fevers (*S. sendai*) including typhoid fever (*S. typhi*, see Section 19.3.1(b)) and the milder paratyphoid fever (*S. paratyphi* A, B and C). Systemic infection may also involve colonization of the viscera, meninges, bones and joints with blood poisoning and abscess formation. These distinctions are not always well defined however, for example *S. typhimurium*, the *Salmonella* with the widest distribution which has been isolated from thirty-seven different animal species, usually causes acute enteritis but occasionally produces a prolonged form of enteric fever.

After infection, convalescents may continue to excrete salmonellae in the faeces and to a lesser extent in urine and pus. They may remain infectious for some time. Identification of the organisms involves pre-enrichment in buffered peptone water or selenite broth and subsequent culture on selective media such as MacConkey agar or bismuth sulphite agar.

Control of salmonellosis is extremely difficult due to the varied and far-ranging reservoirs of infection that exist in domestic and wild animals and birds, many of which are symptomless carriers only. The environment is continually being contaminated by *Salmonellae* in the excreta of such animals. Although the major cause of salmonellosis is contamination of food-stuffs derived from infected animals, water-associated incidence of the disease can be prevented by good sanitation and transmission of infection minimized by observing grazing restrictions on sludged pasture and avoiding sewage contamination of water consumed by domestic animals and fowl.

(d) Shigellosis

Shigella organisms cause bacterial dysentery. They are distributed worldwide with different species predominating in different areas; *S. sonnei* and *S. flexneri* comprise 90% of the human *Shigella* population in the USA, whereas in Central America and the Far East, *S. dysenteriae* is most common. Shigellosis affects young children to the greatest extent and shows a seasonal distribution, peaking in autumn and winter. The only animals which can become infected are primates, but man is the main reservoir.

Shigella bacteria are short, Gram-negative, non-motile rods which grow optimally at 37°C under aerobic conditions. On ingestion, the bacteria penetrate the cells of the epithelial lining of the large intestine and multiply, forming lesions in the ileum and colon. The infectious dose is low, of the order of ten organisms. After an incubation period of one to seven days, an endotoxin is produced which causes dysentery; in *S. dysenteriae* an exotoxin and neurotoxin is also produced which causes convulsions and paralysis. The symptoms of bacterial dysentery are initially fever and abdominal cramps, then diarrhoea which becomes increasingly severe over two to four days, with profuse bloody stools. High temperatures and vomiting also occur. The severity of shigellosis ranges from mild transitory diarrhoea to complete prostration, depending on the species and degree of infection. Excretion of *Shigella* continues for one week after infection. Isolation of the bacterium from faeces can be carried out by culturing on selective media such as desoxycholate citrate agar. Final identification methods include biochemical and agglutination tests.

Shigella do not survive long outside the human body, thus direct person-to-person contact forms one mode of transmission. Foodborne and water-borne infections, however, are both significant. The presence of other bacteria decreases the viability of *Shigella*, thus infections are more likely to be contracted from well water than river water, for example, where the total bacterial count is likely to be higher. No vaccines are available, there-fore control of the disease requires disruption of the anal–oral route of transmission by means of good sanitary practice.

(e) Diarrhoeal disease (enteropathogenic Escherichia coli)

Outbreaks of diarrhoeal illness caused by waterborne enteropathogenic strains of *Escherichia coli* have been traced to sewage-contaminated drinking water. Recorded incidences include one of gastro-enteritis among conference delegates in Washington DC, where the same serotype of enteropathogenic *E. coli* was isolated from the water supply and the faeces of those affected. An outbreak in Uppsala, Sweden in 1965 produced 442 cases in two weeks, with 261 people ill at its peak.

The natural habitat of *E. coli* is the alimentary tract of man and warm-blooded animals; it is one of the most abundant of the intestinal bacteria. Over two hundred specific serological types of *E. coli* exist, determined on the basis of their somatic "O", surface "K" and flagellar "H" antigens. The majority are harmless but a limited number cause disease, either within the gut itself or when accidentally introduced into other parts of the body. Infections outside the gastro-intestinal tract are frequently those of the urinary tract, occurring by transfer of the patient's own faecal flora.

Infections within the gut are caused by three different groups of *E. coli*. The first group, enteropathogenic *E. coli* (EPEC), comprises eleven serogroups: 026, 055, 086, 0111, 0114, 0119, 0125, 0126, 0127, 0128 and 0142. This group causes diarrhoea in infants and children of up to five years old, but rarely affects adults. The second group, enteroinvasive *E. coli* (EIEC), includes the serotypes 0124, 0136, 0143, 0144, 0152 and 0164 and causes diarrhoea in both children and adults. The mechanism of pathogenicity of these serotypes is similar to that of *Shigella* with invasion of the intestinal mucosa, causing extensive inflammation, necrosis and ulceration of the tissues with blood being passed in the faeces as a result. The third group, enterotoxigenic *E. coli* (ETEC), contains the serotypes 06, 08, 015, 025, 027, 078, 0148 and 0159. This group causes watery diarrhoea in infants and adults. Two types of toxin are produced, one heat-stable of low molecular weight and one heat-labile, with a higher molecular weight. The heat-stable toxin (ST) has a relatively short-lived action. The heat-labile toxin (LT) causes a similar response to that of the cholera toxin, resulting in secretion of body fluids and electrolytes into the gut. Production of both toxins is plasmid mediated and may thus be transferred between serotypes, although this appears uncommon.

(f) Legionellosis

Legionellosis was first observed in 1976 when, following an American Legion convention in Philadelphia, 221 people became ill with a pneumoniacal-type infection and 34 died. Symptoms of the disease, termed Legionnaire's disease, become evident in two to ten days and include muscular pain, headache, cough, fever, chills and renal dysfunction. Mortality for the disease ranges from 5% in otherwise healthy adults to 15% in the elderly. Several outbreaks have occurred in Britain since 1976, including one at Corby in 1979 and another at a hospital at Kingston-upon-Thames in 1980.

The aetiologic agent for the disease is *Legionella pneumophila*, a short, tapering, Gram-negative aerobic rod with a single polar flagellum. It is a fastidious microbe which can be isolated only by culture on enriched media such as blood agar supplemented with iron and cysteine, or charcoal yeast agar. Identification for diagnostic purposes is done by indirect fluorescent-antibody tests or by using micro-agglutination tests to detect antibodies to the bacterium.

Legionella normally parasitize free-living amoebae and can survive for periods of up to one year in the thick-walled cysts produced by amoebae under adverse conditions. While encysted, they are very resistant to chemical attack and are thus unaffected by bactericidal water treatment methods such as chlorination. Under favourable conditions the host amoeba becomes debilitated by increasing numbers of Legionellaceae, so it ruptures and releases vesicles containing the bacteria. The mode of transmission and infection of man is thought to be by inhalation of such vesicles in water droplets. Once in the lung legionellae colonize macrophages, amoeba-like defence cells, and continue to multiply. Between a hundred and a thousand *Legionella* cells are required to initiate infection.

L. pneumophila has been found to colonize drinking water, warm water and cooling water systems, adhering by means of a slime layer to parts such as shower heads and rubber washers. Its clumping growth in inaccessible sections of plumbing systems tends to protect it from anti-bacterial treatments that may be administered. Moreover, its ability to survive for long periods in stagnant water causes problems in recently reactivated water systems. Measures that have been proposed to control the colonization of water systems by *Legionella* include the maintenance of water temperatures above 70°C and flushing all taps for fifteen minutes before use or the replacement of all gaskets and washers with those which cannot act as a source of nutrients for the bacteria. Once having been cleared from a water system, recolonization by *Legionella* can be prevented by maintenance of free residual chlorine or by use of chemicals such as quaternary ammonium compounds with tributyltin oxide or calcium hypochlorite. Total prevention of legionellosis will be difficult, however, until the original source(s) of the bacterium and its mode of entry into water supplies are identified.

(g) Leptospirosis

Incidences of leptospirosis (Weil's disease) have been reported in Europe, North and Central America and the East, with an annual incidence rate for the total population of approximately 1%. The disease is contracted by contact with water polluted with the urine of infected animals. Outbreaks in the USA have been variously traced to infected cattle and pigs which had access to slow-flowing streams used downstream for bathing.

The aetiologic agents are *Leptospira*, coiled motile aerobic bacteria which belong to the spirochaete group. Many different serotypes exist, the most common human isolates being *L. icterohaemorrhagiae*, *L. canicola* and *L. pomona*. The portal of entry of the leptospirae is the skin, abrasions or the mucous membranes. The organisms gain access to the bloodstream and are disseminated throughout the body, causing acute infections of the kidneys, liver and central nervous system and giving a characteristic jaundiced effect. The severity of the disease can range from subclinical to fatal.

Leptospirosis can be transmitted to man from a wide range of animals including rodents, domestic animals and wildlife. The leptospirae cause mild, chronic infections in these animals and as a result are continuously shed in the animals' urine. Prevention of the disease thus involves restricting bathing and other water-related recreational pursuits in waterbodies likely to be highly polluted with animal wastes, e.g. near discharge points for farming or animal handling operations.

19.3.2 Viral infections

Viral infections are caused either by enteroviruses which live and reproduce within the gastro-intestinal tract, for example coxsackieviruses, echoviruses and polioviruses, or by those which only occasionally infect the gastro-intestinal tract such as rotaviruses, reoviruses, adenoviruses and the hepatitis A virus. Although the last are not strictly enteric viruses, they are found in the faeces of infected individuals and are transmitted via the faecal contamination of water.

(a) Hepatitis

The largest waterborne epidemic of hepatitis was in New Delhi between 1955 and 1956. Over 35,000 cases of infectious hepatitis (hepatitis A) were reported following major contamination of the city's water supply by sewage. The epidemic peaked six weeks after the contamination incident, despite heavy chlorination of the polluted water supply. Similar incidences of contamination resulted in sixty-eight waterborne outbreaks of hepatitis in the USA between 1946 and 1978. In addition to the direct intake of hepatitis virus in drinking water, consumption of shellfish which have accumulated viruses from polluted water has similarly resulted in hepatitis outbreaks. Outbreaks are rarely fatal.

Following ingestion, the hepatitis A virus passes undamaged through the acid conditions of the stomach and lodges in the small intestine. There it enters the epithelial mucosal cells, multiplies and eventually moves to the liver, which becomes inflamed. Viruses are excreted in the faeces halfway through the incubation period of fifteen to sixty days. Symptoms of hepatitis include fever, tiredness, loss of appetite and abdominal pains, with jaundice. Owing to its stability and high resistance to disinfectants, control of hepatitis is difficult. There is evidence to suggest that water treatment methods adequate for the destruction of bacterial pathogens may not be sufficient to prevent viral contamination. Protection of viruses within particles of flocculated organic matter may reduce the apparent effectiveness of chemical anti-microbial agents such as chlorine. Avoidance of the contamination of drinking water and water-derived foodstuffs with faecal matter are therefore essential to prevent infection in areas where hepatitis is endemic.

(b) Viral gastro-enteritis

Gastro-enteritis is probably the most common waterborne disease in developed countries. In the USA there were 311 outbreaks between 1946 and 1970 involving 69,694 cases, and 132 outbreaks between 1971 and 1978. The causative agents for such incidents are not always known, but the enteric viruses, i.e. poliovirus, coxsackievirus A and B, echovirus and enterovirus types 68–71, are frequently implicated. Of the 311 outbreaks from 1946–70, 178 were of viral origin, affecting 45,250 people.

Enteric viruses are small, with diameters of < 28 nm and a single strand of

RNA. Like the hepatitis A virus, they are resistant over a wide pH range. They usually cause infections of the respiratory system or gastro-intestinal tract, but can additionally cause meningitis, paralysis, fever and diarrhoea to varying degrees of severity. Viruses other than the enteric group which are found in sewage, and which thus have the potential for transmission via faecal pollution, are listed in Table 19.2.

(c) Poliomyelitis

Although the respiratory route is the most common method of transmission for the poliomyelitis virus, the faecal–oral route via food or water is also important. For example, two epidemics of waterborne poliomyelitis were recorded in the USA in 1952–3.

In underdeveloped countries with poor sanitation, infection of children with poliomyelitis is common. The disease is relatively mild in children, who thus may acquire immunity at an early stage before they reach an age where the disease would be crippling. Improved sanitation in the Western world in the nineteenth century, however, reduced this acquisition of natural immunity and poliomyelitis epidemics began to occur. Until vaccines were developed thirty-five years ago, poliomyelitis was one of the most severe infectious diseases of Western countries. Two types of vaccine are used, called Salk and Sabin, and since the introduction of mass infant vaccination cases have been rare. Despite a degree of natural immunity, poliomyelitis still causes considerable problems in less developed areas of Africa and Asia.

The poliovirus is a picornavirus which occurs in three immunological types. On ingestion or inhalation, the virus multiplies and infects target organs of the intestinal mucosa, lymph nodes and central nervous system. The meninges and the motor neurones of the spinal cord and the brainstem are attacked, causing permanent paralysis. During the active phase of the infection, primary transmission of viruses is by droplet exhalation, but the virus continues to be excreted with the faeces for several weeks.

19.3.3 Parasitic protozoan infections

(a) Malaria

Malaria has been responsible for more human deaths than any other disease. Two hundred million people are affected annually and over three million die. One quarter of all African adults suffer from it at various times and a quarter of a million children die from it each year. It is not transmitted by faecal pollution but by an insect vector to which water is essential for maintenance of its aquatic larval life stage.

The disease is caused by the pathogenic protozoan *Plasmodium*, which belongs to the group Sporozoa. *Plasmodium* has primate, dog and rodent reservoirs. Its arthropod vector, which transmits it from host to host, is the female anopheline mosquito. The sexual reproduction of the protozoan occurs within the mosquito and when it bites a victim, forms of *Plasmodium* are injected into the bloodstream. These locate in the liver, multiply asexually and are rereleased into the bloodstream where they invade the red blood cells. After further multiplication they rupture and invade yet more blood cells. When a mosquito bites an infected individual the pathogen enters the mosquito's stomach and begins its sexual reproduction cycle again.

Four species of *Plasmodium* infect man with differing intensity. *P. falciparum* causes malignant malaria which can be fatal, whereas *P. vivax*, *P. ovale* and *P. malarie* cause benign malaria. The periodic rupture and release of the pathogen in the bloodstream causes a cycle of symptoms in the host. Chills, fever, headache and sweating occur every 48 h in *P. vivax* and *P. ovale* infections and every 72 h in *P. malarie*. Additional symptoms are an inflamed spleen, weakness and anaemia. *P. falciparum* causes more persistent symptoms of fever and swelling of the brain and lungs. The very rapid multiplication of this species in the blood vessels round the heart also causes impairment of cardiac function. Without treatment, benign malaria will recur spontaneously, but malignant malaria will kill. Treatment is effected with chloroquine and primaquine.

Prevention of malaria is carried out by eradication of the mosquito vector, by destroying its breeding grounds and the adult and larval stages. This is difficult since in wet areas it is impossible to restrict its breeding sites and in dry areas irrigation canals are required which also provide a suitable habitat for the mosquito larvae. In addition, the emergence of pesticide-resistant strains of mosquito has reduced the efficiency of, for example, DDT or dieldrin in controlling the disease.

(b) Amoebic dysentery

Amoebic dysentery is caused by the parasitic protozoan *Entamoeba histolytica*, which is frequently transmitted via faecally contaminated water. Carrier rates vary with the sanitary conditions prevailing in different areas and the levels of personal hygiene practised. The proportion of the population carrying *E. histolytica* may be 1–5% in the USA but is 50–80% in some areas of the tropics. The amoeba is also carried by animals.

In mild infections, *E. histolytica* colonizes the intestine and feeds on the intestinal flora. In more severe cases, however, the intestinal wall itself is attacked, with the formation of lesions. Infections of the liver, spleen and brain may occur. Symptoms range from abdominal pain with mild diarrhoea to acute dysentery. During the course of an infection, the amoebae are excreted as cysts which can survive outside the body until ingested by a new host. Diagnosis of amoebic dysentery is made by identification of the cysts in the faeces.

(c) Amoebic meningoencephalitis

The aetiologic agent for the fatal disease amoebic meningoencephalitis is the amoeba *Naegleria gruberi*. *Naegleria* species are commonly found in soil, sewage and surface water and occasionally in swimming pools and in drinking water abstracted from rivers. Exposure to the pathogenic strain *N. gruberi* occurs while swimming or diving in infected water. The organism gains access to the body by penetrating the mucous membranes of the nasal cavity, from where it travels to its target organs of the brain and spinal cord. Symptoms are manifested in four to seven days and death occurs four to five days later.

(d) Flagellate diarrhoea

Flagellate diarrhoea is caused by the intestinal flagellated protozoan *Giardia lamblia*. Like *E. histolytica*, *G. lamblia* forms cysts, and can be identified in sewage in this form. Contamination of drinking water by sewage provides the main transmission route. An epidemic of flagellate diarrhoea (giardiasis) in Colorado in 1965 which affected over a hundred people was traced to contamination of well water with sewage. In a further outbreak in New Hampshire, *G. lamblia* cysts were found in treated water; it was later established that raw water was being mixed with the treated water at the treatment plants involved. On ingestion, *G. lamblia* attaches itself to the intestinal wall of the host by means of a suction disc. In severe infections, the resultant coverage of the intestinal surface restricts absorption of, for example, fats into the body. A variety of symptoms are produced, the most frequent being that of protracted, intermittent diarrhoea.

(e) Other protozoan infections

Other waterborne enteric protozoa which parasitize the intestinal tract include the ciliate *Balantidium coli*, which causes balantidial dysentery and intestinal ulcers, and the sporozoans *Isopora belli* and *I. huminis* which cause gastro-intestinal infections. These pathogenic organisms have a growth phase within the intestinal tract but are transmitted via faecal contamination in the form of cysts.

19.3.4 Parasitic worm infections

(a) Tapeworm

The beef tapeworm *Taenia saginata* has cattle as its intermediate host and humans as its final host. Infection rates in humans are highest in E. Africa, at 50% of the population, and are high in Central and Southern America and

Eastern Europe. Elsewhere the incidence of *T. saginata* is generally less than 2%. Cattle become infected by grazing pasture recently spread with infected sewage or by drinking heavily infected water. The tapeworm locates in the muscle tissue of cattle and humans become infected by ingesting the larval stage in raw or poorly cooked beef. The symptoms of infection include abdominal pain, digestive disturbance and weight loss. *Taenia* eggs (ova) are produced in the human intestine and excreted with the faeces. Surveys of raw sewage have shown one to two ova per 10 l, with a considerable number discharged in the effluent. *T. saginata* ova are resistant and can survive for long periods in sewage and sludge under cool, moist conditions. Survival of ova in sludged pasture has been recorded for as long as 365 days; survival tends to be greater when sludge is applied in winter rather than in summer.

Control of tapeworm infections can be made by observing grazing restrictions on sludged pastures until it is considered that ova have been destroyed, storing sludge for at least a year before agricultural use or disinfecting sludge by methods such as pasteurization, irradiation, aerobic thermophilic stabilization, composting or lime treatment. Avoidance of human infection can be implemented by maintaining high standards of meat inspection after slaughter and of meat preparation.

(b) Roundworm

Roundworm or ascariasis is caused by the large intestinal roundworm *Ascaris lumbricoides*. Infection is most common in children but the pathogen is frequently isolated in the faeces of sewage workers and farm workers who spread sludge on land. Infection occurs by the transmission of ova through faecal contamination of water, soil and vegetables. One epidemic of ascariasis in Germany in 1949 was attributed to the application of raw sewage to garden crops. Over 90% of the population were infected and the community's raw sewage contained 540 *Ascaris* eggs per 100 ml. Like *Taenia saginata*, the ova can remain viable for long periods in soil, therefore similar restrictions on sewage and sludge treatment and disposal must be observed to prevent the spread of infection.

(c) Schistosomiasis

Schistosomiasis or Bilharzia disease is a debilitating infection affecting 200 million people annually. Reservoirs for the disease are primates, cattle, dogs, sheep and rodents. It is endemic to Africa, the Middle East and the Far East and was spread to the Caribbean and South America via the slave trade.

Schistosomiasis is caused by the trematode worm *Schistosoma*, notably the species *S. mansoni*, *S. haematobium* and *S. japonica*. The parasite spends part of its life cycle within a water snail, its vector, and part within a human or animal host. Humans become infected when swimming, wading or working, e.g. for rice growing, in freshwater containing infected snails. The larvae leave the snails to penetrate the body and the adult male and female worms subsequently inhabit the blood vessels. Migration of the trematodes to the bladder, bowel, intestines and liver can also occur. The disease is transmitted by faecal pollution of waterbodies; ova excreted by infected individuals hatch and the larvae seek a suitable snail host. Prevention of the disease thus involves both eradication of the snail vector and provision and use of adequate sanitary facilities to avoid the multiple use of waterbodies for agriculture, recreation, ablution and waste disposal.

(d) Other worm infections

Additional water-related tropical diseases caused by parasitic worms include filariasis, guinea worm and hookworm. Filariasis affects 300 million people a year, mainly in Africa. It is caused by a variety of filarial nematode worms which are transmitted by an arthropod vector. The worms colonize the lymphatic system and cause elephantiasis, a massive swelling and out-growth of the affected tissue. Infection of the optic nerve and retina causes glaucoma and cataracts, a condition known as "river blindness" since it is generally contracted near water.

Guinea worms are carried by zooplankton. Infection by immersion in infested water causes disability due to ulceration and abscess formation at the site of entry. It has been estimated that one quarter of Nigerians between fifteen and forty are debilitated for up to ten weeks per year by guinea worm infections.

Hookworm disease or ancylostomiasis is endemic to the south-east USA in addition to the tropics. It is caused by the helminths *Necator americanus* and *Ancylostoma duodenale* which penetrate the skin of those in contact with faecally polluted water. The bloodsucking activity of the helminths gives symptoms of anaemia and malnutrition.

In common with other parasitic diseases, control must be expedited by destroying the vectors or their breeding grounds, e.g. by draining swamps, and by reducing direct human contact with infected waterbodies. Education in good sanitary and hygienic practice and access to unpolluted water are paramount.

Further reading

Bitton, G., Damron, B. L., Edds G. T. and Davidson, J. M. (eds) (1980) *Sludge – Health Risks of Land Application*, Ann Arbor Science, Michigan.

Ford, T. E. (ed.) (1993) *Aquatic Microbiology: An Ecological Approach*, Blackwell Scientific Publications, Oxford.

Linton, A. H. (1982) *Microbes, Man and Animals – The Natural History of Microbial Interactions*, John Wiley and Sons, Chichester.

Mitchell, R. (ed.) (1978) *Water Pollution Microbiology*, John Wiley and Sons, New York.

Pelczar, M. J., Chan, E. C. S. and Krieg, N. R. (1993) *Microbiology: Concepts and Applications*, McGraw-Hill, New York.

Stanier, R. Y., Ingraham, J. L., Wheelis, M. L. and Painter, P. R. (1987) *General Microbiology*, 5th edn, Macmillan Education, London.

White, W. R. and Passmore, S. M. (eds) (1985) *Microbial Aspects of Water Management*, Society for Applied Bacteriology Symposium Series No. 14, Blackwell Scientific Publications, London.

Chapter 20

Water treatment

20.1 Introduction

From a microbiological point of view the treatment of water for public supply primarily involves the removal or destruction of organisms which are pathogenic. Included among such organisms are the bacteria *Salmonella typhi*, *S. paratyphi*, *Bacillus shigella* and *Vibrio cholerae*, the amoebic cyst-forming *Entamoeba histolytica* and several viruses, including coxsackie A2 and polio virus. Treatment of water to render it free of such organisms involves physical processes of removal or chemical processes of destruction. Non-pathogenic organisms which contribute to the suspended solids load, e.g. algae present in eutrophic waters, must also be removed in order to maintain the aesthetic quality of the water.

The range of treatment processes employed includes water storage, flocculation, coagulation, sedimentation, filtration and disinfection. All of these have an effect on the microbiological quality of the treated water, mainly through the removal of suspended matter, with the final process of disinfection ensuring the absence of any pathogens in the water passing into supply.

20.2 Water treatment processes

The types and numbers of stages of treatment employed at any given works will be dependent to a large extent on raw water quality. For example, the treatment of a good quality surface water in the southeast of England may involve storage in a large reservoir, followed by a rapid sand filtration or microstraining, then slow sand filtration and finally disinfection. In other areas or other circumstances sedimentation may be employed. Many fine particles or colloids are fairly stable and difficult to settle out. Therefore coagulation or flocculation are sometimes used to destabilize the suspension and encourage the formation of settleable flocs. Common types of coagulant chemicals which form flocs in water treatment are Al^{3+} and Fe^{3+} salts. The coagulated water is fed upwards through a sludge blanket clarifier

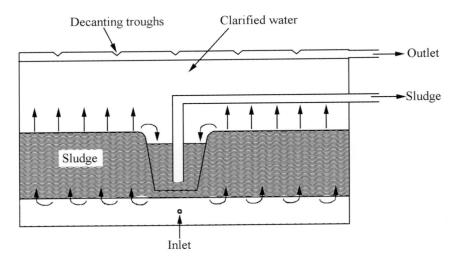

Fig. 20.1 A sludge blanket clarifier (adapted from Hall, 1997).

(Fig. 20.1) which combines the roles of flocculation and floc separation. The floc is held in suspension and forms a "sludge blanket" which results in an efficient flocculation process. Sludge is taken from the top of the blanket which maintains the desired level and the clarified water is taken from decanting channels at the surface of the water. The efficiency depends on the blanket concentration. This is controlled by flow rate, water temperature, floc density and flow distribution.

The process is generally used before rapid filtration as it will reduce the load on the filters. It is also utilized in the removal of precipitates produced as a result of softening lowland waters.

Raw water storage reservoirs are often considered the first line of defence against the transmission of waterborne diseases. Periods of storage can be up to several months during which time counts of enteric bacteria, typhoid and other pathogenic organisms fall to a significant extent. This is partially due to sedimentation, partially due to the lethal effects of both ultraviolet and visible light and partially due to the effects of predation. If large storage reservoirs are used in areas where the rivers acting as sources of abstraction receive sewage or other effluents then should the river become polluted the reservoir can maintain the volume of supply during the time it takes the river to recover. One disadvantage of storage reservoirs is the opportunity presented for the excessive growth of algae (see Chapter 18).

The stages prior to disinfection are designed primarily to remove suspended solids and inasmuch as the micro-organisms themselves constitute part of the suspended solids concentration, they will be removed also. The fairly large organisms, including algae and amoebic cysts, can be removed to a significant degree by coagulation and filtration. Bacterial removals of 99% have also been reported. Some removal of algae can also occur during microstraining. Sometimes, however, phytoplankton can cause problems in sedimentation and filtration. Some plankton do not settle very readily, even in the presence of chemical coagulants, while others may pass through certain types of filter and yet others clog filters very rapidly. The cyanobacteria will often float due to their air vacuoles; in serious cases it may be necessary to employ dissolved air flotation for their removal. A diagram of a dissolved air flotation (DAF) system is shown in Fig. 20.2.

DAF is a separation process whereby solid particles are floated to the surface of the water by attachment to small air bubbles, typically 20–100 μm in size. These particles eventually form a film of sludge on the water's surface which is removed by a mechanical scraper. Around 10–15% of the treated water is recycled through a pressurized air saturation system and is injected into the untreated water entering the tank. Owing to a reduction in pressure on injection of the air-saturated water, fine bubbles are produced. This type of method is employed in the removal of algae in water treatment plants.

Slow sand filtration is particularly of interest because it is fairly effective in the removal of pathogens while, unlike the other unit processes, being dependent upon the activities of other micro-organisms resident within the filter bed for its effectiveness in removing other contaminants including dissolved organic matter. Slow sand filtration was the first of the modern water

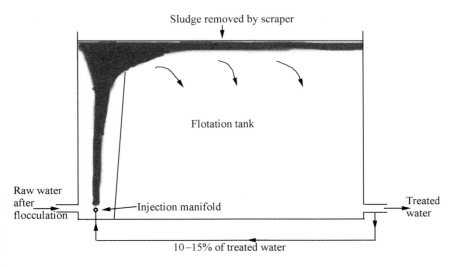

Fig. 20.2 Dissolved air flotation system.

treatment processes to be developed. Its use became almost universal by the middle of the nineteenth century and although it later lost favour to some degree, especially in the United States, as a result of its inability to cope with highly turbid raw waters, it is still widely used in the UK, especially in the London area.

A simplified diagram of a slow sand filter is given in Fig. 20.3. Slow sand filters essentially consist of a bed of sand, perhaps as much as 1.4 m deep initially, but frequently becoming as shallow as 0.45 m as a result of cleaning, overlying a layer of gravel in turn overlying an underdrainage system. Fairly uniform sand of an effective size of 0.15–0.4 mm (equivalent to the mesh size of a sieve that will pass 10% of it) is most frequently used. This will give an effective minimum pore diameter within the bed of about 20 μm and would not, therefore, be expected to retain bacterial cells effectively. However, some sedimentation of particles smaller than 20 μm will occur within the bed.

Slow sand filters are normally operated with a head of water above the bed of 1.0–1.5 m, which gives a typical rate of flow of about 2.5 m^3 per m^2 of bed surface per day, although the range reported is from 0.6 to 12 $m^3\,m^{-2}\,d^{-1}$. Periodically the surface of the bed is skimmed off as the head loss becomes larger. Normally, it is not replaced until a minimum bed depth is reached, when resanding may be necessary.

At the surface of the bed a biologically active layer forms as a result of settling from the water above the filter combined with the straining effect of the sand. It is a complex layer, comprising mud, organic detritus, algae, bacteria and fungi in the form of a mat. The effectiveness of the filter is thought to be partially due to the straining characteristics of this layer and

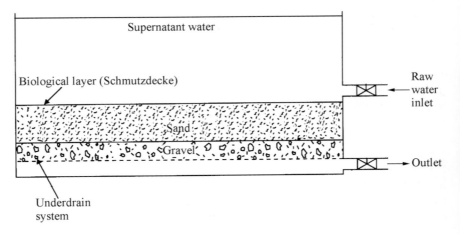

Fig. 20.3 A slow sand filter.

partially due to the biological activity. Nearly all of the suspended matter in the raw water is trapped by this filter skin, sometimes called the Schmutzdecke, some of it acting as substrate for the micro-organisms therein. Leaching of some of this organic material from the filter skin into the subsurface layers occurs, and this encourages bacterial activity at depths of up to 0.7 m. Mature filters will produce a completely nitrified effluent. For this and other reasons, slow sand filters should remain aerobic throughout their depth.

The removal of bacteria in slow sand filters is probably very largely due to predation, particularly by protozoa. Between 10^5 and 10^6 ml^{-1} flagellated protozoa have been found within the top 0.3 m of a slow sand filter. Several reports of bacterial removal efficiencies in excess of 99.9% have been made, although significant decreases often occur during colder weather, suggesting a biological rather than physical mechanism of removal. Similar removal efficiencies for viruses (particularly polioviruses) have been observed. Virus removal seems less dependent on temperature. It has been suggested that virus removal may be enhanced by the period of storage of raw water above the filter.

20.3 Disinfection

Disinfection is normally the final process through which the water passes prior to its entry to the distribution mains. This is important because disinfection should be applied to water which is free of suspended particles and dissolved organic matter, otherwise such materials may physically shield any micro-organisms present from the effects of the disinfectant or they may react directly with the disinfectant and reduce its potency.

The distinction made between sterilization and disinfection in Chapter 5 still applies here to water treatment. It will be recalled that sterilization implied a treatment causing the complete destruction of all living organisms and that such a treatment should be effective permanently. Complete sterilization of a water supply would be extremely difficult to achieve and would not in any case be considered necessary. Thus, disinfection is effective for a limited period and can permit recontamination of the treated water, although some disinfection methods will confer some residual disinfection capability on the treated water as it passes through the distribution system.

Disinfection designed for routine use in water treatment facilities should be cheap, efficient, safe to handle and preferably should confer some residual disinfecting capacity on the water passing into supply. Clearly, the disinfectant should be harmless to consumers of the water while having a significant toxicity to the target organisms. Chlorine, particularly, and ozone to a lesser extent conform largely with these criteria.

20.3.1 Theory of disinfection

For a given concentration of a disinfectant the rate of kill is given by:

$$\frac{dN}{dt} = -KN \tag{20.1}$$

where K is the reaction rate and N the number of viable organisms. Upon integration:

$$K = -\frac{\ln(N_t/N_0)}{t} \tag{20.2}$$

(compare this with equation (10.6)).

K is a constant only for a given system, and is dependent on a range of factors, the most significant of which is the concentration of the disinfectant, $[D]$. Then,

$$\frac{dN}{dt} = -KN[D]^n \tag{20.3}$$

so that:

$$\frac{\ln(N_t/N_0)}{t} = -K[D]^n \tag{20.4}$$

Taking logarithms of both sides gives a straight line relationship which can be solved graphically to give K and n. If $n = 1$ then equation (20.4) is known as Chick's law. Frequently, however, $n \neq 1$, reflecting the complex mechanism of disinfection. Values of K for HOCl (see Section 20.4.2) range from about 10^3 l mol^{-1} min^{-1} for the more resistant amoebic cysts to 10^6 or more for coliform bacteria.

The form of equation (20.4) is such that, theoretically at least, a kill of 100% is not achieved. It is normal, however, to specify a kill rate of, say, 99.9% (i.e. $N_t/N_0 = 0.001$). For such a constant kill rate, $\ln(N_t/N_0)$ is constant, which means that $[D]^n t$ is constant also. If $n = 1$, then the time of exposure is as important as concentration of disinfectant.

20.3.2 Chlorination

Chlorine is a fairly strong oxidizing agent. However, its toxicity to a wide range of micro-organisms does not appear to be due to its oxidation effect, but is more probably due to its interference with the functions of some crucial enzymes. From this point of view, its oxidizing power can be inconvenient because chlorine will be consumed in reaction with reduced compounds such as ammonia, thus decreasing its effectiveness as a disinfectant.

(a) Chlorine compounds

Chlorine (Cl_2), when added to water, forms hypochlorous acid:

$$Cl_2 + H_2O \rightleftharpoons HOCl + HCl$$
$$\text{Hypochlorous acid}$$

Hypochlorous acid is a weak acid, and will dissociate to form H^+ and OCl^- ions. The extent to which this occurs is dependent on the pH. For pH values up to 6.7 at least 90% of the chlorine is in the form of HOCl, but at pH 9 less than 5% remains in the undissociated form. This is important, since HOCl is a far more potent disinfectant than OCl^-. For example, HOCl is about one hundred times more toxic to *E. coli* than OCl^-. Figure 20.4 shows the stages involved in the chlorination process.

If ammonia is present this will be converted successively into chloramide (NH_2Cl), chlorimide ($NHCl_2$) and finally iminotrichloride or nitrogen trichloride (NCl_3). (Trivial names for these compounds are still in widespread use; they are mono-, di- and trichloramine.) The forms of chlorine present may be divided into two categories. The "free chlorine residuals" or "free available chlorine" include Cl_2, HOCl and OCl^-. Combined chlorine residuals or combined available chlorine include the chloramines and any other chlorine compounds (e.g. organochlorine compounds). The chlorine present in the form of chloramines is referred to as combined available chlorine because these compounds still possess some disinfecting capability, albeit less than the

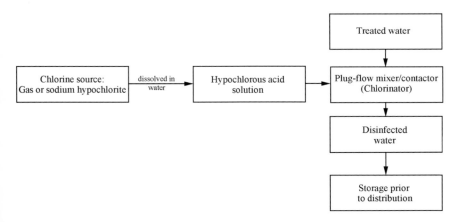

Fig. 20.4 The chlorination process.

free available chlorine compounds. The quantity of chlorine used to disinfect a particular water is therefore dependent to a degree on the concentration of ammonia present.

(b) Methods of chlorination

In fairly clean water containing low concentrations of ammonia and organic matter simple chlorination involving the addition of 0.2–0.5 mg l^{-1} of Cl$_2$ is normally sufficient to give a persistent free available chlorine residual. Where there are significant quantities of ammonia present, breakpoint chlorination is practised. This involves the addition of sufficient chlorine so that after reacting with any ammonia present there is some free available chlorine residual left. Usually, the amount of chlorine added initially is sufficient to form monochloramine. Although dichloramine and nitrogen trichloride could also be formed this does not normally occur if the chlorine to nitrogen ratio is kept fairly low. Nitrogen trichloride is fairly unstable in the presence of monochloramine, and mixtures of mono- and dichloramine are also unstable:

$$NH_2Cl + NHCl_2 \rightleftharpoons N_2 + 3HCl$$

Further addition of chlorine causes the breakdown of the monochloramine:

$$2NH_2Cl + HOCl \rightleftharpoons N_2 + 3HCl + H_2O$$

A breakpoint is reached, where there is no ammonia or monochloramine left in solution nor is there any free available chlorine. Further addition of chlorine produces free available chlorine in solution.

In cases where the quality of the raw water is uncertain, and resistant pathogens may be present, superchlorination may be employed. This involves the addition of a large excess of chlorine for about 30 min. The water may then be dechlorinated using sulphur dioxide.

In certain instances, the presence of combined available chlorine may be more desirable than free available chlorine. Chloramines are preferable to hypochlorous acid where the chlorination of organics may lead to taste and odour problems since they are less prone to react with the organic matter. For this and other reasons, chloramines constitute a more stable residual in the supply system, although they have a reduced activity. The deliberate formation of monochloramine is achieved in a process called chloramination which involves first the addition of ammonia to water followed by chlorination. Dichloramine and nitrogen trichloride formation is avoided if possible, because these compounds are considered a nuisance, producing unpleasant tastes and odours. If the ratio of chlorine to ammonia is kept low and the pH maintained above 7 only small quantities of dichloramine and nitrogen trichloride will be formed.

(c) Susceptibility of micro-organisms to chlorination

Generally, of those organisms of significance, the coliforms are the most susceptible to chlorine, followed by some viruses, *Entamoeba histolytica* and finally bacterial spores. Some typical values of HOCl concentrations required to achieve 99.6% kill in 20 min are shown in Table 20.1. Of the pathogens of greatest concern, *V. cholerae* is at least as susceptible to chlorination as *E. coli*, while the susceptibility of many viruses is more uncertain. In situations where protection against viruses is known or suspected to be necessary a larger chlorine dose and contact time three times that normally considered necessary for *E. coli* removal may be required. The World Health Organisation proposed that 0.05 mg l^{-1} of free chlorine with an exposure of 1 h would be sufficient to inactivate viruses. Although the spore-forming bacteria are the most resistant to chlorination, the occasional presence of a few cells of *Clostridium perfringens* in treated water is not of great significance in terms of immediate or direct risks to health. *C. perfringens* does, however, imply a remote or infrequent source of faecal contamination. In the majority of cases, there is a strong dependence on temperature and pH. Typically, concentrations of free available chlorine five to ten times higher than those at pH 7 would be required to achieve the same kill at pH 8.5. Generally, about 50% more free available chlorine would be required for effective disinfection at <5°C compared with 10°C.

(d) Disinfection levels in practice

Disinfection levels can often be expressed in terms of the product of concentration and time of exposure (*Ct*). For example, a *Ct* value of 30 mg l^{-1} min could imply a 1 mg l^{-1} chlorine concentration for 30 min or 2 mg l^{-1} for 15 min. In practice there are restrictions, however. A contact time of less than 10 min would not be considered desirable for domestic water supply, 30 min being more acceptable and 1 h being desirable. In water at 10°C at a pH of 7 a *Ct* value as low as 8 mg l^{-1} min may be acceptable, but four or five times this value may be required for alkaline waters (pH < 8.5) at lower temperatures in order to inactivate *E. coli*, hepatitis and

Table 20.1 Typical concentrations of hypochlorous acid required for 99.6% kill of various organisms in 20 minutes

Organism	pH	Temperature (°C)	HOCl concentration (mg l^{-1})
E. coli	7	20–29	0.02
Coxsackie virus	7	5	0.25
E. histolytica	7	20–29	3
B. anthracis	7.2	20–29	6

polio viruses. In general terms, chlorine doses of 0.5–2 mg l^{-1} and exposure times of 30–60 min are common.

(e) Chlorine dioxide

Some texts treat chlorine dioxide as a separate disinfectant to chlorine because it is not formed from the direct addition of chlorine to water or from reaction of chlorine with any compound normally found in water. However, it is usually generated at the treatment plant from the addition of normally chlorinated water to sodium chlorite solution:

$$2NaClO_2 + Cl_2 \rightleftharpoons 2ClO_2 + 2NaCl$$

Excess chlorine is added to push this equilibrium over to the right, thus preventing chlorite ion from passing into supply. Thus, the disinfectant is often a mixture of chlorine and chloride dioxide. Chlorine dioxide has a bactericidal efficiency comparable with that of free chlorine and seems to be effective against viruses, although this effect remains largely unquantified. It produces a stable residual in water passing into supply and has the additional advantage of being more effective at high pH values. Moreover, it does not react with ammonia, so that it is particularly useful where breakpoint chlorination would normally be necessary to achieve a stable residual.

20.3.3 Ozone

Ozone (O_3) is a fairly powerful oxidizing agent which has a more rapid effect than chlorine in destroying viruses and bacteria, including spores. Ozone is produced by the effect of a high voltage discharge in dry air. The air containing the ozone is then injected into tanks of the water to be treated. The transfer efficiency into solution is quite high because ozone is considerably more soluble than oxygen in water. Between 1 and 2 mg l^{-1} dissolved ozone and a contact time of 5–15 min is usual practice, although much larger concentrations can be used for disinfecting raw waters of poorer quality. A diagram illustrating the features of an ozonation plant is given in Fig. 20.5.

The major disadvantage of ozone is that it leaves no persistent residual disinfecting capability in the water passing into supply. For this reason a free chlorine residual may be introduced following disinfection with ozone.

20.3.4 Other disinfectants

For large-scale water treatment systems, designed for the production of drinking water, the use of alternative disinfectants is rare. However, there are a number of agents which satisfy the criteria for a disinfectant which find

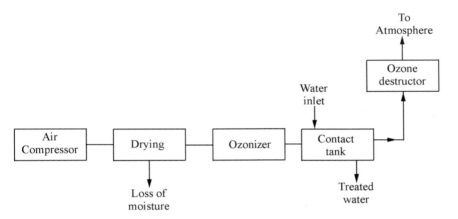

Fig. 20.5 Features of an ozonation plant.

use in other applications. For example, certain disinfectants, such as bromine and ultraviolet radiation, may be useful for swimming pool water.

The other halogens appear logical candidates as disinfectants. However, fluorine is far too reactive and the lower halogens are too costly for general use, although iodine can be used in emergency situations.

The bactericidal effects of ultraviolet (UV) radiation at wavelengths of 250–300 nm are well known. Clearly, the efficiency of UV is severely reduced if the water to be treated is at all turbid or contains any other material which has significant absorbance. Another disadvantage of UV is that it leaves no residual behind.

References and further reading

Barnes, D. and Wilson, F. (1983) *Chemistry and Unit Operations in Water Treatment*, Applied Science Publishers Ltd, Barking.

Ellis, K. V. (1986) Slow sand filters. *Crit. Rev. Environ. Control*, **15**, 315–354.

Hall, T. (1997) *Water Treatment Processes and Practices*, 2nd edn, WRc, Swindon, pp. 33–97.

Tebbutt, T. H. Y. (1998) *Principles of Water Quality Control*, 5th edn, Butterworth-Heinemann, Oxford.

Twort, A. C., Law, F. M., Crowley, F. W. and Ratneyaka, D. P. (1994) *Water Supply*, 4th edn, Edward Arnold, London.

Index